Dynamic Positioning of Offshore Vessels

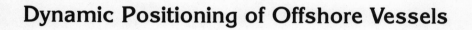

Max J. Morgan

Marine Division
Honeywell Inc.

Dynamic Positioning of Offshore Vessels

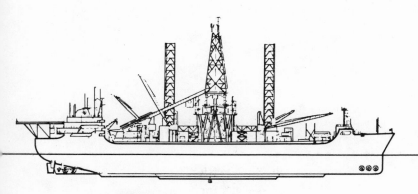

Division of
The Petroleum Publishing Company
Tulsa, Oklahoma

Copyright © 1978
PPC Books Division
The Petroleum Publishing Co.
1421 South Sheridan (P. O. Box 1260)
Tulsa, Oklahoma 74101, U.S.A.

Library of Congress Catalog Card Number: 77-94268
International Standard Book Number: 0-87814-044-1
Printed in U.S.A.

1 2 3 4 5   82 81 80 70 78

This book is dedicated to
R O Y E. P E A R S O N

who died suddenly last year.

Roy was my good friend

and served as an able associate in the design

of several dynamic

positioning systems

Preface

MAN'S LONG PRESENCE IN THE MARINE ENVIRONMENT continues as he turns to the sea to look for the resources required to maintain his way of life. Over the past two decades he has searched the sea for petroleum reserves. In the coming years large deposits of minerals, in addition to petroleum, will continue to draw him to the sea.

Each new offshore endeavor requires development of the equipment used to perform the work and in the platform used to support and transport the equipment. This progression of technology is well illustrated by the search for petroleum. Land-based drilling equipment used to locate oil and gas has been modified to work offshore in deeper and deeper water.

This development is still in progress as petroleum exploration moves into water depths greater than 3,000 ft.

Platforms which transport and support drilling equipment over the well site were initially barges and jack-ups. Now large ship-shape and semisubmersible drilling vessels are used.

To hold these floating vessels in position during drilling, multi-anchor spread mooring systems were first used on the vessels. Multiple thrusters with coordinated manual thruster control were then installed in the hulls of the vessels. The manual control was soon replaced by an automatic feedback control system and dynamic positioning came into being.

Dynamic positioning has since progressed from the single-thread analog controller of the early 1960s to dual redundant digital controllers. Currently the trend is back to single-thread systems with nearly exclusively digital implementation of the controllers. In the future, digitally-based controllers will dominate the dynamic positioning market.

It is likely, however, that the controllers will be implemented in microprocessors with permanent memory programs.

Because of this dominance of digital controllers, I have concentrated on digital control system methods in this book.

I entered the field of dynamic positioning when it was nearly a decade old. From the very beginning it was evident that dynamic positioning was here to stay. Dynamic positioning is essential for deep water applications where mooring systems are impractical. These applications require minimum transit time from one location to another, and more frequent maneuvering than is possible with moored vessels.

The future of dynamic positioning should be bright since we have only begun to develop applications for it. Therefore, I felt compelled to document this important means of controlling the position of an offshore vessel.

A second source of motivation for the book came from my work in the design and technical marketing of several dynamic positioning systems. In the technical marketing role I sensed the need for an explanation of the systems approach to outfitting a dynamically-positioned vessel.

My contacts affirmed the need for a more widespread understanding of at least the basics of feedback control by those responsible for the construction of dynamically positioned vessels.

In order to meet these needs I have tried to first treat the dynamic positioning system as a total system. Second I have focused on feedback control and how it relates to the dynamic positioning system. In each case the treatment is limited. Each element of the system alone could fill a book.

I would like to acknowledge the contributions of the employees of the Honeywell Marine Systems Division. Over the past seven years my work for Honeywell has been concentrated on their ASK (Automatic Station Keeping) dynamic positioning product line. This work gave me the opportunity to acquire the background to write this book.

The publications department of Honeywell provided typing, editing, and illustrating services for this book. Special thanks go to Joyce Haney, Kenso Nagumutzu, Lorna Gibson, Ginger Marshall, and Candace Mckenna.

I also wish to acknowledge the helpful comments of Henry Van Calcar, David Reitz, Ed Swanson, and Ken Prentiss on the technical contents of this book.

Contents

1

Introduction

1.1 Dynamic positioning

When a free-floating body is placed in an offshore marine environment, environmental forces and moments act on the body, tending to cause it to translate and rotate. The environmental elements which generate the forces and moments include wind, current, and waves.

As a result, if the body or vessel is to remain within a given distance of a reference point on the bottom or to follow within a given distance of a path referenced to the bottom or the earth, the capability of generating counterforces and moments must be provided on the vessel.

Historically, the counterforces and moments have been provided by sails, oars, rudders, mooring lines, and propellers. Propellers are the best-known member of a more general classification of counter-force producing devices called thrusters. When vessel is moving from one point to another, all five forms of counterforce generation perform with varying degrees of ease and precision. When the vessel is moving very slowly or is station keeping at a fixed point, only the last two means of generating counterforces and moments perform effectively.

As any mariner knows, offshore environments are known for their variability. Consequently, the environment forces acting on the vessel not only constantly change in magnitude but also in direction. This makes it necessary to provide counterforces and moments in all directions.

In the case of mooring lines, lines are deployed in all directions from the vessel to maintain the vessel at a given position in the changing marine environment. However, as water depth increases, the holding power of the mooring system decreases and the difficulty of deployment increases.

1

In fact, at some water depth the multipoint mooring system is totally impractical.

Unlike the mooring system, the propeller method of furnishing counterforces and moments is not water-depth dependent. Therefore, the cost of the propulsion system will not increase as water depth increases, as opposed to considerable cost increase with increasing water depth for a mooring system.

Like the mooring system, the propeller system must also furnish thrust in all directions. However, in the case of propellers there are several configurations which can be used to satisfy the counterforce requirement.

One example is shown in Fig. 1-1 where the main screws furnish longitudinal counterforce and the lateral thrusters furnish counterforce at right angles to the screws. In addition, the lateral thrusters furnish counter moment. To maximize their counter moment capa-

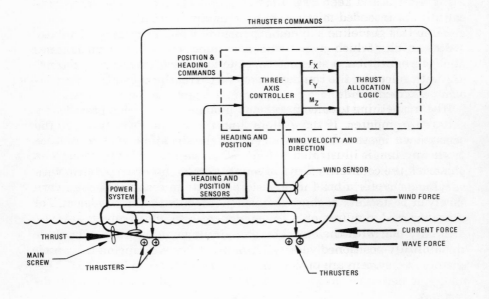

Figure 1-1. A Dynamic Positioning System

bility, the lateral thrusters are located as far forward and aft of the center of rotation of the vessel as possible.

Although the propeller system is a critical element of a dynamic positioning system, a dynamic positioning system is not a dynamic positioning system because the system furnishing counterforces against the environment is a propeller system.*

The distinguishing feature of a dynamic positioning system is the use of measured position to generate counter force commands to maintain the position of the vessel near the desired location, i.e., feedback control. Although not an explicit part of position control, the heading of the vessel must be controlled so that the attitude of the vessel with respect to the environmental elements can be maintained in a controlled manner.

Otherwise, the vessel will assume an attitude in which either the counterforce system cannot furnish sufficient power to counter the environment and keep its position, or the motion of the vessel will not permit the intended mission of the vessel to continue.

From this guideline a dynamic positioning system can be formulated using a sensor which gives position reference of the vessel with respect to a given location, a sensor for measuring vessel heading, and something to calculate the commands to the counterforce devices to implement the commands.

The "something to calculate the commands" could be a human operator or a computer. If the system includes a computer, the dynamic positioning system becomes an automatic feedback control system. Such a system is illustrated in Fig. 1-1 and is the topic of this book. In this book the counterforce system is primarily propellers, or thrusters.

In this chapter a brief introduction to dynamic positioning systems is given to form a background for the remainder of the book. The missions and requirements for dynamic positioning systems are reviewed along with the environmental disturbances which act on the dynamically-positioned vessels. Then, the basic elements of a dynamic positioning system are outlined.

1.2   Historical

Dynamic positioning of floating vessels had its beginning in the early sixties. The first vessels outfitted for dynamic positioning ranged from 450 to approximately 1000 long tons of displacement and were

---

* Dynamic positioning appears throughout the book and is sometimes abbreviated by the letters, DP, where there is no possibility of confusion.

designed for coring, cable laying, or surface support of underwater work.

A list of the early dynamically-positioned vessels in Table 1–1 gives characteristics of the various elements of the dynamic positioning systems.

To the best of the author's knowledge, the first dynamically positioned vessel using automatic feedback was the *Eureka*. This system is thoroughly discussed in Shatto and Dozier's "A Dynamic Stationing System for Floating Vessels."

In Fig. 1–2 the profile view of four of the eight early DP vessels is shown, but it is not obvious that the vessels are dynamically positioned. Likewise, the size or shape of the hull obviously has little to do with whether or not a vessel is dynamically positioned. The best clue that the vessels are dynamically positioned is their multiple thrusters.

Not listed in Table 1–1 is a very well-known vessel that was to be dynamically positioned but was never completed. The vessel was to be built for Project Mohole. The Mohole vessel is shown in Fig. 1–2. From the beginning this project pushed the state-of-the-art and beyond. Perhaps that is why the project was never finished. However, many benefits for future dynamically-positioned vessels came from the Mohole project.

TABLE 1-1. A LIST OF EARLY DYNAMICALLY POSITIONED VESSELS WITH THEIR POSITION SENSOR AND CONTROLLER FEATURES

VESSEL	POSITION SENSOR	CONTROLLER		
		TYPE	REDUNDANT	WIND FEED FORWARD
CUSS I	RADAR	ANALOG (MANUAL)	NO	NO
EUREKA	TAUT WIRE	ANALOG	NO	NO
CALDRILL	TAUT WIRE	ANALOG (MANUAL)	NO	NO
CAPISTRANO	ACOUSTIC	ANALOG	NO	NO
USS NAUBUC	RADIO	DIGITAL	NO	NO
GLOMAR CHALLENGER	ACOUSTIC	DIGITAL	NO	NO
TEREBEL	TAUT WIRE	DIGITAL	NO	NO
DUPLUS	TAUT WIRE	DIGITAL	NO	NO

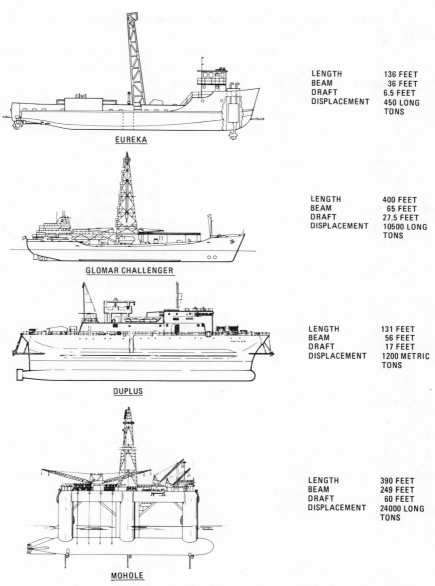

EUREKA

LENGTH	136 FEET
BEAM	36 FEET
DRAFT	6.5 FEET
DISPLACEMENT	450 LONG TONS

GLOMAR CHALLENGER

LENGTH	400 FEET
BEAM	65 FEET
DRAFT	27.5 FEET
DISPLACEMENT	10500 LONG TONS

DUPLUS

LENGTH	131 FEET
BEAM	56 FEET
DRAFT	17 FEET
DISPLACEMENT	1200 METRIC TONS

MOHOLE

LENGTH	390 FEET
BEAM	249 FEET
DRAFT	60 FEET
DISPLACEMENT	24000 LONG TONS

Figure 1-2. Early Dynamically Positioned Vessels

Perhaps the most successful — and by far the best known — of the early dynamically positioned vessels in the world is the *Glomar Challenger*.[6] Its adventures throughout the world, gathering core samples in nearly every ocean on earth in water depths to 20,000 ft, has given the necessary evidence for many geogological findings. For example, the plate-tectonics theory received enormous substantiating proof from the *Challenger's* expeditions.

The next generation of dynamically-positioned vessels is listed in Table 1–2. Heading the list is the SEDCO 445 which went into service in 1971 and whose profile is shown in Fig. 1–3. The main features of the SEDCO 445 dynamic positioning system which set it apart from the early systems given in Table 1–1 are a digital controller contained in a commercial 16-bit minicomputer and redundancy in all elements of the system to make possible extended periods of operation without in-

TABLE 1-2. A LIST OF SECOND-GENERATION DYNAMICALLY POSITIONED VESSELS WITH THEIR POSITION SENSORS AND CONTROLLERS

VESSEL	POSITION SENSORS	CONTROLLER		
		TYPE	REDUNDANT	WIND FEED FORWARD
SEDCO 445	AC, TW, RA	DIGITAL	YES	YES
SAIPEM DUE	AC, TW	DIGITAL	YES	YES
PELICAN	AC, TW	DIGITAL	YES	YES
HAVDRILL	AC, TW, RA	DIGITAL	YES	YES
DISCOVERER 534	AC, TW	DIGITAL	YES	YES
ARCTIC SURVEYOR	AC	DIGITAL	NO	YES
WIMPEY SEALAB	AC, TW	DIGITAL	NO	YES
KATTENTURM	AC	DIGITAL	NO	YES
BEN OCEAN LANCER	AC, TW, RA	DIGITAL	YES	YES
SEVEN SEAS	AC, RA	DIGITAL	YES	YES
PAC NORSE	AC, TW, RA	DIGITAL	YES	YES
SEDCO 709	AC, TW, RA	DIGITAL	YES	YES
SEDCO 472	AC, TW, RA	DIGITAL	YES	YES
SEDCO 471	AC, TW, RA	DIGITAL	YES	YES
BIG JOHN	AC, TW	DIGITAL	YES	YES
PETREL	AC, TW	DIGITAL	YES	YES
PELERIN	AC, TW	DIGITAL	YES	YES

AC = ACOUSTIC   TW = TAUT WIRE   RA = RISER ANGLE

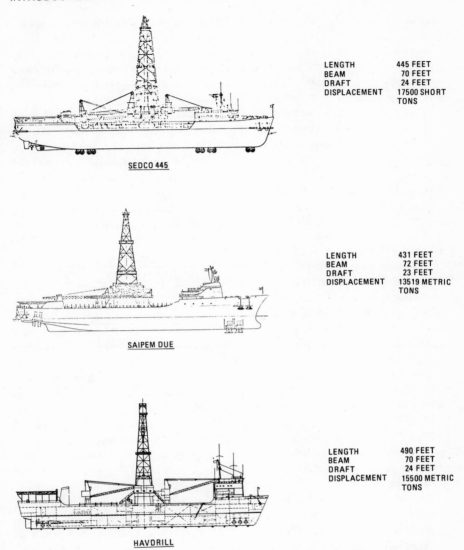

LENGTH   445 FEET
BEAM   70 FEET
DRAFT   24 FEET
DISPLACEMENT   17500 SHORT
                TONS

SEDCO 445

LENGTH   431 FEET
BEAM   72 FEET
DRAFT   23 FEET
DISPLACEMENT   13519 METRIC
                TONS

SAIPEM DUE

LENGTH   490 FEET
BEAM   70 FEET
DRAFT   24 FEET
DISPLACEMENT   15500 METRIC
                TONS

HAVDRILL

Figure 1-3. The First Generation Exploratory Dynamically Positioned
Drillships

terruptions. The system design called for 150 days of continuous operation.

Also, the SEDCO 445 had a large number of thrust-producing units, 11 lateral screws and two main screws. Many very good papers exist regarding the SEDCO 445 and other second-generation vessels. The interested reader is referred to References 3, 4, and 5.

Two other members of the second-generation dynamically-positioned drillships are also shown in Fig. 1.3. Each has its unique features. However, they all use approximately the same sensor suite and digital computer controller system. The greatest DP-related difference among the three drillships is their thruster system. The SEDCO 445 has 13 controllable speed propellers, the *Havdrill* has seven controllable pitch propellers, and *Saipem Due* has four cycloidal propellers. In the original installation all four thrusters of the *Saipem Due* were cycloidal. Later, the most forward thruster was replaced by a ducted, steerable, controllable pitch propeller.

Another less significant DP-related difference in the three drillships in Fig. 1–3 is that they all have had, at one time, a multipoint mooring system. The SEDCO 445 started with an eight-point wire-rope mooring system which was removed in 1975. In contrast, *Havdrill* did not have a mooring system in the beginning, but when it was outfitted to drill in the ice of the Northern Canadian waters, an eight-point mooring system was installed.

From the beginning, *Saipem Due* has had an eight-point chain mooring system which has been used with and without the dynamic-positioning system to position the vessel during drilling.

Another class of the second generation of dynamically-positioned offshore vessels is shown in Fig. 1–4. Unlike the previously-mentioned second generation DP vessels, this class of offshore vessels is much smaller and does not have the redundancy in the DP system that the drillships have.

The reason is that their mission is to support offshore surface and subsea work, such as pipeline inspection. More discussion of system requirements as they relate to vessel mission appears in Section 1.4.

Another class of the second-generation DP vessel is shown in Fig. 1–5. The Hughes *Glomar Explorer* is the largest vessel to be dynamically positioned to date.[9] It is also the first dynamically positioned mining ship. However, the DP system is very similar to the systems on other drillships.

The major difference is in the primary position reference system of

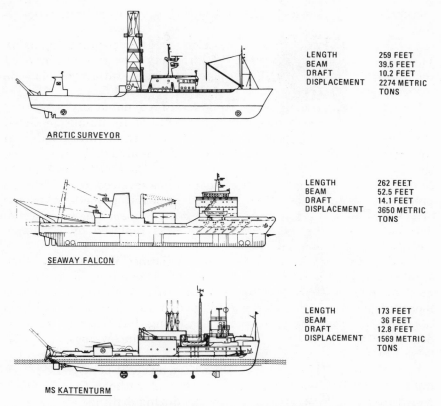

LENGTH	259 FEET	
BEAM	39.5 FEET	
DRAFT	10.2 FEET	
DISPLACEMENT	2274 METRIC TONS	

ARCTIC SURVEYOR

LENGTH	262 FEET	
BEAM	52.5 FEET	
DRAFT	14.1 FEET	
DISPLACEMENT	3650 METRIC TONS	

SEAWAY FALCON

LENGTH	173 FEET	
BEAM	36 FEET	
DRAFT	12.8 FEET	
DISPLACEMENT	1569 METRIC TONS	

MS KATTENTURM

Figure 1-4. First Generation Dynamically Positioned Support Vessels

the *Glomar Explorer*. It is a long baseline acoustic system, whereas for drillships the primary system is a short baseline acoustic system. The *Glomar Explorer* does have a short-baseline system which is used in a secondary manner.

The future of dynamic positioning holds many fascinating possibilities. The first is not actually in the future and it is shown in Fig. 1–5. The SEDCO 709 is the first semisubmersible to be dynamically positioned for any purpose.[7][8] Furthermore, its 24,000 HP thruster system has the largest capacity over the full azimuthal range.

The future challenge of drilling for oil in areas which are under ice likewise presents the next challenge for dynamically positioned

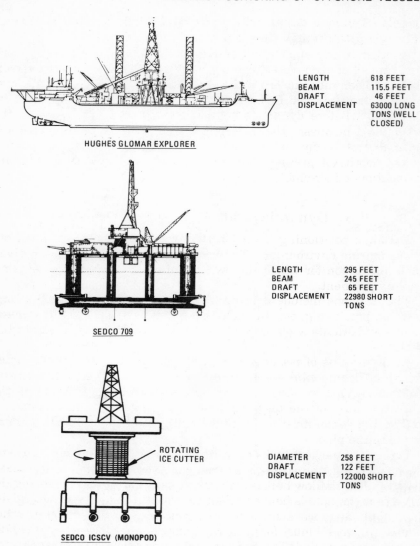

HUGHES GLOMAR EXPLORER

LENGTH	618 FEET
BEAM	115.5 FEET
DRAFT	46 FEET
DISPLACEMENT	63000 LONG TONS (WELL CLOSED)

SEDCO 709

LENGTH	295 FEET
BEAM	245 FEET
DRAFT	65 FEET
DISPLACEMENT	22980 SHORT TONS

ROTATING ICE CUTTER

DIAMETER	258 FEET
DRAFT	122 FEET
DISPLACEMENT	122000 SHORT TONS

SEDCO ICSCV (MONOPOD)

Figure 1-5. Advanced Dynamic Positioning Systems

vessels. A unique design now under study (Fig. 1-5) has the vessel cutting a path through the ice pack as it drills for oil.

Another future application of dynamic positioning will be for position control of deep-water pipe-laying vessels. To date pipelay vessels have all been moored. The first step toward a dynamically positioned pipelay vessel is the *Castoro VI* which is owned and operated by SAIPEM.[10] Before dynamically positioning will find a place in the pipelaying business, the mooring system will necessarily become incapable of accomplishing the positioning task. This will require water depths of greater than several thousand feet or a new pipe launching and assembly method.

1.3   Dynamic positioning system missions

Dynamic positioning begins with a free-floating vessel in an offshore marine environment. The vessel is there to accomplish a given task or mission for which the owner and operator of the vessel is receiving payment.

The missions include drilling for oil, diving support, fire fighting, mining, pipelaying, etc. In each case the manner in which the vessel moves and orients itself to accomplish its designated task is somewhat different.

In the process of performing its designated task the vessel is acted on by environmental elements including wind, waves, and current. Additionally, there may be other forces and moments acting on the vessel because of the task the vessel is performing. For example, a pipelaying vessel must maintain proper tension in the pipe to prevent bending the pipe.

One of the earliest tasks for a DP system was drilling for core samples. In coring, the vessel positions itself over one point on the ocean floor while a shallow hole is drilled to obtain core samples. During the time that the hole is being drilled, the vessel can safely move a given horizontal distance away from the hole in any direction. In other words, the vessel must maintain station within a circle about the hole that is being drilled. This circle is sometimes referred to as a watch circle.

The maximum permissible radius of the watch circle is generally specified in percent of water depth. This measure is used because the drill string can be tilted to an angular limit before drilling must be stopped as Fig. 1–6 shows. The angular limit does not depend on water

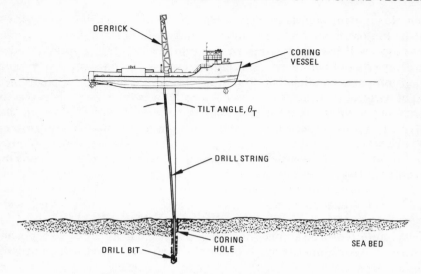

Figure 1-6. The Coring Vessel Task

depth and can be easily translated into the horizontal measure, i.e., percent of water depth, which is likewise not dependent on water depth. The conversion from angular tilt to percent of water depth is:

$$\text{Percent water depth} = 100\% \tan \theta_T \qquad (1.1)$$

Generally, the coring vessel is expected to perform its task during and up to certain environmental limits. These limits can include a specification of maximum mean wind speed, wave height, and current speed. With such a specification, the vessel is expected to maintain itself within the specified watch circle up to the specified environmental limits. When the environmental conditions are less than the specified limits, the vessel usually remains within a much smaller watch circle.

In the coring application the heading of the vessel is not critical. This can be used to advantage. Since most coring vessels use ship-shape hulls, there is a preferential heading to the environment. This preferred heading will not only reduce the horsepower required to hold the vessel in the specified watch circle, but also reduce wave-induced motion of the drilling platform and provide for improved crew comfort.

Once the coring vessel completes the hole or disconnects from the hole, the watch circle radius becomes nearly arbitrary so far as the basic task or mission is involved. However, an environmental specification may still exist which applies to defining the conditions in which the vessel will remain in the area of the hole waiting to re-establish the drilling process before the vessel can break off and head for milder environmental conditions or shelter.

The task of drilling for oil is not radically different than coring in terms of the vessel remaining within a given watch circle around the wellhead up to a specified environmental condition as shown in Fig. 1-7. However, the oil drilling task introduces unusually strict requirements for reliable position keeping for many consecutive days. The reason for the reliability of position keeping is because disconnection from the subsea equipment and reconnection is time consuming, and if not properly executed, costly damage to the drilling equipment can occur.

The time required to drill the well can reach 150 days. This places additional requirements on the dynamic positioning system which will be discussed later.

Although disconnection from the subsea equipment is undesirable during severe environmental conditions, the drilling operation must be suspended and the marine riser disconnected from the subsea

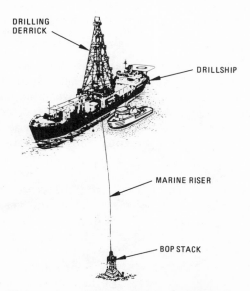

DRILLING DERRICK

DRILLSHIP

MARINE RISER

BOP STACK

Figure 1-7. The Drillship Application

equipment. In order to resume the drilling operation, re-entry into the well must be performed. To perform a re-entry, the wellhead and subsea equipment left behind during the disconnection must be relocated. Then the drilling vessel must be repositioned over the subsea equipment so that the marine riser can be reconnected.[4]

The process of re-entry places additional positioning requirements on the dynamic positioning system. First, the dynamic positioning system must be capable of moving the vessel to relocate the wellhead. Once back over the wellhead, the dynamic positioning system must be capable of making very fine positioning adjustments to assist in the reentry process.

Another category of tasks for offshore vessels covers those in support of some other offshore operation. These tasks include the support of diving operations and supply of other offshore vessels or sites. In the latter case the supply vessel must position itself sufficiently close to the point being supplied for the easy transfer of goods as illustrated in Fig. 1–8.

If the point being supplied is fixed, the position requirements are simpler than if the point being supplied is moving. When the point being supplied is moving, the relative position of the supply vessel to point being supplied must be measured with sufficient accuracy to

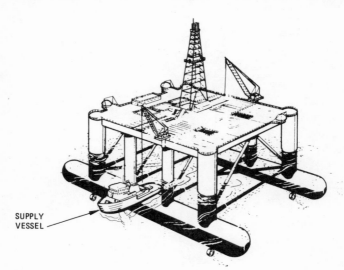

SUPPLY
VESSEL

Figure 1-8. The Supply Application

ensure that the supply vessel can be safely positioned near the vessel being supplied. Additionally, the dynamic positioning system must be sufficiently responsive to hold the supply vessel apart from the vessel being supplied by a safe distance.[11]

In the diving support role (Fig. 1–9) the dynamically positioned vessel must be capable of holding position to the accuracy required by the diving operation. In some cases, such as during deployment and recovery, the accuracy of positioning is significantly more demanding than during the times when the diving operation is in full progress.

However, even during the less accurate operational periods, the vessel must be ready for immediate reaction to support an event requiring quick responses. Likewise, the dynamic positioning system must be able to stay sufficiently close to the diving operation so that the diving operation can be monitored from the surface and solid communications can be maintained.

Always a danger in an oil or gas drilling or producing process is a blowout which results in uncontrolled fire on the structure over the wellhead. To assist in the control of such fires, special fire-fighting vessels (Fig. 1–10) are required which can quickly reach the site of fire, and then concentrate large volumes of water on the structure to cool it until the flow feeding the fire is controlled.

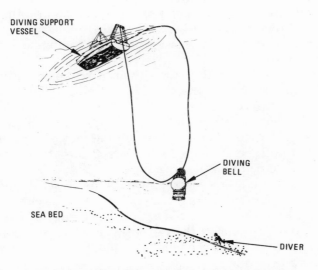

Figure 1-9. The Diving Support Application

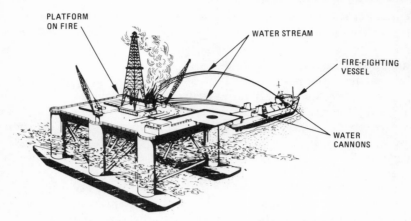

Figure 1-10. The Fire-Fighting Application

The positioning requirements for the fire-fighting vessel are such that the vessel can be held a reasonable distance from the burning structure so that the water cannons on the vessel can be accurately trained on the fire.

Generally the fire-fighting strategy calls for the vessel to be positioned upwind from the burning structure to avoid the smoke and flames of the fire. In positioning upwind from the fire, the dynamic positioning system must be either sufficiently reliable or have fallback modes to avert the obvious disaster of a positioning-holding failure upwind from the fire.

Likewise, since fires will occur without any consideration for the environmental conditions, the fire-fighting vessel must be able to perform in very heavy weather.

With such heavy weather requirements the vessel motions might restrict the performance of the fire-fighting process. However, the water cannons will be gimballed with servocontrol to maintain their stream on a fixed point on the structure.

Another class of position control involves the vessel moving along a given track. The specific tasks being performed in the class of position control are pipeline support, cable laying, pipelaying, pipe burial, and mining. For pipeline support the vessel tracks a subsea pipeline, some object attached to or inside the pipe, or a submersible following a pipeline, as illustrated in Fig. 1-11.

In the cable laying and pipe laying application shown in Fig. 1-12

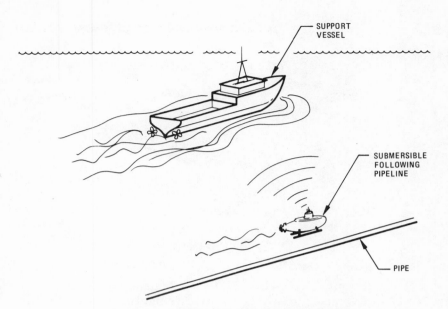

Figure 1-11. The Pipeline Support Application

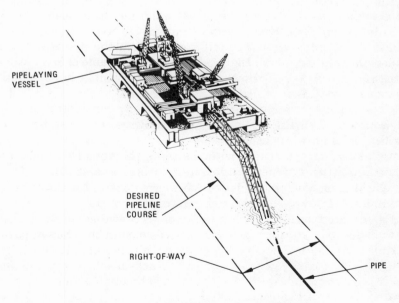

Figure 1-12. The Pipelaying Application

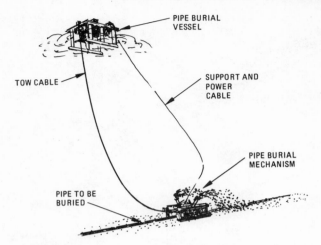

Figure 1-13. The Pipe Burial Application

the vessel moves along a preplanned course laying a cable or pipe behind the vessel. The positioning requirements on the dynamic positioning system are twofold. First, the vessel must move along a designated right-of-way. Second, the vessel must maintain station so that the cable or pipe is not damaged during the process of laying.

Pipe burial differs from the laying process in that the pipe is already laid on the sea bed. Then, the pipe burial vessel moves along over the pipe (Fig. 1–13) to pull the burial mechanism along the pipe as it buries the pipe. The positioning requirements of the burial vessel are based on the burial mechanism requirements.

The last of the track-following position-control missions is mining. For this mission the surface vessel which is dynamically positioned moves along a given course with a mining mechanism below on the bottom following a predetermined pattern as illustrated in Fig. 1–14. The pattern is designed to move the mining mechanism so that the subsea area is thoroughly covered to recover as much bottom material in a given area as possible. If the positioning of the surface vessel is not sufficiently accurate, the mining process will not be as optimum as possible.

1.4   The disturbance elements

An offshore vessel at sea is acted on by disturbance forces which will cause the vessel to move and change its heading. The most common

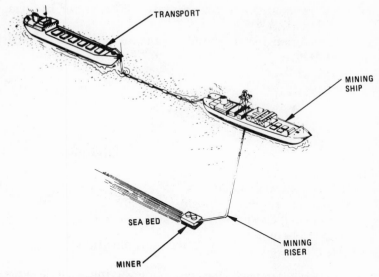

TRANSPORT

MINING
SHIP

SEA BED

MINING
RISER

MINER

Figure 1-14. The Mining Application

environmental disturbances acting on the vessel are the wind, the waves, and the sea current. Each has its own characteristics and will be discussed briefly in this section.

The extent of the discussion is to develop the descriptions of the disturbances which are used in the design and testing of dynamic positioning systems. The reader interested in further details is referred to the comprehensive articles and texts listed at the end of this chapter.

In addition to the environmental disturbances discussed in this section, there are often forces acting on the vessel associated with the tasks the vessel is performing. For example, the pipe-laying vessel has the force acting on it to properly shape the pipe as it is launched from the stinger. These task-related forces are not a subject of this section.

The most familiar of the environmental disturbances is the wind. At a given point above the ground the wind can be described by a speed and a direction, both of which are constantly changing. Furthermore, the wind speed and direction can each be described by an average value or steady component plus a time-varying component. The time-varying component is random in nature. For purposes of analysis and convenience the randomness of the wind can be described in terms of power spectral densities.[18][19][20]

One such description for the wind speed was proposed in 1961 by Davenport and is:[14]

$$\frac{fS(f)}{K\overline{v}^2} = 4\left\{\frac{x^2}{(1 + x^2)^{4/3}}\right\} \tag{1.2}$$

where:

$$x = \frac{fL}{\overline{v}}$$

f = frequency, hertz
L = reference length = 4,000 ft
$\overline{v}$ = the mean hourly wind velocity at height Z, ft/second
K = the drag coefficient for the surface (referred to the mean velocity at height Z)
Z = the standard reference height = 10 m

The Davenport spectrum for increasing mean wind speed is illustrated in Fig. 1–15. Another wind speed spectrum is given in Reference 24.

The variance of any random variable can be determined from its spectral density by integrating the spectral density function over the entire range of frequencies.[19] In the case of the wind speed, the variance can be determined from the Davenport spectrum as follows:

$$\sigma^2(v_A) = \int_0^\infty S(f)df \tag{1.3}$$

$$= 6K\overline{v}^2$$

where $\sigma^2(v_A)$ is the variance of the wind speed. The drag coefficient, K, used in the spectral density and the equation for the variance varies as a function of the surface over which the wind is moving. The value given by Davenport for open, unobstructed country is 0.005.

The wind direction can also be characterized by a random variation about a mean value, ψ_A. However, to date the author has not found a published spectral density for wind direction.

As the wind moves over the surface of the ocean, the wind interacts with the water surface to form waves. The higher the wind speed, the longer the wind blows (duration); and the larger the area over which the wind blows (fetch), the larger the waves.[17] The waves formed by the wind, which is generally referred to as the sea state, are described

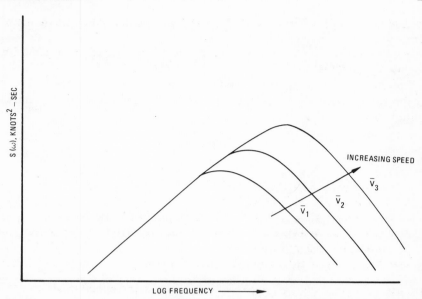

Figure 1-15. The Davenport Wind Speed Spectrum for Various Mean
Wind Speeds

by numerical descriptors which can be correlated to the observed characteristics of the waves (see Appendix D).

They also can be described mathematically by a higher frequency
random spectrum plus a lower frequency random spectrum. One higher
frequency random spectrum which has widespread usage is given as
follows:[15]

$$S_w(\omega) = \frac{A}{\omega^5} \exp\left[-\frac{B}{\omega^4}\right] \tag{1.4}$$

where ω equals radian frequency. The values of the coefficients, A and
B, are different for two formulations of wave spectrum. The first, referred to as the Bretschneider spectrum, has the following coefficient
values:[15]

$$A = 4200\,\frac{H_s^2}{\widetilde{T}_s^4}$$

$$B = \frac{1050}{\widetilde{T}_s^4} \tag{1.5}$$

where:

H_s = significant wave height
$\widetilde{T}_s$ = average period of the significant waves

The second is the International Ship Structures Congress' modification to the Bretschneider spectrum and has the following coefficient values:[15]

$$A = 2760\, \frac{H_s^{\,2}}{\widetilde{T}_s^{\,4}}$$

$$B = \frac{690}{\widetilde{T}_s^{\,4}} \tag{1.6}$$

In Fig. 1–16 the spectrum given in Equation 1.4 is shown for three significant wave heights and average periods. Significantly, as Fig. 1–16 shows, as the significant wave height and average period increase, the peak of the spectral density increases and moves to lower frequencies. This effect is significant to the design of the controller of the dynamic positioning system.

The significant wave height, H_s, is defined as the average of the one-third highest waves. If the sea is fully developed, ("fully developed" is defined as the state at which there is an energy balance between the

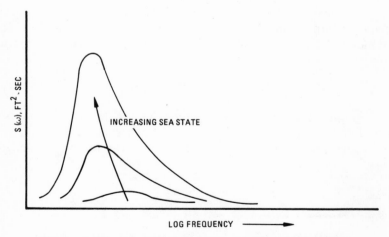

Figure 1-16. The High-Frequency Wave Spectrum

waves and the wind generating the waves) then the significant wave height is functionally related to the mean wind speed as follows:[15]

$$H_s, \text{ft} = 0.025 \; \bar{v}_A{}^2 \tag{1.7}$$

where:

$$\bar{v}_A = \text{the average wind speed in knots}$$

Similarly, for a fully-developed sea the average wave period is related to the average wind speed as follows:[15]

$$\widetilde{T}_s, \text{seconds} = 0.64 \; \bar{v}_A \tag{1.8}$$

To be fully developed, the wind must blow over the surface of the ocean for a given length of time over a certain area.[17] The length of time is referred to as the duration and the area is referred to as the fetch. The manner in which the wave spectrum changes as the sea develops is important in the discussion of the controller of the dynamic positioning system. In Fig. 1–17 the manner in which the spectrum changes as the sea develops and decays is shown.

The spectrum of wave energy is not only a function of frequency but also a function of the direction of propagation.[17] The effect of the direction of propagation is that as the angular deviation from the dominant direction of the travel of wave energy increased, the wave spectrum as a function of frequency decreases. The functional decrease has

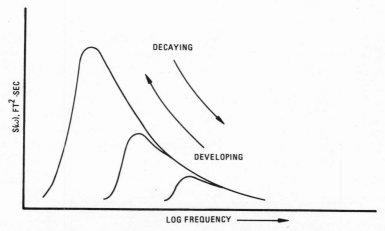

Figure 1-17. The Wave Spectrum of a Developing and Decaying Sea

been formulated as a cosine square function, but there is no generally agreed upon angular function at this time.

The high-frequency wave spectrum represents the orbital or oscillatory motion of water particles. As these waves strike an offshore vessel, they interact with the vessel structure, causing the vessel to slowly drift. The prediction of the force which causes the vessel to drift has been thoroughly studied and the study is not yet complete because the results to data have only met with limited success.

Furthermore, the mathematical techniques used to predict the wave drift forces are quite complex.[22][23][24]

To date, in the sea states experienced by most operating dynamic positioning systems, there is no accurate information available regarding the magnitude of the wave drift forces. At the same time, the environmental loading information available from operating dynamic positioning systems indicates that the magnitude of wave drift force is not generally very large in comparison to the wind and current forces. However, there is no doubt that such forces exist.

The final environmental element which acts on the dynamically positioned vessel is the natural movement of the water over the sea bed, commonly referred to as sea current. Currents can be classed as either slope, wind-driven, or thermohaline.[16]

In the slope class, tidal effects caused by the motion of the moon around the earth and by storm surges along a coastline are common sources of this current. Outflows from rivers can, also, produce current out into the ocean.

The wind-driven sea current is, of course, caused by an interaction between the wind and the sea. The direction of wind-driven currents, however, is not in the same direction as the wind, but is tens of degrees from the mean wind direction because of the rotation of the earth.[16]

The final class of current has as its source the differences in heating and cooling of the surface of the ocean and the salinity changes at the surface caused by evaporation, ice formation and melting, and precipitation. Generally speaking, these currents can be characterized by magnitudes of less than a few knots which vary over periods of hours and days. The slowly varying aspect of the current is significant in the design of the controller of the dynamic positioning system.

Forces resulting from the environmental disturbances are a function of the drag characteristics of the vessel upon which the disturbances are acting. The drag characteristics are a function of the physical characteristics of the vessel and are discussed in Chapter 8 on dynamic models and simulation techniques.

1.5   Basic system elements

The system which is used to solve the dynamic positioning problem includes, as shown in Fig. 1–1, the following basic elements:

1. Sensor system
2. Controller
3. Thruster system
4. Power system

The function of the sensor system is to measure the required information with sufficient speed and accuracy to enable the controller to calculate the thruster commands needed to counter the environmental elements. Since the purpose of the DP system is to control position and heading, the sensor system must include a means of determining position and heading.

In the case of position, several systems are available which have sufficient speed and accuracy. For example, there are several position reference systems which use the transmission of acoustic signals to determine position. This accuracy is at least as good as 1% of water depth. More discussion of the position reference sensors is given in Chapters 2 and 3.

A unanimous choice for the heading sensor is the gyrocompass. With its long life and proven marine experience it has the perfect attributes for a DP system of an offshore vessel in a drilling operation.

Another sensor which has become as common to a DP system as the position sensor is the wind sensor. The purpose of the wind sensor is to furnish wind speed and direction measurements for the controller to calculate thruster commands in a classical feed-forward manner as indicated in Fig. 1–1.

The controller for a DP system can be either an analog or digital computer. The latter has found more widespread acceptance because of its high degree of flexibility and reliability. The main functions of the controller are to:

1. Process the sensor information to obtain actual position and heading
2. Compare the actual position and heading to the commanded values to generate error signals

3. Calculate force and moment commands in the three control axes (two for position and one for heading) to reduce the errors in the average sense to zero

4. Calculate the counter wind forces and moment to give the desired feed-forward anticipation of wind variations

5. Sum the feed-forward wind forces and moment and error-generated forces and moment to form the total force and moment command

6. Distribute the force and moment commands among the active thrusters in a logical manner

7. Convert thrust commands to thruster command variables for each active thruster, i.e., rpm, pitch, etc.

All of these functions must be performed once or twice per second. As a result, the computer has to have a high-speed capability.

Beyond furnishing the proper thruster commands to counter the environment, the controller must perform the following important functions:

• Compensate for inherent DP system delays which produce unstable closed-loop action (stability compensation)
• Eliminate corrupting sensor signals which tend to unnecessarily exercise the thruster (thruster modulation)

The system delays include computational lags, vessel inertia, thruster lags, and sensor lags. Included in the corrupting sensor signals are wave-induced motions and electronic noise.

Combining stability compensation and thruster modulation with the necessary response time of the DP system to counter the environmental elements constitutes a critical design tradeoff for the controller. The tradeoff involves obtaining the "best" response time with adequate stability margins to compensate for system uncertainties and nonlinearities within a guideline of a tolerable amount of thruster modulation.

This design tradeoff must be a part of the production process which leads to the final delivery of the DP system and must involve sufficient verification of predicted results prior to delivery. Otherwise the system stands little chance of achieving satisfactory performance when installed, and considerable time may be required after installation for "tuning" the system. In fact, the performance of the system may never reach its desired goals with the on-board design procedure.

The next element of the DP system is the thrusters, of which there are many types and manufacturers. The thrusters must provide the basic function of furnishing counterforces and moment against the environmental elements to hold the vessel within the specified operating watch circle. The choice of thrusters is a tradeoff involving many factors, some of which are based on experience with a particular manufacturer. In short, there is not a clearly superior thruster for the dynamic positioning application. In a later discussion some practical aspects of thruster selection are covered.

There is a tendency to overlook an important supporting system to the DP system. The supporting system is the power system, as indicated in Fig. 1–1. In fact, the power system can enter into the choice of the thruster system. The power system can influence the choice of the thrusters through the prime movers of the thrusters.

For example, if a power system is to be configured with a maximum ac capability, then the thrusters will be driven by ac motors, which means that the thrusters will have a controllable pitch with a constant speed of rotation. However, if a power system is to be configured with a maximum dc capability to give a high degree of commonality between the drilling and the thruster system, then the thrusters will be driven by dc motors, which means that the thrusters will be fixed-pitch with controllable speed.

1.6   Control system nomenclature

A dynamic positioning system is a multiloop feedback control system and in this book a certain amount of discussion is included regarding the DP system as a control system. As a result, at this early point, introductory information is given to assist the noncontrol system reader through the first chapters until more mathematical details are required and given.

In the following discussions of the dynamic positioning system many explanations will be supported by block diagrams. The basic elements of block diagrams are shown in Fig. 1–18. The primary element is the block itself. The output, Y, is functionally related to the inputs to the block. The relationships can take many forms. Generally, in control system, however, the relationships are mathematical and are referred to as transfer functions. The transfer functions can be linear, nonlinear, a logical relation, etc.

As a general practice, the name of the system element or the trans-

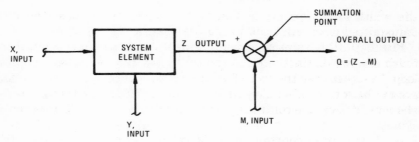

Figure 1-18. Block Diagram Elements

fer function is written in the block diagram. Inputs and outputs to the block are identified by arrows and are labelled by a symbol as demonstrated in Fig. 1–18.

A second element of a block diagram is summation point. The output of the summation point is simply the sum of the inputs to the summation point, using the sign given in the diagram.

To illustrate a complete block diagram, a simplified block diagram of a DP system is shown in Fig. 1–19. The block diagram includes the basic elements shown in Fig. 1–18 except the arrows interconnecting the blocks have width as opposed to being a line. The significance of

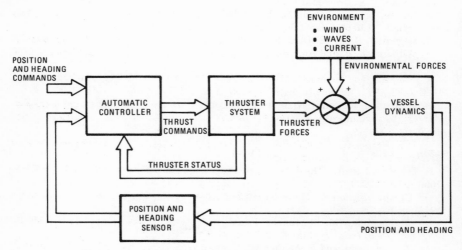

Figure 1-19. A Block Diagram of a DP System

the width to the arrows is that there are multiple quantities being transmitted from one block to the next.

Fig. 1–19 illustrates several other aspects of control systems. First, there are two distinct loops to the control system, an inner and outer loop. In either case the loops feed information from one element of the system back to another element which influences the output of the first element. Hence, the control system is referred to as a feedback control system.

Another type of control system is an open-loop control system (Fig. 1–19). If the thruster system outputs have no influence on the thruster commands, then the thruster control system is called open-loop. The name obviously refers to the absence of feedback. Later the benefits of feedback or closed-loop control over open-loop control are discussed.

Block diagrams can become extremely complex, especially in highly coupled processes. In fact, the complexity can easily make understanding and analysis of the system difficult, if not impossible. To overcome this difficulty there are rules of manipulation of block diagram where simplifications can be achieved. Reference 12 gives further details.

REFERENCES

1. John J. D'Azzo and Constantine H. Houpis, *Feedback Control System Analysis and Synthesis,* McGraw-Hill Book Company, New York, 1966.
2. Shatto, H. L. and Dozier, J. R., "A Dynamic Stationing System for Floating Vessels," Shell Oil Report dated September 1961, revised in September 1962.
3. Hammett, D. S., "SEDCO 445 – Dynamic Stationed Drill Ship," OTC 1626, 1972.
4. Williford, F. B. and Anderson, A, "Dynamic Stationed-Drilling SEDCO 445," OTC 1882, 1973.
5. Sjouke, J. and Lagers, G., "Development of Dynamic Positioning for IHC Drillship," OTC 1498, 1971, Volume II, page 797–805.
6. Graham, J. R., Jones, K. M., G. D. Knorr, and Dixon, T. F., "Design and Construction of the Dynamically Positioned *Glomar Challenger,*" Marine Technology, April 1970, pp. 159–179.
7. D. S. Hammett, "SEDCO 709 – First dynamically stationed semi-submersible, Part 5: Sea Trials," *Ocean Industry,* April, 1977.
8. D. S. Hammett, "The First Dynamically Stationed Semi-Submersible – SEDCO 709," OTC 2972, 1977.
9. *"Glomar Explorer's* Many Technical Innovations," *Ocean Industry.* December, 1976, pp. 67–73.
10. J. Cranfield, "Mediterranean Trials Prove Feasibility of Pipelaying in 2000 Ft. Waters," *Ocean Industry,* March 1977, pp. 28.
11. J. S. Sargent and P. N. Cowgill, "Design Considerations for Dynamically Positioned Utility Vessels," OTC 2633, 1976.

12. K. Ogata, *Modern Control Engineering,* Prentice-Hall, Englewood Cliffs, NJ, 1970.

13. J. S. Sargent and J. J. Eldred, "Adaptive Control of Thruster Modulation for a Dynamically Positioned Drillship," OTC 2036, 1974.

14. A. G. Davenport, "The Spectrum of Horizontal Gustiness Near the Ground in High Wind," Quart, J. R. Met. Soc., Volume 87, April 1961, pp 194–211.

15. Walter H. Michel, "Sea Spectra Simplified," Marine Technology, January 1968, pp 17–30.

16. Gerhard Neumann and Willard J. Pierson, Jr., *Principles of Physical Oceanography,* Prentice-Hall, Inc., New Jersey, 1966.

17. B. V. Krovin-Kroukovsky, *Theory of Sea Keeping,* Soc. of Naval Architects and Marine Engineers, New York, 1961.

18. R. B. Blackman and J. W. Tukey, *The Measurement of Power Spectra,* Dover Publications, 1958, New York.

19. W. B. Davenport, Jr. and W. L. Root, *An Introduction to the Theory of Random Signals and Noise,* McGraw-Hill, 1958.

20. Wilbur Marks, *The Application of Spectral Analysis and Statistics to Sea Keeping,* Soc. of Naval Architects and Marine Engineers, 1963, New York.

21. C. H. Kim and F. Chou, "Prediction of Drifting Force and Moment on an Ocean Platform Floating in Oblique Waves," International Shipbuilding Progress, October, 1973, pp 388–401.

22. Verhagen, J. H. G., "The Drifting Force on a Floating Body in Irregular Waves," Office of Naval Research 8th Symposium on Naval Hydrodynamics, August, 1970, pp 955–979.

23. Hajime Maruo, "The Drift of a Body Floating in Waves," Journal of Ship Research, December 1960, pp 1–10.

24. I. Van der Hoven, "Power Spectrum of Horizontal Wind Speed in the Frequency Range from 0.0007 to 900 cycles per Hour," Journal of Meteorology, Volume 14, April 1957, pp 160–164.

2

Acoustic Position Reference Sensors

2.1 Introduction

A dynamic positioning system is composed of several basic elements, each important to the success of the overall system. The next two chapters consider the first element of the DP system, the sensors. Quite naturally, the task or tasks that the particular vessel is designed to perform bears heavily on the sensors to be used with the DP system.

The purpose of the sensors is to gather the required information with sufficient speed and accuracy for the controller to calculate the thruster commands so that the vessel performs the desired task. The kinds of information the controller requires include vessel position, vessel heading, and wind speed and direction. In addition, certain offshore tasks may require other sensors, such as vessel motion and environmental sensors.

Position reference is not a new problem to the offshore environment. Since man has put to sea in ships, there has been a need to know where he was and which way he was going. However, the speed and accuracy with which the position was determined was not critical on the open sea. Even today the determination of position reference for nonmilitary purposes of navigation between two widely separated points does not require the accuracy that certain offshore dynamic positioning tasks do.

The offshore position reference needs to include not only navigation but also an accurate, repeatable local position reference. The former is to locate the vessel at the correct location on the earth to perform its assigned task. The local position reference is primarily for thruster control. In most cases the accuracy and speed of measurement for the local position reference system is higher than the navigation system.

However, if a navigation system is capable of doing both navigation and local position measurements, a much simpler system can be achieved.

The remainder of this chapter is devoted to acoustic position reference systems. They are strictly a local position reference system restricted to a small coverage area. Therefore, they require a separate navigation system to locate the vessel geographically. However, they are the primary position reference system now in use in dynamic positioning systems.

There are many types of acoustic position reference systems. For example, there are short baseline, long baseline, time-of-arrival, phase comparison, pinger, and transponder acoustic systems. However, they all depend on the propagation of an acoustic signal from one point to another through the water. Therefore, the propagation characteristics of acoustic energy in water influence, to a great extent, the performance of the acoustic system.

An acoustic position reference system is composed of a subsea acoustic unit, a vessel-mounted acoustic unit, and a signal processor/position

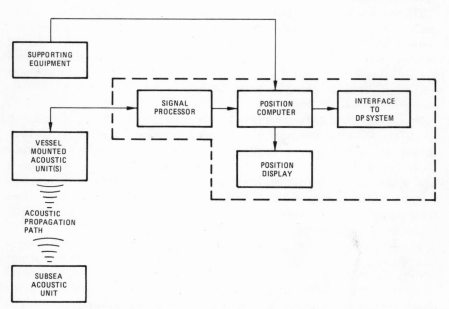

Figure 2-1. Basic Acoustic Position Reference System

computer unit which is also vessel mounted. These units are interconnected acoustically and electrically as indicated in Fig. 2–1. For each
type of acoustic position reference system, the three units are somewhat different in the function they perform and their complexity.

Before discussing the various types of acoustic systems, it is easier
to understand the pros and cons of each system if the properties of
sound propagation through water are first discussed. Basic to certain
acoustic position reference systems is the speed of propagation of sound
in water.

2.2   Acoustic propagation

Acoustic systems operate by projecting acoustic energy into the fluid
medium. In the simplest system the acoustic energy travels only from
a subsea beacon to receivers on the vessel. In the more complicated
acoustic systems the acoustic energy is transmitted from the vessel
to a subsea transponder, then the transponder transmits an acoustic
pulse back to the vessel. In either case acoustic energy is projected into
the fluid medium for propagation toward the receiver portion of the
acoustic system.

Not unlike any transmission phenomena, the medium through
which the signal is propagated affects the transmitted signal. First, the
acoustic signal experiences a transmission loss or reduction in intensity as it travels out from the projector because of the spreading
effect. Spreading is best understood by considering the very simple example of a small sound source in a lossless, homogenous, infinite medium. As sound is radiated from the source, the sound travels in all
directions, equally distributed over the surface of a sphere. With the
assumption of a lossless medium the power in the signal remains constant.

However, the area over which the power is spread is constantly increasing as the square of the distance from the source. As a result, the
intensity of the sound must decrease with increasing distance from the
source as demonstrated by the following expression:

$$P = 4 \pi R_1^2 I_1 = 4 \pi R_2^2 I_2 = \ldots \qquad (2.1)$$

where:

        $P =$ the acoustic power emitted by the source
        $I =$ the signal intensity
        $R =$ distance from the source

The rate of decrease of signal intensity with distance for the ideal case is square-law. Another commonly-used technique of expressing the decrease of signal intensity or increase in transmission loss is to use a logarithmic ratio of intensities with a reference intensity at a reference distance of one yard. In equation form, the expression for transmission loss becomes:

$$\text{Transmission loss, } \delta_{TL}, = 10 \log \frac{I_1}{I_2} = -10 \log R_2{}^2 = -20 \log R_2$$

$$(2.2)$$

where δ_{TL} has units of decibels, dB, and R_2 is expressed in yards.

A second transmission loss that an acoustic signal experiences in water is attenuation. Attenuation can be caused by absorption, scattering, or interference. Absorption is caused by some acoustic energy being converted into heat as the energy is propagated through the medium. Scattering is a loss resulting from the signal being diverted from its original path by something in the medium, such as fish, sea weed, air bubbles, etc. Interference can be either reinforcing or canceling and is caused by the acoustic signals reflecting from a surface and interacting with the nonreflected signal.

Each of these effects has received extensive coverage (see References 4 and 5). Therefore, only the results of these effects are given here and the interested reader is referred to other sources for greater detail.

In the case of absorption the transmission loss is commonly expressed through a logarithmic absorption coefficient, α, which relates to signal intensity and range as follows:[4]

$$\alpha, \frac{dB}{\text{Kiloyard}} = \frac{10 \log I_1 - 10 \log I_2}{R_2 - R_1} \qquad (2.3)$$

The absorption coefficient for sea water is greater than for distilled water and varies as a function of frequency and temperature. The variations are illustrated in Fig. 2–2 for sea water with a salinity of 35 parts per thousand. Absorption also varies with water depth. The variation with depth is a decrease of about 2% for an increase of every 1,000 ft in depth.

The combined transmission loss of spherical spreading and absorption can be expressed approximately as:[4]

$$\text{Transmission loss, decibels} = -\left(20 \log R + \frac{\alpha R}{1000}\right) \qquad (2.4)$$

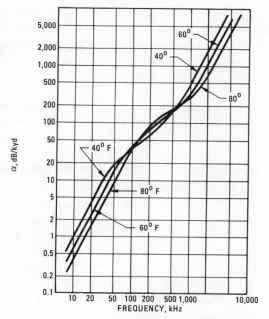

Figure 2-2. The Absorption Coefficient in Sea Water as a Function of
Frequency [5]

where R is in yards. Using Equation 2.4, the transmission loss for an
acoustic signal of 20 kilohertz in 60°F sea water is 62.5 decibels at
1,000 yards (3,000 ft).

Attenuation of an acoustic signal caused by scattering and interference depends on many factors which are totally variable from one
situation to the next. In the case of scattering, the amount of matter in
the water which makes the mass of water inhomogenous must be taken
into account. The kinds of foreign matter are numerous and include
everything from tiny particles of dust and marine organisms to geo-
logical formations rising from the sea bottom.

Because of the unpredictability of the kinds and concentration of
scatterers, a transmission loss caused by scattering is not readily
formulated as it is for spherical divergence and absorption.

One form of inhomogenity in water is gas bubbles. When the gas is
dissolved in the water, the gas has no effect on the transmission characteristics of the acoustic signal. However, if the gas is in the form of

bubbles, the effect can be very significant both in terms of decreasing the local velocity of propagation and in increasing transmission attenuation. The effect of the velocity variation influences the accuracy of the acoustic system and the transmission attenuation influences the maximum range of the system or the amount of power required to reach a certain depth.

In either case the resonant frequencies of the bubbles with respect to the frequency of the transmitted signal are important to the degree of influence the bubbles will have on the acoustic signal. Air is the most common gas or mixture of gases in the water surrounding a dynamically-positioned vessel. The resonant frequency of an air bubble is dependent on its size and its depth as shown in the following expression:

$$f_r, \text{hertz} = \frac{256{,}700}{d_a} \sqrt{1 + 0.03\,h} \qquad (2.5)$$

where:

d_a = the diameter of the air bubble, mils
h = the depth of the bubble, ft

For bubble sizes much smaller than the resonant size, the velocity of propagation is given as follows for β much less than one[5]

$$v = v_w \left[\frac{1}{1 + 2.5 \times 10^4 \beta} \right]^{1/2} \qquad (2.6)$$

where:

β = the fraction of air by volume that the water contains
v_w = the velocity of sound in air-free water

An example value of velocity reduction resulting from air bubbles as predicted by Equation 2.6 for bubble/water mixture of 0.01% (β = 0.0001) is 47%.

At frequencies much higher than the resonant frequency of the smallest bubble in the volume of water, the air has insignificant effect. In the region of resonance the sound velocity will change sharply between the value predicted by Equation 2.6 and the value for water without air.

Attenuation of the acoustic signal by air bubbles in the water can be expressed in a compact form:

$$\text{Attenuation, dB/yard} = 4.34\,\bar{n}\,\sigma_e \qquad (2.7)$$

where:

$\bar{n}$ = the average number of resonant bubbles per cubic yard of water
σ_e = the extinction coefficient for the resonant bubble

The extinction coefficient is a function of a damping constant of the bubbles, the resonant frequency and diameter of the resonant bubbles, the velocity of sound in air-free water, and the frequency under consideration. References 4 and 5 give further details.

Likewise, the formulation of a transmission loss caused by interference is impractical for the general case because interference depends on the reflection of acoustic energy. As a result, the geometry of the acoustic projector with respect to the water surface, the bottom, and other reflecting objects must be taken into account. Additionally, the reflective characteristics of the surface which the acoustic signal strikes and the angle of incidence affects the degree of interference.

An important physical property of acoustic energy in the determination of position is its speed of propagation in water. This parameter has been measured by various investigators and an empirical expression has been refined several times. One empirical expression is as follows:[5]

$$c, \text{ft/sec} = 4{,}625 + 7.68 \ (T - 32) - 0.0376 \ (T - 32)^2 + 3.35 \ S + 0.018 \ h \tag{2.8}$$

where:

T = the temperature, °F.
S = the salinity of water, parts per thousand
h = the depth of the water, ft

As the empirical expression given in Equation 2.8 shows, the speed of sound in water varies as a function of temperature, salinity, and depth. As a result the speed of sound varies with geographic location, time-of-day, season, and water depth. A typical velocity profile is shown in Fig. 2–3.

One effect of velocity variations as function of depth, salinity, and temperature is refraction or ray bending. If the acoustic signal travels through distinct layers of water with different sound velocities, the path of the sound obeys Snell's law:

$$k_j = \frac{\cos \theta_i}{c_i} \tag{2.9}$$

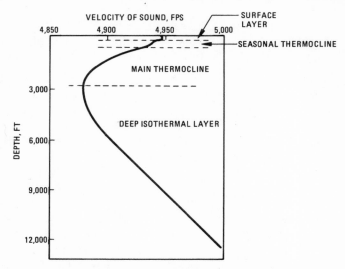

Figure 2-3. A Typical Velocity Profile [5]

where:

θ_i = the grazing angle of the jth ray at the ith layer boundary

k_j = a constant for the jth ray

A refraction according to Snell's law in a layered medium is illustrated in Fig. 2-4a. However, layering does not normally occur. Instead changes occur gradually which result in the path of the acoustic signal being warped into sweeping curvatures.

The effect of ray bending is to degrade the accuracy of the acoustic position reference system. In addition, ray bending can actually warp the path of the acoustic signal to the point that the transmitted acoustic energy will not reach beyond a certain range, creating what is referred to as a shadow zone (Fig. 2-4b). This effect, of course, limits the operating range of the acoustic system to an absolute value.

A property of the medium in which the acoustic energy is being transmitted which influences the reception of the acoustic signal is the ambient noise present in the medium. Ambient noise includes all acoustic sounds at an acoustic receiver which are not the desired signal. The sources of noise which make up ambient noise include surface water motion, rain, ship, man-made, and marine life. In the case of

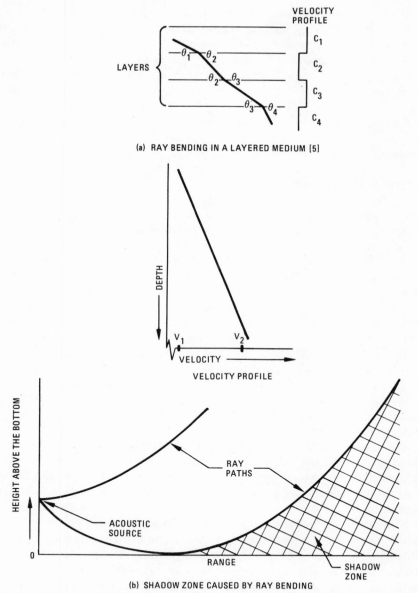

(a) RAY BENDING IN A LAYERED MEDIUM [5]

VELOCITY PROFILE

(b) SHADOW ZONE CAUSED BY RAY BENDING

Figure 2-4. Acoustic Path Refraction Caused by Variation in Sound Velocity (Ray Bending)

surface water motion noise, the exact generating mechanism is not known, but breaking waves are suspected because of a strong correlation between this noise and sea state and wind velocity, making this noise weather dependent.

Significant increases in acoustic noise have been noted as a result of falling rain. The level of noise is apparently a function of rate of rainfall and perhaps the rainfall area.[4] Increases of nearly 30 dB over the 5- to 10-kHz frequency range have been recorded in "heavy" rain.[4] Above 10-kHz the noise spectrum caused by rain drops decrease depending on the rainfall rate.

The noise generated by marine life is restricted to shallow water areas where the sea creatures live. Most ship and man-made noises are restricted to shallow-water areas near a busy port or active offshore area. In both cases there is a large variation in their magnitude.

Another source of ship noise is the noise generated by the ship to which the acoustic devices are attached or vessels alongside the same vessel. For a dynamically-positioned vessel the main source of self-noise generated by the ship is the thrusters. In addition, on-board machinery generates noise which is transmitted through the hull into the water.

The hull also interacts with the sea to generate acoustic noise and carry air into the region surrounding the hull. This air may find its way into the acoustic transmission path from the subsea unit and the vessel mounted unit, causing serious transmission loss.

2.3   Geometrical considerations

In any acoustic position reference system a geometrical pattern or array of either transmitters or receivers is located either on the vessel or the reference frame to which the vessel is dynamically positioning, e.g., the sea bottom. When the array is mounted on the vessel, the dimensions of the array are limited and, thus, the name short-baseline results.

Conversely, when the array is on the bottom, the dimensions of the array are sized by transmission limitations of the acoustic system. However, a pattern covering an area with sides equal to the water depth is easily achievable, making the name "long-baseline" applicable, especially in deep water.

Whether the array is vessel-mounted or bottom-mounted, the geometrical considerations involved in the determination of the location of the vessel within the area of coverage of the acoustic system are similar. In this section the measurement geometry is developed. The development uses certain simplifying assumptions. The purpose of the development is not to give exact solutions as given in Reference 7, but to develop sufficient detail so that the performance of existing acoustic systems can be evaluated.

In Fig. 2–5 the basic geometry for a surface-mounted array is shown. The following simplifying assumptions are made for the geometry shown in Fig. 2–5 over the generalized case:

1. The array elements are in the same plane.

2. The array elements are arranged in a rectangular pattern centered and with sides parallel to the acoustic system coordinate frame, designated as X_A, Y_A, Z_A.

The ranges from the array elements to the reference point in Fig. 2–5 are expressed in equation form as follows:

$$R_1{}^2 = (x_A - d_x)^2 + (y_A - d_y)^2 + z_A{}^2 \qquad (2.10)$$
$$R_2{}^2 = (x_A - d_x)^2 + (y_A + d_y)^2 + z_A{}^2 \qquad (2.11)$$
$$R_3{}^2 = (x_A + d_x)^2 + (y_A + d_y)^2 + z_A{}^2 \qquad (2.12)$$
$$R_4{}^2 = (x_A + d_x)^2 + (y_A - d_y)^2 + z_A{}^2 \qquad (2.13)$$

where:

R_i = the range from the ith array element to the reference point

x_A, y_A, z_A = the coordinates of the reference point

$2d_x$, $2d_y$ = the known length of the sides of the rectangular array

The ranges given in Equations 2.10 through 2.13 are the quantities which can be measured by the acoustic system; whereas the coordinates of the reference point are the quantities of interest. Therefore, Equations 2.10 through 2.13 must be solved for the coordinates of the reference point. The results of such an algebraic exercise are:

$$x_A = \left[\frac{R_3 - R_2}{2d_x}\right]\left[\frac{R_3 + R_2}{2}\right] = \left[\frac{R_4 - R_1}{2d_x}\right]\left[\frac{R_4 + R_1}{2}\right] \qquad (2.14)$$

$$y_A = \left[\frac{R_2 - R_1}{2d_y}\right]\left[\frac{R_1 + R_2}{2}\right] = \left[\frac{R_3 - R_4}{2d_y}\right]\left[\frac{R_3 + R_4}{2}\right] \qquad (2.15)$$

$$z_A = [R_2{}^2 - (x_A - d_x)^2 - (y_A - d_y)^2]^{1/2} \qquad (2.16)$$

Only one solution is given in Equation 2.16 for the vertical coordinate, z_A, when in fact z_A can be determined from any of the Equations

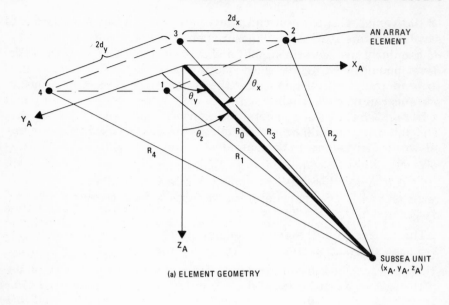

(a) ELEMENT GEOMETRY

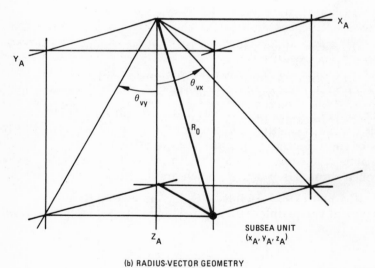

(b) RADIUS-VECTOR GEOMETRY

Figure 2-5. Basic Surface-Mounted Array Geometry

2.10 through 2.13 once x_A and y_A are determined from Equations 2.14 and 2.15.

Similarly, as Equations 2.14 and 2.15 show, x_A and y_A can be determined in two ways. Further examination of Equations 2.14 and 2.15 reveals that from the ranges for any three array elements, x_A and y_A can be determined. Thus, four array elements provide measurement redundancy.

Then, by using Equation 2.16 with a range from one of the array elements being used to determine x_A and y_A, the vertical distance z_A can be computed. Generally, z_A is not used in DP systems, but its computation is included for information purposes.

To compute the coordinates of the reference point using Equations 2.14, 2.15, and 2.16 the ranges are needed, which means that the time required for the acoustic signal to travel from the reference point to the array elements must be measured. Unless the time of departure of the acoustic signal from the reference point is known, the round-trip time needs to be measured so that the departure time as well as the return times at each array element are known. This round-trip time system is referred to as a transponder system and is discussed in a later section.

The resulting coordinates for the reference point from Equations 2.14, 2.15, and 2.16 are determined in the array coordinate system. To transform the resulting coordinates into horizontal frame, coordinate transformations for the angular variations between the horizontal frame and the array-coordinate frame are required. Usually when the array is mounted on the surface vessel, the angular variations between reference frames are the roll, pitch, and yaw of the vessel.

There is a technique of determining position without knowing the time of departure of the acoustic signal from the reference point or the round-trip time. The technique involves forming differences in the ranges. The range difference for use in the x-axis are $(R_3 - R_2)$ and $(R_4 - R_1)$, and in the y-axis are $(R_2 - R_1)$ and $(R_3 - R_4)$.

A typical range difference formed from Equations 2.10 through 2.13 is:

$$(R_3 - R_2) = z_A \left[\left(\frac{x_A + d_x}{z_A} \right)^2 + \left(\frac{y_A + d_y}{z_A} \right)^2 + 1 \right]^{1/2} \tag{2.17}$$

$$- z_A \left[\left(\frac{x_A - d_x}{z_A} \right)^2 + \left(\frac{y_A + d_y}{z_A} \right)^2 + 1 \right]^{1/2}$$

If $(x_A \pm d_x)$ and $(y_A + d_y)$ are much smaller than z_A, then Equation 2.17 can be approximately expressed as follows:

$$R_3 - R_2 = z_A \left[1 + \frac{1}{2}\left(\frac{x_A + d_x}{z_A}\right)^2 + \frac{1}{2}\left(\frac{y_A + d_y}{z_A}\right)^2 \right]$$

$$- z_A \left[1 + \frac{1}{2}\left(\frac{x_A - d_x}{z_A}\right)^2 + \frac{1}{2}\left(\frac{y_A + d_y}{z_A}\right)^2 \right] \quad (2.18)$$

which simplifies to:

$$R_3 - R_2 = \frac{2 x_A d_x}{z_A} \quad (2.19)$$

or

$$\tan \theta_{vx} = \frac{x_A}{z_A} \approx \frac{R_3 - R_2}{2 d_x} \approx \sin \theta_{vx} \quad (2.20)$$

The same result is obtained from Equation 2.14 if $(R_3 + R_2)$ is approximately equal to $2 z_A$.

Equation 2.20 is better understood by referring to Fig. 2–6 where a two-dimensional array is shown. As z_A becomes larger and larger with respect to x_A and d_x, R_3 and R_2 become more nearly parallel. If R_3 and R_2 were parallel, then the ratio of $(R_3 - R_2)$ over $2 d_x$ would be equal to the $\sin \theta_{vx}$. Since z_A is much greater than x_A, the angle θ_{vx} is small which implies that the sine of the angle and the tangent of the angle are approximately equal. In addition, they are both approximately equal to their argument, the angle θ_{vx}, which is exactly what Equation 2.20 implies.

To measure the difference in the two ranges, time of departure is not necessary as is shown in the following expression:

$$R_3 - R_2 = c_3(t_3 - t_0) - c_2(t_2 - t_0)$$
$$= c(t_3 - t_2) \text{ if } c_2 = c_3 \quad (2.21)$$

where:

$c_3 = $ the velocity of sound in the media along the path to array element 3

$c_2 = $ the velocity of sound in the media along the path to array element 2

$t_3 = $ the time-of-arrival at the third array element

$t_2 = $ the time-of-arrival at the second array element

$t_0 = $ the time-of-departure from reference point

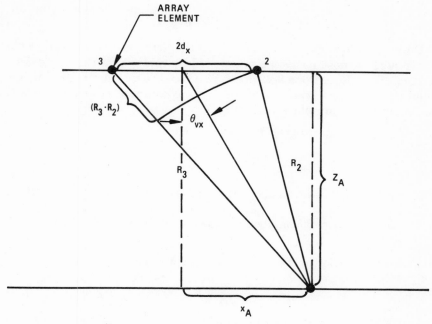

Figure 2-6. Two-Dimensional Array Configuration

Thus, the difference in two ranges can be determined simply by measuring the difference in time of arrival and knowing the velocity of sound along the path. This technique of measuring is discussed more thoroughly in the following section.

Equation 2.21 is true even if the velocity of the sound in the media is not constant along the path to the array elements. The only requirement of the velocity of sound for the two paths is that it is equal and known. Since the array elements for the ship-mounted system are quite near to one another, the velocity of sound to two separate array elements will be nearly equal, especially when the subsea reference point is directly beneath the vessel-mounted array.

If the direction angles, θ_x and θ_y, shown in Fig. 2–5 can be measured and if the vertical array coordinate is known, then the other two array coordinates, x_A and y_A can be expressed as:

$$\frac{x_A}{z_A} = \frac{\cos \theta_x}{(1 - \cos^2 \theta_x - \cos^2 \theta_y)^{1/2}} = \tan \theta_{vx} \qquad (2.22)$$

$$\frac{y_A}{z_A} = \frac{\cos \theta_y}{(1 - \cos^2 \theta_x - \cos^2 \theta_y)^{1/2}} = \tan \theta_{vy} \tag{2.23}$$

Correction for angular variations in the array coordinates from the horizontal is most easily accomplished by computing the angles, θ_{vx} and θ_{vy}, from Equations 2.22 and 2.23. Once the two angles are known, then the angular variations of the array-coordinate system can easily be computed by simple subtraction:

$$\theta_{vx}' = \theta_{vx} - \theta_p \tag{2.24}$$
$$\theta_{vy}' = \theta_{vy} - \theta_r \tag{2.25}$$

where:

$\theta_p =$ the angular rotation about the y-axis of the horizontal coordinate system

$\theta_r =$ the angular rotation about the x-axis of the horizontal coordinate system

Then the position of the vessel can be computed as follows:

$$\frac{x}{h} = \tan \theta_{vx}' \tag{2.26}$$

$$\frac{y}{h} = \tan \theta_{vy}' \tag{2.27}$$

where:

$h =$ the vertical distance from the origin of horizontal plane and the corresponding parallel plane through the beacon

These expressions are in array coordinates. If the array is mounted to a floating vessel, certain coordinate transformations, angular and translational, will be necessary to properly solve the control problem. In the control problem the reference coordinate frame is centered on a subsea reference point at mean stillwater conditions. Here the reference coordinate system is defined as a right-hand coordinate system with the x-axis aligned to north, the y-axis to the east and the z-axis downward.

A second coordinate system is used in the control problem shown in Fig. 2–7. The second system is referenced to the point on the dynamically positioned vessel which is to be aligned over the subsea reference point. The vertical coordinate of the vessel coordinate system, i.e. z_B, is equal to the corresponding reference coordinate, z_R, which is the

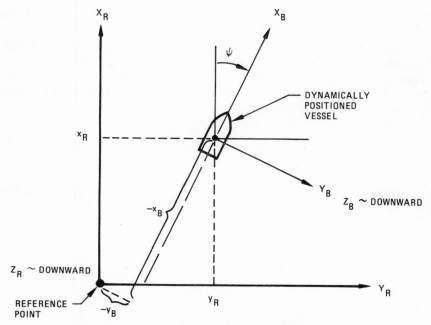

Figure 2-7. The Control Problem Geometry

vertical distance from the subsea reference point and the reference point on the DP vessel.

In this book the x-axis is aligned to the bow of the vessel, the y-axis to starboard, and the z-axis downward. As Fig. 2–7 shows, the rotation between the body coordinate system of the vessel and the reference system becomes the heading of the vessel measured with respect to north.

If the array coordinate system is coincident with the body coordinate system, then the offsets from the reference point measured in the array system can be used to solve the control system. However, rarely will be the case when the array is mounted coincident to the body-coordinate system. Likewise, the vessel will generally not be in the stillwater condition. Instead the roll, pitch, and yaw of the vessel will induce changes in the array-measured coordinates that must be corrected for transforming the array measurements into either body or reference coordinates.

Multiple angular and translational coordinate transformations when encountered in one diagram can be more baffling than enlightening. As a result, in the following the transformations from array measurements to body coordinates will be illustrated in two dimensions before the generalized case is considered. From Fig. 2–8 the measurements made in array coordinates are x_A and z_A. The first coordinate transformation required to translate the measurements into body coordinates is a linear translation equal to the distances the array coordinate system is from the origin of the body coordinate system, i.e.:

$$\begin{bmatrix} x_A' \\ z_A' \end{bmatrix} = \begin{bmatrix} x_A \\ z_A \end{bmatrix} - \begin{bmatrix} -L_{AX} \\ -L_{AZ} \end{bmatrix} \tag{2.28}$$

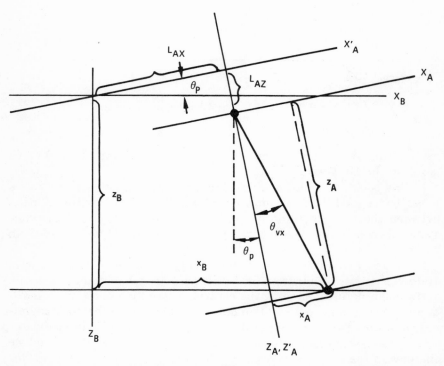

Figure 2-8. Two Dimensional Coordinate Transformation from Array Coordinates to Body Coordinates

where the equation form is according to matrix notation. The final transformation of the array measurements into body coordinates is the angular rotation and can be expressed as follows:

$$\begin{bmatrix} x_B \\ z_B \end{bmatrix} = \begin{bmatrix} \cos\theta_p & \sin\theta_p \\ -\sin\theta_p & \cos\theta_p \end{bmatrix} \begin{bmatrix} x_A + L_{AX} \\ z_A + L_{AZ} \end{bmatrix} \tag{2.29}$$

where θ_p equals the angular rotation of array system with respect to the body system.

Equation 2.29 is derived with the assumption that z_A is measured. If z_A is not measured and, instead, $x_A/z_A = \tan\theta_{yx}$ is measured, and z_B is known by measurement of the water depth, then the horizontal offset in body coordinates can be calculated using the following expression:

$$\begin{aligned} x_B = z_B \tan(\theta_p + \theta_{vx}) + L_{AX}\cos\theta_p + L_{AZ}\sin\theta_p \\ + (L_{AX}\sin\theta_p - L_{AZ}\cos\theta_p)\tan(\theta_p + \theta_{vx}) \end{aligned} \tag{2.30}$$

The quantity $\tan(\theta_p + \theta_{vx})$ is, of course, computed from the measurement of $\tan\theta_{vx}$ and the angle, θ_p. Then, the angle θ_{vx} is computed from the arctangent function, the pitch angle, θ_p, is added to θ_{vx}, and the tangent of the sum is recomputed.

For the more generalized case the transformations include translations in all three coordinates and rotations about all three vessel axes. The manner in which the body coordinate system is selected to align with the bow of the vessel eliminates the consideration of one of the angular rotations, yaw. So only roll and pitch angular transformations are required as shown in Fig. 2–9. If the angular rotations shown in Fig. 2–9 are performed, the following expression for the location of reference point in body coordinates results:

$$\begin{bmatrix} x_B \\ y_B \\ z_B \end{bmatrix} = \begin{bmatrix} \cos\theta_p & 0 & \sin\theta_p \\ 0 & 1 & 0 \\ -\sin\theta_p & 0 & \cos\theta_p \end{bmatrix} \begin{bmatrix} 1 & 0 & 0 \\ 0 & \cos\theta_r & \sin\theta_r \\ 0 & -\sin\theta_r & \cos\theta_r \end{bmatrix} \begin{bmatrix} x_A + L_{AX} \\ y_A + L_{AY} \\ z_A + L_{AZ} \end{bmatrix} \tag{2.31}$$

where θ_r is the roll angle and θ_p is the pitch angle.

If in Equation 2.33 z_A is not measured and, instead $x_A/z_A = \tan\theta_{vx}$ and $y_Z/z_A = \tan\theta_{vy}$ are measured and z_B is known by the measurement of the water depth, then the two horizontal offsets in body coordinates can be computed as follows:

$$x_B = (z_B - R)\left[\frac{\tan\theta_p\cos\theta_r + \tan\theta_{vx} - \sin\theta_r\tan\theta_{vy}}{\cos\theta_r - \tan\theta_p\tan\theta_{vx} - \sin\theta_r\tan\theta_{vy}}\right] + Q \tag{2.32}$$

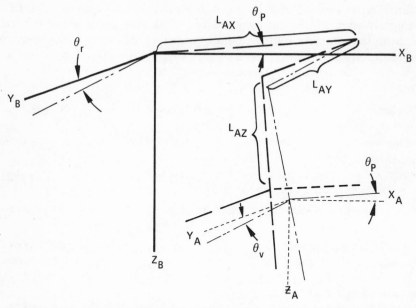

Figure 2-9. Angular Transformations for Array Offset From the
Center of Rotation

$$y_B = (z_B - R)\left[\frac{\tan\theta_r + \tan\theta_{vy}}{\cos\theta_p - \dfrac{\sin\theta_p}{\cos\theta_r}\tan\theta_{vx} - \cos\theta_p \tan\theta_r \tan\theta_{vy}}\right]$$
$$+ P \quad (2.33)$$

where:

$$Q = L_{AX}\cos\theta_p - L_{AY}\sin\theta_r\sin\theta_p + L_{AZ}\sin\theta_p\cos\theta_r$$
$$P = L_{AY}\cos\theta_r + L_{AZ}\sin\theta_r$$
$$R = -L_{AX}\sin\theta_p - L_{AY}\sin\theta_r\cos\theta_p + L_{AZ}\cos\theta_r\cos\theta_p$$

For small angles Equations 2.32 and 2.33 simplify to:

$$x_B = (z_B - L_{AZ})\tan(\theta_p + \theta_{vx}) + L_{AX}$$
$$y_B = (z_B - L_{AZ})\tan(\theta_r + \theta_{vy}) + L_{AY}$$
$$(2.34)$$

The two angular summations are computed by the technique given
previously for Equation 2.30.

The horizontal offsets computed by Equations 2.32 and 2.33 are between the reference point on the vessel and the subsea unit. If the sub-

sea unit is not located at the subsea reference point, then another off-set must be added to Equations 2.32 and 2.33. If the offset of the subsea unit from the subsea reference point is expressed in earth coordinates as a distance and a direction, then the offset of the reference point on the vessel with respect to the subsea reference point is expressed as follows:

$$x_{BT} = x_B + R_B \cos (\theta_B - \psi_s)$$
$$y_{BT} = y_B + R_B \sin (\theta_B - \psi_s) \qquad (2.35)$$

where:

$R_B =$ the distance from the subsea unit to the subsea reference point

$\theta_B =$ the angle from north to the distance between the subsea unit and the subsea reference point

$\psi_s =$ the heading of the reference frame on the vessel with respect to north

In order to implement Equation 2.35 not only must the horizontal distance and direction from the subsea unit to the subsea reference point be known, but also the heading of the reference on the vessel with respect to north must be known or measured.

Another acoustic system geometry exchanges the array geometry with the reference point on the bottom in a manner similar to that shown in Fig. 2–10. Unlike the acoustic system with the reference point on the bottom, in this reverse system, the reference point moves relative to the array elements. Likewise the array elements are nearly never in the same plane and their location with respect to one another is only approximately known at best.

The ranges from the array elements on the bottom to the reference point shown in Fig. 2–10 on the kth interrogation are expressed mathematically as follows:

$$R_{ik}^2 = (x_i - x_{vk})^2 + (y_i - y_{vk})^2 + (z_i - z_{vk})^2 \qquad (2.36)$$

where:

$R_{ik} =$ the range from the ith array elements to the reference point on the kth interrogation

$x_i, y_i, z_i =$ the Cartesian coordinates of the ith array elements

$x_{vk}, y_{vk}, z_{vk} =$ the Cartesian coordinates of the reference point on the kth interrogation

The only quantities which are known in Equation 2.36 are the ranges which can be measured. Thus, some method must be employed

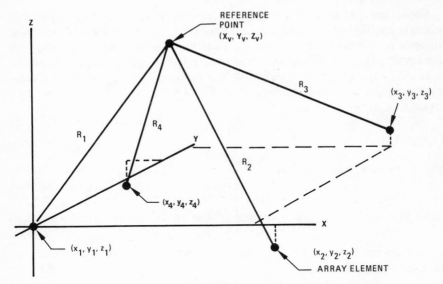

Figure 2-10. Long Baseline Geometry

to use the range measurements to determine not only the coordinates of the reference point but also the coordinates of the array elements.

In a paper given by Young[8] an exact solution is given to determine the coordinates of the array elements and the reference point using range measurements taken at different points relative to the array. The results of Young's paper are beyond the scope of this discussion.

Once the coordinates of array elements and the water depth, z_v, are known, then the method of determining the coordinates of the reference point is the same as used for the array at the surface where the differences in range squared are computed. In equation form the range squared appears as follows for the ith and jth range:

$$R_i^2 - R_j^2 = [(x_v - x_i)^2 + (y_v - y_i)^2 + (z_v - z_i)^2] \\ - [(x_v - x_j)^2 + (y_v - y_j)^2 + (z_v - z_j)^2] \quad (2.37)$$

which reduces to:

$$(R_i - R_j)(R_i + R_j) = 2\,x_v(x_j - x_i) + 2\,y_v(y_j - y_i) \\ + 2\,z_v(z_j - z_i) + D_{ij}^2 \quad (2.38)$$

where:

$$D_{ij}^2 = (x_j^2 - x_i^2) + (y_j^2 - y_i^2) + (z_j^2 - z_i^2)$$

Since the locations of the array elements and the water depth are known by the array calibration process given by Young[8] and the ranges are measured, the only unknowns in Equation 2.38 are the horizontal coordinates of the reference point, x_v and y_v. As a result, if a third range measurement is made or two additional range measurements are made, the following two linear equations with two unknowns can be written.

$$
\begin{bmatrix} (x_1 - x_3) \ (y_1 - y_3) \\ (x_2 - x_4) \ (y_2 - y_4) \end{bmatrix} \begin{bmatrix} x_v \\ y_v \end{bmatrix}
$$
$$
= \begin{bmatrix} -\dfrac{(R_3 - R_1)(R_3 + R_1)}{2} + \dfrac{D_{13}{}^2}{2} + z_v \ (z_1 - z_3) \\[2ex] -\dfrac{(R_4 - R_2)(R_2 + R_4)}{2} + \dfrac{D_{24}{}^2}{2} + z_v \ (z_2 - z_4) \end{bmatrix} \quad (2.39)
$$

or in a matrix form:

$$ \underline{A}\overline{X}_v = \overline{R} + \overline{b} \quad (2.40) $$

The solution of Equation 2.40 for the coordinates of the reference point is given in matrix form as:

$$ \overline{X}_v = \underline{A}^{-1} [\overline{R} + \overline{b}] \quad (2.41) $$

where A^{-1} equals the inverse of the matrix A. The inverse of A can be expressed as follows:

$$
\underline{A}^{-1} = \begin{bmatrix} \dfrac{x_1 - x_3}{\Delta} & -\dfrac{(y_1 - y_3)}{\Delta} \\[2ex] -\dfrac{(x_2 - x_4)}{\Delta} & \dfrac{(y_2 - y_4)}{\Delta} \end{bmatrix} \quad (2.42)
$$

where $\Delta = (x_1 - x_3)(y_2 - y_4) - (x_2 - x_4)(y_1 - y_3)$. For Equation 2.42 to exist, Δ must be non-zero which is true if all the subsea array elements are not colocated.

2.4 Short-baseline systems (SBS)

The short-baseline type of acoustic system is characterized by a single subsea acoustic device and an array of acoustic receivers mounted on the lower side of the hull of the vessel.

There are several variations of the short-baseline system, each with different hardware. However, they all estimate range to calculate position as shown in the previous section. In this section three variations of short-baseline systems are discussed: the time-of-arrival pinger system, the time-of-arrival transponder system, and the phase comparison pinger system.

The pinger acoustic system is the simplest. In this system the subsea acoustic unit periodically emits a short burst of acoustic energy in the form of a sinusoidal pulse. The periodicity and frequency of the sinusoid are controlled internally to the subsea unit and are preset before the subsea unit is deployed.

In order to compute position from the periodic burst of acoustic energy, the vessel-mounted acoustic unit uses an array of acoustic receiving elements. As shown in the previous section, if the elements are separated by known distances, their planar orientation to the vessel is known, and their planar orientation to local vertical is known, then the relative position of the array with respect to the pinger can be computed from the difference in time-of-arrival of the acoustic pulses from the pinger. The basic geometry of the receiving elements from which the position calculation is made is shown in Fig. 2–5.

For this simple acoustic system the times-of-arrival of the same acoustic pulse at three receiving array elements mounted on the vessel are detected. Then, the differences in time-of-arrival between the x-axis and y-axis pairs of array elements are computed. The corner array element is used to compute the difference in the time-of-arrival for both the x-axis and the y-axis and is often referred to as the reference element. Once the difference in the time-of-arrival is computed for each vessel coordinate axis, the difference in the ranges from the array elements to the pinger can be estimated as shown in Equation 2.21.

Since the distances between array elements are known, the estimated differences in the ranges can be used to estimate the tangents of the direction angles from the centroid of the array to the subsea pinger (Fig. 2–5). The equations for calculating the direction angles from the difference in the time-of-arrival of the acoustic pulses is shown in Equation 2.20 and is repeated here for convenience:

$$\theta_{vx} \approx \arcsin\left[\frac{c(t_3 - t_1)}{2d_x}\right] = \arcsin\left[\frac{c\,\Delta t_x}{2d_x}\right] \qquad (2.43)$$

$$\theta_{vy} \approx \arcsin\left[\frac{c(t_2 - t_1)}{2d_y}\right] = \arcsin\left[\frac{c\,\Delta t_y}{2d_y}\right] \qquad (2.44)$$

These equations are exact when the paths of the acoustic pulses to the hydrophones are parallel which occurs when $x_A + d_x \ll z_A$ and $y_A + d_y \ll z_A$.

Equations 2.43 and 2.44 use the second array element as the reference element. Any of the other array elements can be used as reference elements as long as the other two elements form an orthogonal set of coordinates with the reference element at the origin. In this book it has been assumed that the array is rectangular and aligned parallel to the vessel coordinate system. Another array geometry could be used as long as three elements are used, but the computations will become more complicated and cross-coupled. Also, the loss of one array element will affect both x-axis and y-axis information.

Naturally the subscripting of the times-of-arrival in Equations 2.43 and 2.44 will be modified to correlate the array elements used and sign convention noted so that proper offset direction is computed from the times of arrival.

The direction angles computed by Equations 2.43 and 2.44 are in array coordinates as Fig. 2–5 shows. To transform them into a horizontal coordinate system referenced to some point on the vessel requires angular and translational transformations like that derived for Equations 2.32 and 2.33. If, in addition, the direction angles and angular motions of the vessel are small, then the vessel offsets in the horizontal reference plane can be expressed in percent of water depth as follows:

$100\% \dfrac{x_B}{h_w}$, percent water depth

$$= \left[\frac{(z_B - L_{AZ})}{h_w} \tan (\theta_{vx} + \theta_p) + \frac{L_{AX}}{h_w} \right] 100\% \quad (2.45)$$

$100\% \dfrac{y_B}{h_w}$, percent water depth

$$= \left[\frac{(z_B - L_{AZ})}{h_w} \tan (\theta_{vy} + \theta_r) + \frac{L_{AY}}{h_w} \right] 100\% \quad (2.46)$$

where:

L_{AX}, L_{AY}, L_{AZ} = the translational distance of the array centroid to the center of rotation of the vessel

θ_r, θ_p = the roll and pitch angles of the vessel, respectively

x_B, y_B = the horizontal coordinates of the subsea pinger

z_B = the vertical distance from the horizontal reference frame to the horizontal plane through the subsea pinger

h_w = the water depth

If additional horizontal translations are desired to reference some other point on the vessel than the center of rotation of the vessel, additional terms must be added to Equation 2.45 and 2.46.

As Equations 2.43 through 2.46 show, the calculation of position of the vessel with the SBS pinger system involves the measurement of the difference in the time-of-arrival and the measurement of the angular orientation of the vessel to the vertical. The measurement of the angular orientation can be performed with a vertical reference sensor such as an inclinometer or a vertical gyro. The measurement of the difference in time-of-arrival is easily performed using an electronic timer which starts when the first pulse arrives and stops when the second pulse arrives. Then with proper scaling the position offset can easily be computed.

The determination of the arrival of the acoustic pulse is a classical problem of detection of a signal of a known frequency in a background of noise.[3] Detection does not occur if the amplitudes of the acoustic pulses diminish because of transmission path losses or attenuation, or if the noise amplitude increases.

In either case detection is not possible because the signal-to-noise ratio decreases below some minimum required by the detection electronics.

As the expression for the calculation of position shows, the error sources which will affect the accuracy of the position calculations are not a function of the subsea pinger. This will remain true if the pinger continues to produce the acoustic pulse with approximately constant pulse width, frequency, and pulse repetition rate.

Actually the tolerances on these quantities and the amplitude of the pulse are not in terms of initial values or stability because the critical measurement parameter is the arrival of the pulse, not its signal characteristics.

These factors contribute to the pinger being a simple device consisting of an electromechanical transducer, an electronic pulse generator, an oscillator, a self-contained power supply, and a pressure case. An example of a pinger is shown in Fig. 2–11.

The electromechanical transducer converts an electrical pulse of

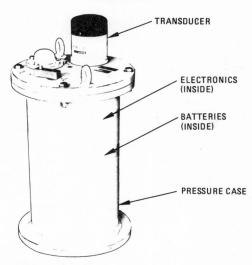

Figure 2-11. A Short-Baseline Acoustic Position Reference System Pinger (Courtesy Honeywell Marine Systems Division)

frequency into an acoustic pulse for projection into the surrounding water. To extend the life of the self-contained power supply, the transducer can be designed to project the acoustic energy in a beam pattern with a favored direction, or over a favored beam width to increase its efficiency.

The pinger electronics are standard and the self-contained power supply is generally batteries. Various kinds of batteries are used. The batteries used in the pinger shown in Fig. 2-11 are sealed lead acid which can be recharged each time the pinger is retrieved. The length of time the pinger can operate depends on the size of the battery pack, the transmitting source level or maximum operating range (depth), and the maximum ping rate.

The pressure case has to be large enough to hold all the electronics and batteries. In addition, the case must be strong enough to withstand the pressure for the design water depth. For example, the case shown in Fig. 2-11 is designed for water depths of over 3,000 ft.

The size and shape of the pinger case will affect its deployment, implanting, and retrieval. For example, to maintain the pinger above the mud line of the sea bed in an approximately vertical attitude, the pinger is made to float with the transducer at the top with some form of anchor below and a deployment/retrieval line back to the surface. The element of pinger handling in an operation which uses an acoustic position reference system either alone or as a part of a DP system cannot be neglected.

The receiver of the acoustic energy on the vessel consists of an electromechanical transducer which transforms the acoustic pulse into an electrical pulse with the least amount of distortion possible. The receiver also contains a preamplifier to amplify the transformed acoustic pulse so that it will be more resistant to any noise along the electrical path from the receiver to the signal processor. An acoustic receiver is generally called a hydrophone. Two examples are shown in Fig. 2–12.

To make the hydrophone more resistant to unwanted noise and reflected acoustic pulses, the transducer elements that convert the acoustic pulses to electrical pulses are designed and arranged in the hydrophone and surrounded by special baffling material to give a preferred direction of reception. In fact, as Fig. 2–13 shows, the directivity pattern for the time-of-arrival hydrophone is 35 decibels higher (or

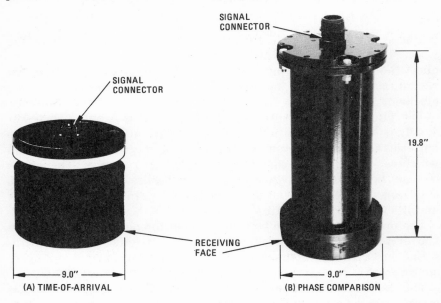

Figure 2-12. Acoustic Hydrophones (Courtesy Honeywell Marine Systems Division)

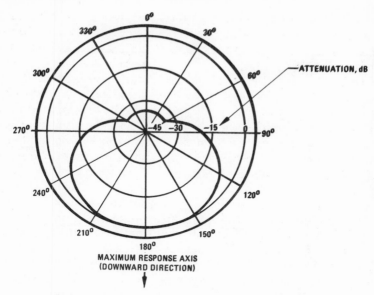

Figure 2-13. Directivity Pattern for a Time-of-Arrival Hydrophone as Shown in Figure 2-12. (Courtesy Honeywell Marine Systems Division)

87.5 times larger in terms of voltage) in forward direction than in the back direction.

This, of course, means that any acoustic pulses reflected from the hul! are going to be severely attenuated compared with the pulses directly from the pinger.

Mounting of hydrophones on the vessel can be done in many ways. A preferred method for a ship-shaped vessel is shown in Fig. 2-14. The stem to which the hydrophone is mounted is usually pipe to which additional lengths can be added once the well through the vessel has been made. With the ability to extend the hydrophone below the keel of the vessel, the reception of acoustic pulses by the hydrophone from the pinger can be improved.

Likewise, the hydrophone can often be moved to an area below the aeration that occurs around the hull of the vessel because of waves and other vessels which come alongside. However, the design of the means of securing the stem in the hydrophone well and the size of the stem become increasingly important the farther the stem is extended below the hull.

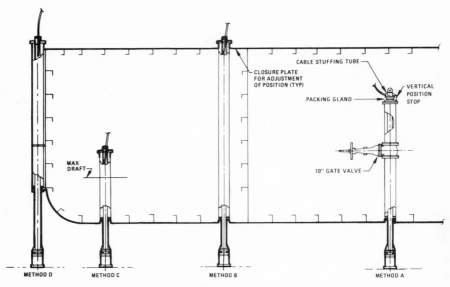

Figure 2-14. Typical Installation Methods for Hydrophones
(Courtesy Honeywell Marine Systems Division)

Location of the hydrophones on the vessel should follow several basic guidelines so that the system is as accurate and as noise resistant is possible. Naturally, for the time-of-arrival system the best policy is to mount the three necessary hydrophones in the same horizontal plane in a rectangular array aligned with the position-reference frame of the vessel as shown in Fig. 2–15.

Generally, for the drilling application the hydrophone array is around the drilling area or moonpool and at least four hydrophones are installed as shown in Fig. 2–15 for redundancy and to improve acoustic reception if only one hydrophone is being interferred with. The hydrophone array is best located around the moonpool of a drillship because heading changes do not rapidly alter the position of the hydrophones to the subsea pinger, and the moonpool area is generally as far from all the thrusters as is possible.

Hydrophone location on a semisubmersible can be more complicated than on the ship-shaped vessel because of lack of access to the hydrophone wells when the vessel is at working draft and the routing of the cable from the hydrophone to the upper deck. An example of a hydro-

TOP VIEW

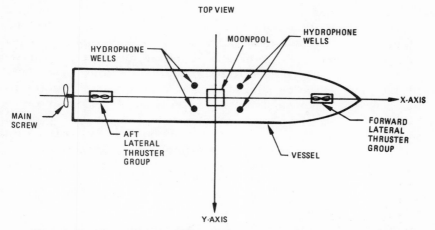

Figure 2-15. Orthogonal Hydrophone Array Configuration
for a Time-of-Arrival System

phone installation on a semisubmersible drilling vessel is shown in
Fig. 2–16.

The array does not need to be square or exactly centered around the
moonpool of the vessel because the position computer will adjust the
location of the hydrophone array from any point on the vessel to any

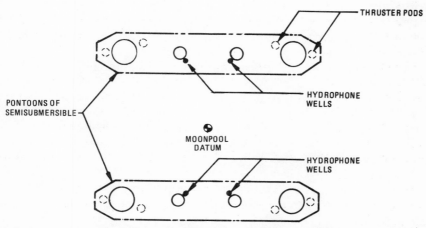

Figure 2-16. Semisubmersible Drilling Rig Hydrophone Configuration

reference point on the vessel. Likewise the subsea pinger does not need to be located over the reference point on the sea bed. In fact, locating the pinger on the subsea reference point can be impossible.

For example, in the case of a drillship the reference point is the wellhead through which the drill string must enter from above. Therefore, the pinger location is also adjusted computationally in the position computer when the system is initially put in service at a work site.

One technique of processing the acoustic signals from two hydrophones to obtain the difference in the time-of-arrival is shown in Fig. 2–17. The acoustic pulses are first transmitted from the hydrophones to narrow bandpass filters tuned to the frequency of the pinger. The filtered pulses are then shaped and differentiated so that the center of the acoustic pulse can be detected.

As Fig. 2–17 illustrates, the first acoustic pulse detected starts an integrator integrating at a constant rate until the second acoustic pulse is detected, which stops the integrator. The rate of integration is scaled according to the factor of the speed of sound divided by the hydrophone baseline as shown in Equation 2.43 and 2.44 and a factor used in the computer for fullscale output for a given percent of water depth.

For example, in an analog system with 10 v equal to 20% of water depth and a 50-ft hydrophone baseline, the integrator scale factor is 5 v per millisecond (1/1,000 of a second). With this scale factor the integrator gives the sine of the angle approximately equal to 20% of water depth. The polarity of the output of the integrator is determined by which pulse arrives first and, therefore, gives the direction the hydrophones are offset from directly over the pinger.

Once the angular direction of the pinger is computed by the integra-

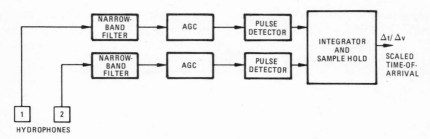

Figure 2-17. One Technique of Generating the Difference in the Time-of-Arrival

tor, then the position computer can compensate for the angular variation of the vessel from the horizontal plane and any offsets in arrays
and pinger locations. Since the pinger location with respect to the
subsea reference point is in earth-reference coordinates, the corrections for pinger location in the position computer requires a coordinate
transformation using the heading of the vessel and a scale change to
obtain percent of water depth.

Measurement of the angular variation from the horizontal plane is
very important to the accuracy of the position calculation of the acoustic position reference sensor as Equations 2.43 and 2.44 show. The
accuracy of the measurement in dynamic and static terms is important.
Consequently, the sensor used to measure the angular variations becomes important. This sensor is covered in Section 3.5.

The position computer takes in the bearing information from the
signal processor, the angular variation from the vertical reference
system, and the offset and bias information properly coordinate-
transformed by the heading angle furnished by the gyrocompass to
calculate the position offset of the pinger from the vessel. The calculation can be accomplished either by an analog or digital computer.
The earlier systems, of course, used an analog computer. However,
with the advent of very low-cost digital components and microprocessors, digital computers are now being used.

With the digital computers not only are more precise calculations
possible at no great increase of hardware but also the electronic drift
present in analog systems are avoided. However, additional input and
output devices are necessary to prepare the information from external
sources for use in the digital computers. Fig. 2–18 shows a comparison
of an analog versus a digital position computer for the time-of-arrival
pinger-type system.

A second type of short baseline system uses a simple pinger on the
sea bed, but instead of measuring difference in time-of-arrival, it
measures phase difference as illustrated in Fig. 2–19. Phase difference
can be related to the mechanical angle of incidence of the acoustic
pulse to the array-receiving elements by the following expression:

$$\Delta \phi = \frac{4\pi \ f_0 \ d \ \cos \theta}{c} \tag{2.47}$$

where:

f_0 = the frequency of the acoustic pulse in hertz
$2d$ = the distance between array elements in feet

$\Delta\phi$ = the difference in phase in radians

θ = the mechanical angle of incidence of the acoustic wave

c = the velocity of propagation of the acoustic pulse

Solving Equation 2.47 for the cosine of mechanical angle of incidence and substituting the result into Equations 2.22 and 2.23 yields expressions for the tangents of the direction angles, θ_{vx} and θ_{vy}, as follows:

$$\tan \theta_{vx} = \frac{\Delta\phi_x}{[k_s^2 - \Delta\phi_x^2 - \Delta\phi_y^2]^{1/2}} \qquad (2.48)$$

$$\tan \theta_{vy} = \frac{\Delta\phi_y}{[k_s^2 - \Delta\phi_x^2 - \Delta\phi_y^2]^{1/2}} \qquad (2.49)$$

where

$$k_s = \frac{4\pi f_o d}{c}$$

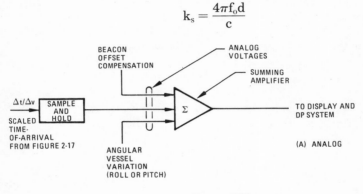

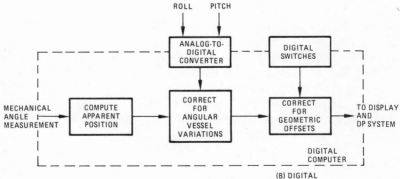

Figure 2-18. A Comparison of the Analog Versus Digital Position Computers

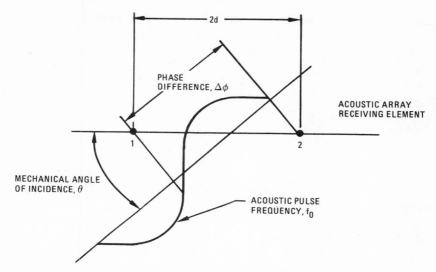

Figure 2-19. Detection of Phase Delay Between Receiver Elements

and which for small phase difference (the array nearly over the pinger) can be approximated as:

$$\theta_{vx} = \frac{c}{4 \pi f_0 d} \Delta \phi_x \qquad (2.50)$$

$$\theta_{vy} = \frac{c}{4 \pi f_0 d} \Delta \phi_y \qquad (2.51)$$

From the tangents of the direction angles the vessel offsets from the pinger can be computed from Equation 2.45 and 2.46.

For the time-of-arrival system the minimum distance between hydrophones is generally quoted as 40 ft. For the phase difference system a system is now available which has a 2-in. separation of the receiving elements.[2] With this small array dimension the entire array can be packaged in one hydrophone assembly as shown in Fig. 2–12.

With only one hydrophone several advantages are obtained. First, there are fewer hydrophone wells and stems to be constructed and installed and, therefore, less cost. Second, the single hydrophone can be independently located on the vessel while the multiple hydrophone arrangement requires the three hydrophones to be mounted in the same horizontal plane in a measured-array pattern. Third, the geometrical dimensional tolerances of the array elements can be compensated for with calibration methods used during hydrophone manufacturing which is easier than calibrating multiple hydrophones in a shipyard. Fourth, with only one hydrophone the vertical-reference assembly can be integrated with the hydrophone assembly in one unit so that vessel motion compensation can be performed more accurately and installation can be further simplified. To date, this advantage has not been practically realized because of available vertical reference sensor assemblies.

The remainder of the phase difference system is exactly like the time-of-arrival system except for the signal processing to measure phase difference. For example, the pinger and the vertical reference system are exactly the same as used for the previously described time-of-arrival system. Therefore, they will not be discussed again. Likewise, the phase measurement technique will not be discussed.[2]

There is a third type of short-baseline system which can use time-of-arrival or phase difference to measure the direction angles of the subsea unit and which can use analog or digital processing to compute position offset. This system differs from the previous short-baseline systems in that it also measures the range to the subsea unit.

To do this, the portion of the acoustic system on board the vessel transmits an acoustic pulse downward toward the subsea unit. At the same time a timer is started. The vessel-transmitted pulse, when received by the subsea unit, causes the subsea unit to transmit an acoustic pulse. When the acoustic pulse from the subsea unit is detected on the vessel, the timer that was started when the vessel transmitted its acoustic pulse is stopped. The accumulated time of the timer is a measure of the round-trip transmission time from the vessel to the subsea unit. With the round-trip time the range can be estimated using the velocity of sound in water times one half the round-trip time.

However, to determine the position offset from the vessel to the subsea acoustic unit, either at least three range measurements or a range plus bearing measurement is required. The implementation of such a system is shown in Fig. 2–20.

As a result of the required multiple measurements, an array of receiving elements is mounted on the vessel along with an acoustic transmitter. The receiving elements are called hydrophones and are generally geometrically arranged the same as for the pinger system. The transmitter is called an interrogator and is usually located near,

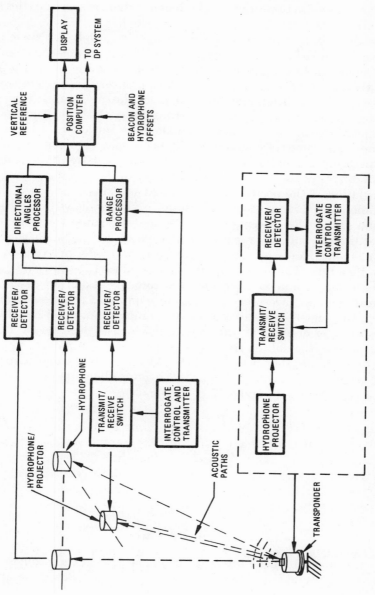

Figure 2-20. Short-Baseline, Time-of-Arrival Transponder System

and in the same horizontal plane, as the receiving elements to simplify the calculations of position.

Since the subsea unit must be able to receive as well as transmit, it is called a transponder. Physically, however, the subsea unit can be almost identical to the pinger subsea unit because the receiving portion of the unit does not require a significant battery pack or electronic package. In fact, if the receiving and transmitting functions can be combined in one electromechanical transducer, the transponder and pinger can be physically indistinguishable. However, because of differing requirements in power and beam patterns and different manufacturer's experiences, each function may be better implemented with its own transducer.

Calculation of the position of the subsea unit in array coordinates for a time-of-arrival transponder system can be made with Equations 2.14 and 2.15 if the interrogation pulse for each range measurement originates from the same geometrical location as the corresponding array element. However, normally there is only one interrogator for each subsea unit. Therefore, Equations 2.14 and 2.15 must be modified to account for a common interrogator at some point on the vessel. The modification must necessarily result in expressions for the location of the transponder in array coordinates which contains either differences in ranges or sums of ranges, one of which is the range from the interrogator to the subsea transponder.

For an array as shown in Fig. 2–5 and the interrogator located at the midpoint between the first and second array elements, the modified form of Equations 2.14 and 2.15 is as follows:[9]

$$x_A = K_A (R_4 - R_1) \tag{2.52}$$

$$y_A = (R_3 - R_4) \left[\frac{d_x}{d_y} K_A + \frac{1}{4d_y} (R_3 - R_1) \right] \tag{2.53}$$

where:

$$K_A = \frac{1}{2} \left\{ \frac{d_y{}^2 + (R_1 + R)(R_4 + R) - \frac{1}{2} (R_3 - R_4)(R_3 - R_1)}{d_x [(R_3 - R_4) + 2(R + R_1)]} \right\}$$

R = the range from the interrogator to the subsea transponder

In terms of round-trip times the equations for the transponder given in Equations 2.52 and 2.53 are as follows:

$$x_A = K_A \, c\left(\frac{t_4 - t_1}{2}\right) \tag{2.54}$$

$$y_A = c(t_3 - t_4)\left[\frac{d_x}{d_y} K_A + \frac{1}{4d_y} c(t_3 - t_1)\right] \tag{2.55}$$

where:

$$K_A = \frac{1}{2} \frac{d_y^2 + c^2 t_1 t_4 - \frac{1}{2} c^2 (t_3 - t_4)(t_3 - t_1)}{d_x c[(t_3 - t_4) + 2t_1]}$$

t_i = the round-trip time for the transmission from the interrogator to the transponder and back to the ith array element

c = the sound velocity

The round-trip time actually measured in the position computer includes the time for the command to travel to the interrogator, the time for the interrogator electronics to issue an acoustical pulse into the water, the time for the transponder to receive, process, and retransmit a return acoustical pulse, and the time for the acoustical pulse to be processed by the hydrophone and transmitted to the position computer. Fortunately, these times remain reasonably constant and are determinable so that they can be compensated for in the position computer.

Unlike the pinger system the transponder system measures the position of the subsea unit in absolute units of distance as opposed to percent of water depth (angular offsets). In addition, the transponder system can measure the vertical separation between the array elements and the subsea units. For tracking subsea objects such as submersibles, and for compensating for ray bending, measurement of vertical separation is useful. However, for applications where the subsea units remain at a constant depth at small horizontal offsets from the hydrophone such as drilling, the vertical separation is not useful.

In the case of the phase difference system with a single hydrophone the absolute position of the subsea transponder is calculated as follows if the interrogator is coresident with the receiving array:

$$x_A = R \cos \theta_x = \frac{c^2}{8 \pi f_o d_x} \Delta T_{RT} \, \Delta \phi_x \tag{2.56}$$

$$y_A = R \cos \theta_y = \frac{c^2}{8 \pi f_o d_y} \Delta T_{RT} \, \Delta \phi_x \tag{2.57}$$

where:

ΔT_{RT} = measured round-trip time
$\Delta \phi_x$ = measured phase difference in the x-axis
$\Delta \phi_y$ = measured phase difference in the y-axis
c = velocity of sound in water
f_0 = frequency of the acoustic pulse from the subsea transponder
$2 d_x$, $2 d_y$ = distance between receiving array elements

With the single hydrophone configuration the interrogator can easily be made coresident with the hydrophone. Therefore, Equations 2.56 and 2.57 will not require modification.

Like the time-of-arrival transponder system, the phase difference system can measure the vertical distance from receiving array to the subsea transponder. The equation to calculate the vertical distance is:

$$z_A = R \cos \theta_z$$
$$= (R^2 - x_A{}^2 - y_A{}^2)^{1/2} \qquad (2.58)$$

In a similar manner to the time-of-arrival system, the position computer must compensate the measured round-trip time for transmission times in the system not related to the transmission time in the acoustic medium. For both types of transponder systems, the angular movements of the vessel and the offsets in the array and subsea unit from the vessel and subsea reference points, respectively, must be compensated for in the position computer.

One distinct difference between the pinger and transponder systems is the data rate, or the frequency with which information is available to make position calculations. In the case of the pinger, the data rate is the rate at which the subsea unit emits acoustic pulses or the ping rate. As a result, the data rate is preset when the subsea unit is deployed and not generally controlled from the surface. It is possible to have a surface commandable pinger which will change its ping rate and its frequency. However, the cost of the commandable units increase rapidly as they become more sophisticated.

By contrast, in the transponder system the subsea transponder will not respond until it is interrogated from the surface. Thus, the data rate in the transponder system is entirely surface controlled.

Having complete surface control is not without its complications. First, if the system is designed to interrogate only after a return pulse from the transponder has been received, the data rate in deep water can become large. For example, if the slant range from the vessel to the transponder is 2,500 ft, the round-trip time is approximately one second which sets the data rate at approximately one sample per second.

However, higher data rates are possible by using multiple interrogations in the period of one round-trip time. The difficulty with this scheme is the spacing of the interrogation pulses so that returning pulses from the subsea unit can be reliably detected. With a digital computer as the position computer, logic can be designed to control such complex timing problems.

Four short-baseline acoustic position reference systems have been discussed. Two use a simple subsea pinger, two use a subsea transponder, two use time-of-arrival, and two use phase difference to compute position. They all assume a value of the velocity of sound in water to compute position; they all must compensate for angular motions of the vessel to compute position; and they all use a single subsea unit.

Likewise, two measure position in terms of percent of water depth (angular offsets) and two measure position in terms of absolute distance units. The same two that measure in percent of water depth do not measure vertical distance, whereas the other two do.

As a result of these similarities and differences a valid question is: which system should a prospective buyer select? The answer depends primarily on the buyer's intended application.

In the selection of an acoustic short-baseline system several factors should be considered, including:

1. Accuracy
2. Area of coverage
3. Data rate
4. Operating convenience
5. Reliability
6. Cost of ownership

Accuracy is of course a basic question, especially for a system to be used as a basic part of a dynamic positioning system. As Table 2–1 shows, the four acoustic systems have the same accuracy for small horizontal offsets. Thus, there is no clear-cut choice based on accuracy among the four systems for small horizontal offsets, such as required in the station keeping application.

However, if the application calls for a large area of coverage, the pinger system is not recommended. In choosing between the two types of transponder systems, selection is not possible based on accuracy and area coverage.

The third factor affecting system selection—data rate—is different for the pinger and the transponder systems. However, in the manner in which data rate affects the dynamic positioning system, namely sampling rate and stability, there is no clear choice among the four systems unless the application calls for intermittent use of the subsea unit. Then the transponder system is superior to the basic pinger system because the transponder data rate is entirely controlled by the surface unit (Table 2-1). A pinger can be made commandable, however, so that it can be started and stopped from the surface.

Operating convenience is a selection factor which depends more on the engineering of a product for operator use, ease of installation, maintenance and repair, and training. As a result, operating convenience depends primarily on the manufactured product. As a general rule, the simpler a system is, the easier it is to operate. Based on simplicity, the pinger system is the preferred choice because it does not have an interrogator and its associated parts.

Among the pinger systems the phase difference system is slightly simpler than the time-of-arrival system because there is only one hydrophone. However, the pinger system does not perform for all applications. In those applications where the pinger system is not usable, the criterion of simplicity favors the phase difference system with its single hydrophone.

Reliability is, like operating convenience, strongly a function of product engineering. However, given equal product engineering, the simpler system is the most reliable because fewer parts can fail. Thus, the pinger system should be more reliable than the transponder system. In the cases where the pinger system will perform the given application, the more reliable system choice should be the phase difference system, given equal product engineering.

Cost of ownership is always a very important selection factor and in the competitive market that exists today, cost may rate supreme. Cost of ownership includes more than just the original purchase of equipment. Cost of ownership includes the cost of installation, repair, maintenance, and training. Unfortunately each of these costs are strongly a function of product engineering. However, they also depend on the complexity of the system, especially installation costs. The phase difference systems are clearly less expensive to install because they have only one hull-penetrating unit.

TABLE 2-1. A COMPARISON OF SHORT-BASELINE ACOUSTIC POSITION REFERENCE SYSTEM CHARACTERISTICS

	PINGER		TRANSPONDER	
	TIME-OF-ARRIVAL	PHASE COMPARISON	TIME-OF-ARRIVAL	PHASE COMPARISON
PERFORMANCE				
ACCURACY				
• SMALL HORIZONTAL OFFSETS	BETTER THAN 1% OF WATER DEPTH; ACCURACY INDEPENDENT OF WATER DEPTH.	BETTER THAN 1% OF WATER DEPTH; ACCURACY INDEPENDENT OF WATER DEPTH.	SAME AS PINGER.	SAME AS PINGER.
• LARGE HORIZONTAL OFFSETS	NOT RECOMMENDED.	NOT RECOMMENDED.	1% TO 3% OF RANGE.	1% TO 3% OF RANGE.
DATA RATE	SET BY PINGER; CAN BE BETTER THAN 10 SAMPLES PER SECOND: NORMALLY ONCE OR TWICE PER SECOND TO CONSERVE BATTERY POWER.	SAME AS TIME-OF-ARRIVAL PINGER.	CONTROLLED BY POSITION COMPUTER; RATE DEPENDS ON SLANT RANGE AND GEOMETRY OF VESSEL TO TRANSPONDER.	SAME AS TIME-OF-ARRIVAL SYSTEM.
AREA OF COVERAGE	MOST ACCURATE AT LESS THAN 20% OF WATER DEPTH, ACCURACY AT GREATER THAN 100% OF WATER DEPTH LIMITED.	SAME AS TIME-OF-ARRIVAL PINGER.	LIMITED BY TRANSPONDER POWER OR BY ACOUSTIC TRANSMISSION EFFECTS; GENERALLY NO GREATER THAN 300 TO 500 PERCENT OF WATER DEPTH.	SAME AS TIME-OF-ARRIVAL SYSTEM.
COMPONENTS*				
SUBSEA UNITS	1	1	1	1
HYDROPHONES	3	1	3	1
INTERROGATORS	0	0	1	1
POSITION COMPUTER/ SIGNAL PROCESSORS	1	1	1	1
VERTICAL REFERENCE UNITS	1	1	1	1
INSTALLATION				
HYDROPHONES	THREE HYDROPHONE WELLS WITH STEMS OF ADJUSTABLE LENGTH; HYDROPHONES SHOULD BE REPLACEABLE AT SEA.	ONE HYDROPHONE WELL WITH STEMS OF ADJUSTABLE LENGTH; HYDROPHONE SHOULD BE REPLACEABLE AT SEA.	THREE HYDROPHONE WELLS WITH STEMS OF ADJUSTABLE LENGTH; HYDROPHONES SHOULD BE REPLACEABLE AT SEA.	ONE HYDROPHONE WELL WITH STEM OF ADJUSTABLE LENGTH; HYDROPHONE SHOULD BE REPLACEABLE AT SEA.
INTERROGATORS	NOT APPLICABLE	NOT APPLICABLE	ONE INTERROGATOR WELL WITH STEM OF ADJUSTABLE LENGTH; INTERROGATOR SHOULD BE REPLACEABLE AT SEA.	INTERROGATOR CAN CORESIDE IN SAME WELL AS HYDROPHONE.
OPERATION				
SUBSEA UNIT DEPLOYMENT AND RECOVERY	SINGLE UNIT ON A RETRIEVAL BASE OR EXPENDABLE FREE-DROPPED.	SAME AS TIME-OF-ARRIVAL.	SAME AS PINGER.	SAME AS PINGER.
INITIALIZATION	NONE REQUIRED.	NONE REQUIRED.	NONE REQUIRED.	NONE REQUIRED.

* THE NUMBER OF COMPONENTS REPRESENTS MINIMUMS FOR BASIC OPERATION; LARGER NUMBERS OF ACOUSTIC ELEMENTS ARE RECOMMENDED TO ACHIEVE THE MOST RELIABLE PERFORMANCE.

If these selection factors are tabulated and the assumption of equal product engineering is made, then for the applications requiring small horizontal offsets (less than 100% of water depth) the preferred short-baseline acoustic system is the phase difference pinger system. If the application calls for intermittent operation, the pinger system will necessarily be converted to a commandable system which makes it equivalent to the phase difference transponder system.

If the application requires large horizontal offsets (a large coverage area), then the phase difference transponder system is preferred.

2.5   Long-baseline systems (LBS)

An alternative configuration to the short-baseline system is one with an array of subsea units and a reference point on the vessel as shown in Fig. 2–10. This alternative system is a purely range-measuring system where the reference unit on the vessel interrogates multiple subsea units in a transponder mode like the transponder short-baseline system. Unlike the short-baseline system, only one receiver on the vessel is required to receive the multiple responses from the subsea units. Once all the subsea responses are received, i.e., ranges measured, then the position computer can calculate the position of the vessel within a coordinate system initially chosen within the position computer.

Since the size of the array is no longer limited by the dimensions of the vessel and the system operates in a transponding mode, the subsea units can be separated by distances limited only by transponder power and acoustic propagation factors. These distances can easily be greater than 100% of water depth. Thus, the name "long baseline" is derived.

Position computations with the long-baseline system can be performed with the previously derived expression given in Equation 2.41:

$$\overline{X}_v = \underline{A}^{-1}[\overline{R} + \overline{b}] \qquad (2.59)$$

where:

$$\overline{X}_v = \begin{bmatrix} x_v \\ y_v \end{bmatrix}$$

$$\underline{A}^{-1} = \begin{bmatrix} \dfrac{x_1 - x_3}{\Delta} & -\dfrac{(y_1 - y_3)}{\Delta} \\ -\dfrac{x_2 - x_4}{\Delta} & \dfrac{y_2 - y_4}{\Delta} \end{bmatrix}$$

$$\overline{R} = -\frac{1}{2}\begin{bmatrix} R_3{}^2 - R_1{}^2 \\ R_4{}^2 - R_2{}^2 \end{bmatrix}$$

$$\overline{b} = \begin{bmatrix} D_{13}{}^2 + z_v \,(z_1 - z_3) \\ D_{24}{}^2 + z_v \,(z_2 - z_4) \end{bmatrix}$$

$$\Delta = (x_1 - x_3)(y_2 - y_4) - (x_2 - x_4)(y_1 - y_3)$$

$$D_{ij}{}^2 = (x_j{}^2 - x_i{}^2) + (y_j{}^2 - y_i{}^2) + (z_j{}^2 - z_i{}^2)$$

In order to use Equation 2.59 to compute the position of the vessel, i.e., (x_v, y_v), no roll and pitch compensation is required. The translational offset of the interrogator and hydrophone from the reference point on the vessel must be included in the position computations. Likewise, the position of the transponders must be either known or determined. Generally, the locations of the transponders are unknown. Consequently, a calibration technique to determine the locations of the transponders is required.

One method of calibrating the transponder system or grid is to sail within the grid making multiple interrogations, recording the resulting ranges, and then solving the following system of quadratic equations when sufficient interrogations are made:

$$R_{ik}{}^2 = (x_i - x_{vk})^2 + (y_i - y_{vk})^2 + (z_i - z_{vk})^2 \qquad (2.60)$$

where k indicates the kth interrogation. At each interrogation there is one range measured for each transponder for a total of m ranges which yields mp quadratic equations if p interrogations are made. The total number of unknowns in the mp equations is $3m + 2p - 3$ since the vertical coordinate, z_{vk}, is the same for all equations and the coordinate reference frame can be specified without any loss of generality as shown in Fig. 2–10.

Thus, to obtain a solution to the mp quadratic equations for the case of three transponders, six interrogations are required or for the case of four transponders, five interrogations are required. If additional interrogations are taken for a given number of transponders, then the accuracy of the calibrations can be improved. Naturally additional interrogations require more time, which is an important factor in any commercial marine application.

Once the calibration interrogations are made, the mp ranges measured and stored for computation of the transponder locations and depth of the coordinate reference frame, a computational technique to solve the mp quadratic equation is needed. Such a technique is given in a paper by Young,[8] and is in use on the Hughes *Glomar Explorer* mining ship.

In addition to giving a computational technique for the mp quadratic equations, Young shows that the sailing pattern must meet certain rules to avoid computational problems with his method. First the pattern must include points not all in the same vertical plane or conic surface. Thus, a straight line or circular course must not be used. Likewise, the pattern must not be too small or too large to avoid computational sensitivities. From a practical point of view the length of the calibration course is important from the standpoint of time.

As a result of these calibration considerations, the sailing course used for the Hughes *Glomar Explorer* was to start near the center of the grid with one interrogation and then sail outward, curving onto a course to form a closed circuit either circular or triangular as shown in Fig. 2–21.[10] The size of the most efficient closed circuit used for the Hughes *Glomar Explorer* was found to be approximately ³⁄₄ the water depth in diameter.[10]

An alternate method of using a long-baseline system which is particularly suited for the station-keeping or slowly-moving track-following applications computes position changes from range differences. Unlike Equation 2.59, which gives the coordinates of the vessel in the transponder coordinate reference system, the alternate method computes changes in vessel position about an operating point or a set point

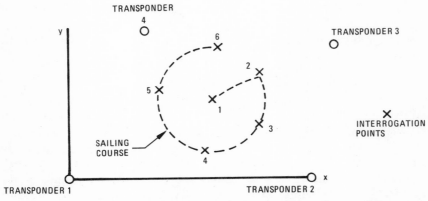

Figure 2-21. An Efficient Calibration Sailing Course for a
Long-Baseline System

in the grid. To compute these changes the following equation is expressed in differential form:

$$F_{ij} = R_{ij} - f_{ij}(x_v, y_v) = 0 \qquad (2.61)$$

where:

$$R_{ij} = R_i - R_j$$
$$f_{ij}(x_v, y_v) = [(x_i - x_v)^2 + (y_i - y_v)^2 + (z_i - z_v)^2]^{1/2}$$
$$- [(x_j - x_v)^2 + (y_j - y_v)^2 + (z_j - z_v)^2]^{1/2}$$

Equation 2.61 expressed in differential form for a transponder grid is:

$$\begin{bmatrix} d\,\Delta\,R_{13} \\ \\ d\,\Delta\,R_{24} \end{bmatrix} = \begin{bmatrix} \dfrac{\partial F_{13}}{\partial x_v} & \dfrac{\partial F_{13}}{\partial y_v} \\ \\ \dfrac{\partial F_{24}}{\partial x_v} & \dfrac{\partial F_{24}}{\partial y_v} \end{bmatrix} \begin{bmatrix} d\,x_v \\ \\ d\,y_v \end{bmatrix} \Bigg|_{(x_v,\,y_v)} \qquad (2.62)$$

Equation 2.62 can be solved for the changes in the vessel position about a given operating point as follows:

$$\begin{bmatrix} d\,x_v \\ \\ d\,y_v \end{bmatrix} \Bigg|_{(x_v,\,y_v)} = \begin{bmatrix} \dfrac{\partial F_{13}}{\partial x_v} & \dfrac{\partial F_{13}}{\partial y_v} \\ \\ \dfrac{\partial F_{24}}{\partial x_v} & \dfrac{\partial F_{24}}{\partial y_v} \end{bmatrix}^{-1} \begin{bmatrix} d\,\Delta\,R_{13} \\ \\ d\,\Delta\,R_{24} \end{bmatrix} \qquad (2.63)$$

where:

$$\frac{\partial F_{ij}}{\partial x_v} = \frac{x_i - x_v}{R_i} - \frac{x_j - x_v}{R_j} = \cos\theta_{xi} - \cos\theta_{xj}$$

$$\frac{\partial F_{ij}}{\partial y_v} = \frac{y_i - y_v}{R_i} - \frac{y_j - y_v}{R_j} = \cos\theta_{yi} - \cos\theta_{yj}$$

$$R_k = [(x_k - x_v)^2 + (y_k - y_v)^2 + (z_k - z_v)^2]^{1/2}$$

Once the grid is calibrated and the vessel is moved near its desired operating point or the operating point is set equal to the present position of the vessel, then the partial derivatives or gradients of the function F_{ij} can be computed and the matrix inverted as indicated in Equation 2.63.

With the coefficient matrix of Equation 2.63 calculated and stored in memory, the changes in the vessel position about the commanded operating point can be computed from the changes in the range differences. The changes in the range differences are expressed as follows:

$$d \, \Delta \, R_{ij} = (R_i - R_j)\Big|_{\substack{\text{at the} \\ \text{measurement} \\ \text{point}}} - (R_i - R_j)\Big|_{\substack{\text{at the} \\ \text{operating} \\ \text{point}}} \qquad (2.64)$$

The range differences at the operating point can be computed and stored in memory at the same time as the gradients. Then, if the long-baseline system measures on-range differences or provides range differences, position changes can be quickly calculated for direct use in the automatic controller.

The implementation of a long-baseline system is somewhat like the short-baseline transponder system shown in Fig. 2–20 in that the position computer initiates the interrogation pulse. Once the interrogation pulse is transmitted, the position computer transfers the system to a receive mode, waiting for pulses to return from the transponders. The transmit/receive feature of the long-baseline system is illustrated in Fig. 2–22. Fig. 2–22 also illustrates the circuitry involved in a long-baseline subsea transponder.

A comparison of Fig. 2–20 and 2–22 shows the subsea transponders to be alike. The only difference in the long-baseline and short-baseline subsea system is that the former uses a minimum of two transponders and generally three or four, whereas the short-baseline uses only one transponder. With the two transponder long-baseline system the general location of the transponder with respect to the hydrophone must be known.

As Fig. 2–22 shows, the receiving portion of the long-baseline system is exactly like its counterpart in the short-baseline system. Again the major difference between the two systems is that the long-baseline system requires three or more range counters and the short-baseline only one. The long-baseline requires no measurement of the bearing of the returning acoustic signals or vessel motion compensation.

For the long-baseline system, data rate is controlled by the equipment on board the vessel. However, the geometry of the transponder grid and the location of the vessel in the transponder grid influence when interrogations can be made. This influence of the grid on the data rate is because multiple acoustic pulses are received for each interrogation.

Thus, unless the vessel is equidistant from the transponders when

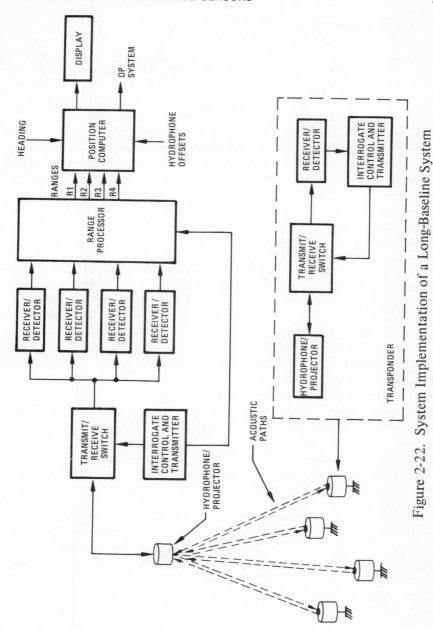

Figure 2-22. System Implementation of a Long-Baseline System

an interrogation is made, the return acoustic pulses will not occur at the same time. If the system does not interrogate until all the return pulses are received, then the maximum data rate is determined by the slant range to the most distance transponder. This means that for the normal long-baseline system where the distance between transponders is equal to the water depth and the water depth is 15,000 ft or deeper, the maximum time between interrogations inside the grid can be over 10 seconds.

As previously mentioned for the short-baseline system, higher data rates can be achieved by interrogating before the return pulses from the prior interrogation are received, i.e., the use of multiple pulses in the water. The problem of properly timing the interrogation pulses and receiving the return pulses is not difficult when the vessel is nearly equidistant from the transponders because all the return pulses arrive at approximately the same time.

However, when the vessel is away from the center of the grid, the arrival time of the return pulses is spread over several seconds. Then, the timing of multiple interrogations requires the sophistication of a modern digital computer.

In the previous section a comparison of the various short-baseline systems is made to assist the prospective buyer in selecting among the various types of short-baseline systems. The characteristics included in the comparison are summarized in Table 2–1. In Table 2–2 the same characteristics for the long-baseline system are listed so that they can be compared to the characteristics for the short-baseline system.

The basic differences between the long-baseline and short-baseline transponder systems can be summarized as follows:

• Number of transponders to deploy
• Calibration of the transponder grid
• Absolute distance accuracy

In terms of the first two differences the short-baseline transponder system is much easier to operate, has a lower cost of ownership, and can be put in operation much quicker than the long-baseline system. However, if the application calls for operations in water depths greater than 1000 ft with an absolute distance accuracy of better than 1% of water depth, the long-baseline system is required in spite of its cost and operational difficulties.

TABLE 2–2

CHARACTERISTICS OF A TYPICAL LONG-BASELINE ACOUSTIC
POSITION REFERENCE SYSTEM

Performance
Accuracy

- Small horizontal offsets (100% of water depth) with proper grid calibration

Better than 0.5% of range offset from the grid center

- Large horizontal offsets (greater than 100% of water depth)

1 or 2% of range offset from the grid center

Data Rate

Controlled by position computer; rate depends on slant range to all transponders and the geometry of the vessel within the transponder grid.

Area of Coverage

Limited by transponder power or by acoustic transmission effects; with sufficient transponder power several hundred percent of water depth is achievable.

Components

Subsea units	4
Hydrophones	1
Interrogators	1
Position computers/ signal processors	1
Vertical reference units	0

Installation

Hydrophone

One hydrophone well with stem of adjustable length; hydrophone should be replaceable at sea.

Interrogators

Interrogator can coreside with hydrophone.

Operation
Subsea unit deployment and recovery

Four units; generally expendable, free dropped because of water depth.

Initialization

Transponder grid calibration required.

2.6   System accuracy and reliability

Elements of the acoustic system have been mentioned which affect the position accuracy of the system. Likewise, the accuracy of each acoustic system has been given in Tables 2-1 and 2-2. The errors which determine the accuracy of the systems include constant and random components. The constant terms can be minimized or eliminated by calibration, but random errors can only be reduced to a minimum.

Each acoustic system is composed of many components: some are electrical, some are electromechanical, and some are mechanical. As for any system, the total system error is a composite of the error contributions of each component. A complete discussion of all the error sources in each acoustic system is beyond the scope of this book. So only the following error sources are discussed: geometric, measurement, and computational.

Geometric errors include errors in the geometrical parameters used to calculate the position of the reference point on the vessel relative to the subsea-reference point, including:

1. The location of the vessel-mounted units relative to the reference point on the vessel.

2. The location of the subsea unit relative to subsea-reference point.

3. The vertical separation between the vessel-mounted units and the subsea units.

The errors in the location of the vessel-mounted units relative to the reference point on the vessel are a function of the survey techniques used during the installation of the units. Therefore, these errors can be reduced to the accuracy of the survey techniques, so that the errors are in inches, not feet.

Errors in the location of the subsea unit relative to the subsea reference point depends on the method of deployment of the unit. If the unit is mounted physically to some structure connected to the subsea reference point, such as the BOP stack used in drilling, then the geometrical relationship to the reference point can be accurately measured. However, if the subsea unit is free dropped or lowered on a line, its location with respect to a subsea reference point can only be approximated until some other measurement correlated to the reference point is made.

The accuracy of the determination of the vertical separation between the vessel-mounted units and the subsea units depends on the installation accuracy of the hydrophones, the accuracy of the height of the sub-sea unit above the bottom, and the accuracy of the measurement of the water depth.

Measurement errors include the errors directly related to the measured quantities from which position is calculated. In all cases either time or phase are the primary measured quantities. The measured time can be round-trip time or difference in time-of-arrival for adjacent hydrophones. Round-trip time is used in systems which measure range and can be measured to a timing accuracy in the millisecond range, which is approximately a 2.5 ft error in range.

For differences in time-of-arrival, the time difference is used to compute angular offset (see Equations 2.43 and 2.44). Timing accuracy in computing angular offset is more critical than for range measurement as the following example shows. If the error in the time difference is one millisecond, then the error in the position offset for a 50-ft hydrophone separation is 10% of water depth.

Thus, to reduce the error in the position measurement in the difference in time-of-arrival system to less than a fraction of 1% of water depth, the accuracy of the timing difference must be better than 100 microseconds. Fortunately this kind of accuracy with modern electronic circuits is not difficult to achieve.

The accuracy of the short-baseline system is also dependent on the measurement accuracy of roll, pitch, and vessel-heading sensors which are used to compensate for the angular motions of the vessel. In the case of roll and pitch, the accuracy of the measurement will carry over directly to the accuracy of the system as shown in Equations 2.45 and 2.46.

For heading, the effect on accuracy depends on the offset of the sub-sea unit from the subsea reference point and the angle between the heading of the vessel and line between the subsea unit and subsea reference point. The accuracy of these sensors is discussed in more detail later.

As shown earlier in this section, each acoustic system uses a set of equations to compute position. The equations include geometric and measured parameters. In addition, the key parameter of the velocity of sound in water is used in the equations to convert time into distance or angle. Except in cases where a high degree of accuracy is required, the velocity of sound is not a measured parameter.

Therefore, a value is assumed which is typical or average for most velocity profiles. The error in the measured angular offset caused by the error in the assumed speed of sound is directly proportional to the magnitude of speed of velocity error (see Equation 2.43 and 2.44).

Accuracy of the data from a position reference system is important to the performance of a dynamic positioning system. The dynamic positioning system cannot hold position any more accurately than the accuracy of the position sensor. The rate at which the acoustic position reference system furnishes data also affects the performance of the dynamic positioning system. Under nominal performance all the currently produced acoustic reference systems are sufficiently accurate and furnish data at a sufficiently high rate to yield a dynamic positioning system with sufficient responsiveness, stability, and accuracy.

Data rate can be interpreted in another way from the repetition rate of a pinger or the interrogator. The data can also be noncontinuous because the acoustic signal cannot be properly detected. The degree of intermittency will depend on the cause of the lack of detection. At any rate, the controller of the dynamic positioning system should be notified by the acoustic system that the acoustic signal has been lost so that some other course of action can be taken. The operator should also be notified of the data loss if it is for an extended length of time.

The course of action taken by the controller will depend on the duration of the interruption and the alternatives the controller has available. For example, if there are other channels of position reference data, the controller can automatically switch to a channel that is not being interrupted. This strategy effectively avoids the total loss of position reference information to the controller for any length of time.

The position reference system to which the controller switches to can be either another acoustic position reference system or another type of position reference system. In any case the controller should not switch to another source of position data unless the new source is valid and operational.

For an acoustic system, the phenomena which causes a failure in signal detection in a properly operating system are either increases in acoustic noise being picked up by the acoustic reveivers or increased attenuation of the acoustic signal in the transmission path between the subsea unit and the vessel-mounted units. For most acoustic systems operating with separation distances between the subsea units and the vessel equal to or less than specified values, the acoustic noise will not be as likely to cause signal detection losses as frequently as increased attenuation in the transmission path. A major source of increased attenuation is air.

There are many sources of air. Sources of air in the acoustic transmission path include:

1. The dumping of material containing air in the region of or upstream of the vessel-mounted units.

2. Propeller-generated either by suction or cavitation by the vessel with the acoustic system, or a vessel near the vessel with the acoustic system.

3. Wave action against the vessel.

The likelihood of air being in the transmission path of the acoustic signal can be minimized by mounting the vessel units on extensions below the hull. Obviously, there are practical limits to this solution.

Also, multiple or redundant units can be mounted on the vessel at different locations to give alternate transmission paths (spatial diversity).[11] For this method the short-baseline system which has only one hydrophone is particularly well-suited. This method can also be used effectively with the long-baseline system.

R E F E R E N C E S

1. Van Calcar, Henry, "Acoustic Position Reference Methods for Offshore Drilling Operations," OTC paper 1141, 1969, Volume II pp 567–582.

2. Roberts, J. L., "An Advanced Acoustic Position Reference System," OTC paper 2173, 1975, pp 265–276.

3. Helstrom, Carl W., *Statistical Theory of Signal Detection,* Pergamon Press, 1960, New York.

4. Urick, Robert J., *Principles of Underwater Sound for Engineers,* McGraw-Hill, 1967, New York.

5. Albers, Vernon M., *Underwater Acoustic Handbook-II,* Pennsylvania State University Press, University Park, Pennsylvania, 1965.

6. Andrews, Frank A., "The Meaning of Error in Short-Base Underwater Navigation System," Catholic University of America, 1 September 1966.

7. Vaishnav, Ramesh and Magrab, Edward, "Exact Theory of Tracking of Moving Underwater Object by Short-Base Navigation System Attached to Imperfectly Stabilized Moving Ship," The Catholic University of America, Washington, D.C., April 1969.

8. Young, Myron C., "An Exact Solution to a Problem in Positioning," Institute of Navigation Conference, St. Louis, Missouri, June 18–21, 1973.

9. Leroy, C. C., DiGiacomo, C., and Prost, J., "Acoustic Measuring System and its Performance," OTC paper 2026, 1974, pp 849–863.

10. Henry Van Calcar, "Verbal Conversations Regarding Transponder Calibration Techniques on the Hughes *Glomar Explorer,*" January 1976.

11. J. L. Roberts, "More Reliable Acoustics Through Spatial Diversity," OTC 2868, 1977.

3

Nonacoustic Position and Other DP Sensors

3.1 Introduction

The remaining position reference sensors include the taut wire, marine riser, radio, satellite, and inertial systems. These systems do not depend on the transmission of acoustic energy through the water to measure position. Instead they use inclinometers, electromagnetic energy, and accelerometers to perform their position measurement. Each of these techniques has its own advantages and disadvantages.

Of the five additional position reference systems, the taut wire system is the most frequently used. In fact, as Table 1–1 indicates, the taut wire was the most frequently-used position reference sensor for the early dynamic positioning system and was the position reference sensor on the first automatically-controlled dynamically-positioned vessel, the *Eureka*.

However, as acoustic technology improved and operating water depths increased, the taut wire system was replaced by the acoustic position reference system as the prime position reference system.

Today, the taut wire system is used as a back-up position reference system on drillships as highlighted in Table 1–2. Another back-up position reference sensor used on drillships is the marine riser system. Like the taut wire, the marine riser position sensor uses a two-axis set of orthogonally-mounted inclinometers to perform its position measurement. The inclinometers in the marine riser system are typically mounted near the sea bed, not on the surface as in the case of the taut wire.

The application of radio position reference systems to dynamically-

86

positioned vessels has been limited previously to the USS *Naubuc*. In the future, radio position reference systems will be used more frequently in dynamic-positioning systems. The type of radio position reference system to be used in these applications will be high-frequency (microwave-range) line-of-sight, active transponder systems.

To date, satellite navigation and inertial position sensor systems have not been used as position reference systems for dynamic positioning systems. Both, however, may become future position reference systems for DP systems, especially in applications where navigation is an important requirement in the operational mission, e.g., mining and pipelaying.

In addition to position sensing, the measurement of the heading of the vessel is essential for a dynamic positioning system. The primary sensor measuring the heading of marine vessels since around the beginning of this century is the gyrocompass. It has been adopted as the heading sensor for dynamic positioning systems.

The measurement of the vertical reference is also essential to dynamic positioning systems using short-baseline acoustic, taut wire, or marine riser position reference systems. Secondly, the attitude of the vessel to the vertical is of interest to operating personnel or to some other subsystem on the vessel.

For example, the subsea handling equipment on mining vessels is motion-compensated so that it remains vertical as the vessel rolls and pitches.

The most critical of the vessel motion and environmental sensors is the wind sensor which is used as a part of the dynamic positioning control system. The manner in which the wind sensor measurements are used in the DP control system is anticipatory so that the thrusters can be commanded to react to sudden changes in the wind at the earliest possible moment.

The effect of using wind measurements in the DP control system has been shown to improve the position-holding accuracy of the DP system. The actual implementation of the wind compensation in the DP control system is given in Chapter 5.

Other vessel motion and environmental sensors described in this chapter are heave (vertical displacement), wave height, and current sensors. The measurements of these sensors are usually used only for performance information in the operating log of the vessel. As a result, they are usually only displayed to operating personnel and logged on a permanent recording system such as a strip chart recorder.

3.2   Taut wire systems

The Taut Wire Position Reference System (TWS) is basically a mechanical system in which the inclination of a line under constant tension is measured and transmitted electrically to the system using the position reference system. There are several varieties of TWS and they are mounted at various locations on the vessel.

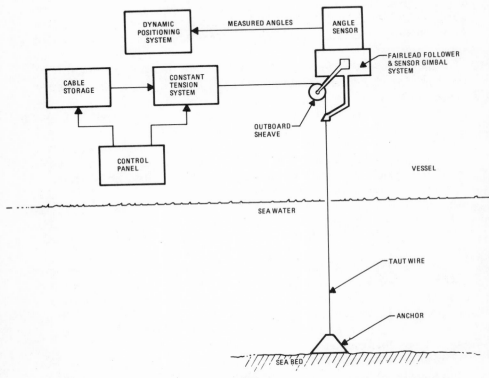

Figure 3-1. The Basic System Elements of a Taut Wire System

In Fig. 3-1 the basic elements of a TWS are illustrated; in Fig. 3-2 an example of a TWS installation is shown. Some of the elements are passive once the system is deployed and operating. These elements are not cirtical so long as they are properly sized to not affect the performance of the active elements. Included in the list of passive elements are:

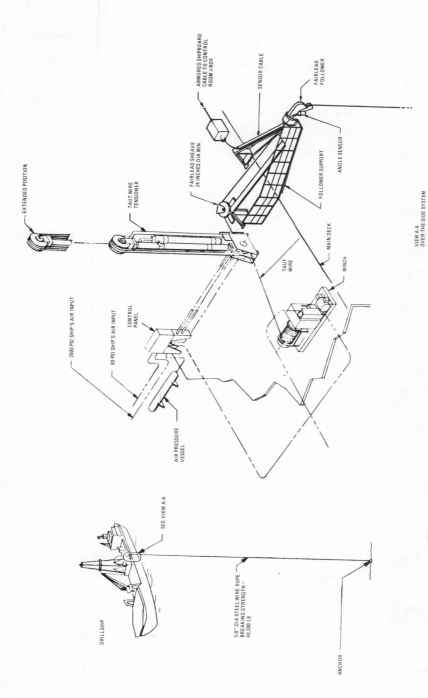

Figure 3-2. Over-the-Side Taut Wire Reference System Assembly

- Anchor: sized to maintain proper tension in the taut wire without dragging, yet easy to deploy and recover.
- Cable storage: sized to hold the line required for the limiting water depth plus offset.
- Outboard sheave: sized to minimize wear-and-tear on the taut wire.

The control panel and constant tension system will be of secondary importance because it is assumed that the tension in the taut wire is maintained within a sufficiently close tolerance of a constant value. As a result, only three elements of the taut wire system are of primary importance to the position reference performance of the TWS:

- Taut wire
- Fairlead follower and sensor gimbal mechanism
- Angle sensor (the inclinometer)

The taut wire is generally a steel cable with a diameter of $3/8$ to $3/4$ in. With these cable sizes a tension level with sufficient safety margin can be used for reasonable accuracy.

The fairlead follower and sensor gimbal mechanism is a portion of the TWS which interfaces the cable to the angle sensor. Consequently, this mechanism should ideally transfer the angle of the cable to the angle sensor without influencing the angle in any manner. Likewise, the gimbal portion of this mechanism should decouple the angle sensor from the roll and pitch of the vessel. Important to the proper operation of each mechanism is balanced, mechanical freedom. They also should be rugged and easy to maintain.

The TWS can be located anywhere on vessels. However, on drillships most installations are amidship to reduce the effect of vessel motion on the constant tension system. The structure that supports the fairlead follower and sensor gimbal can be configured either above or below water. The installation shown in Fig. 3–2 is more convenient for deployment/recovery and maintenance, but it can be in the way of any vessel that comes alongside the vessel on which it is installed.

The angle sensor is typically composed of two orthogonally-mounted, pendulous-type inclinometers which give electrical signals which are generally routed from the dynamic positioning room to the sensor by a multiconductor, shielded electrical cable as indicated in Fig. 3–2. The sensor is very similar to the vertical reference pendulum pot sensors discussed in Section 3.6.

Ideally, the taut wire runs in a straight line between its attachment point to the vessel and the sea-floor anchor. Fig. 3–3 illustrates the geometry of the ideal taut wire from which the offsets of the taut wire attachment point to the vessel from the anchor can be computed using the following equations:

$$\frac{XTW^*}{h_{TW}} = \tan \theta_x = \frac{\sin \theta_{xm}}{\sqrt{1 - \sin^2 \theta_{xm} - \sin^2 \theta_{ym}}} \tag{3.1}$$

$$\frac{YTW^*}{h_{TW}} = \tan \theta_y = \frac{\sin \theta_{ym}}{\sqrt{1 - \sin^2 \theta_{xm} - \sin^2 \theta_{ym}}} \tag{3.2}$$

where:

h_{TW} = the vertical separation between the attachment point of the taut wire to the vessel and the anchor

θ_{xm} = the measured angle in the plane containing the x-axis of the sensor

θ_{ym} = the measured angle in the plane containing the y-axis of the sensor.

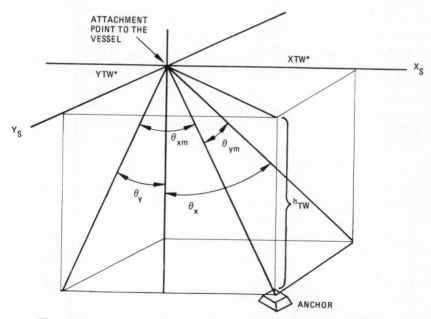

Figure 3-3. The Ideal Taut Wire Measurement Geometry

If the taut wire is attached to the vessel in the horizontal plane as shown in Fig. 3–4 with an angular misalignment with the vessel, α_{TWS}, then the offsets can be expressed in matrix form as follows:

$$\frac{1}{h_{TW}}\begin{bmatrix} XTW \\ YTW \end{bmatrix} = \frac{1}{h_{TW}}\begin{bmatrix} \cos \alpha_{TWS} & -\sin \alpha_{TWS} \\ \sin \alpha_{TWS} & \cos \alpha_{TWS} \end{bmatrix}\begin{bmatrix} XTW^* \\ YTW^* \end{bmatrix} \qquad (3.3)$$

Transformation of the offsets from the attachment point of the taut wire on the vessel to the anchor into the vessel coordinates with reference to a subsea reference point is given as follows:

$$\begin{bmatrix} PXTW \\ PYTW \end{bmatrix} = \begin{bmatrix} XTWAB \\ TYWAB \end{bmatrix} - \begin{bmatrix} XTWO \\ YTWO \end{bmatrix} + \begin{bmatrix} XTW \\ YTW \end{bmatrix} \qquad (3.4)$$

where:

XTWAB, YTWAB = the coordinates of the currently used locations of the anchor in vessel-fixed coordinates referenced to the subsea reference point

XTWO, YTWO = the offsets of the attachment point of the taut wire from the reference point on the vessel in vessel-fixed coordinates

PXTW, PYTW = the indicated location of the reference point on the vessel to the subsea reference point.

So that the anchor remains fixed when vessel heading changes occur, the anchor location is expressed in north-east earth coordinates. The conversion of the anchor location into vessel-fixed axes is as follows:

$$\begin{bmatrix} XTWAB \\ YTWAB \end{bmatrix} = \begin{bmatrix} \cos \psi & \sin \psi \\ -\sin \psi & \cos \psi \end{bmatrix}\begin{bmatrix} XTWAN \\ YTWAN \end{bmatrix} \qquad (3.5)$$

where:

XTWAN, YTWAN = the currently used locations in N-E coordinates
ψ = the heading of the vessel

Substitution of Equation 3.5 into Equation 3.4 yields:

$$h_{TW}\begin{bmatrix} x_{TW} \\ y_{TW} \end{bmatrix} = \begin{bmatrix} XTW \\ YTW \end{bmatrix} - \begin{bmatrix} XTWO \\ YTWO \end{bmatrix} + \begin{bmatrix} \cos \psi & \sin \psi \\ -\sin \psi & \cos \psi \end{bmatrix}\begin{bmatrix} XTWAN \\ YTWAN \end{bmatrix} \qquad (3.6)$$

where x_{TW} and y_{TW} equal $PXTW/h_{TW}$ and $PYTW/h_{TW}$, respectively, and are the indicated taut wire position in per unit water depth. If the misalignment of the angle sensor is included in Equation 3.6, then:

$$h_{TW}\begin{bmatrix} x_{TW} \\ y_{TW} \end{bmatrix} = \begin{bmatrix} \cos \alpha_{TWS} & -\sin \alpha_{TWS} \\ \sin \alpha_{TWS} & \cos \alpha_{TWS} \end{bmatrix} \begin{bmatrix} \dfrac{h_{TW} \sin \theta_{xm}}{(1 - \sin^2 \theta_{xm} - \sin^2 \theta_{ym})^{1/2}} \\ \dfrac{h_{TW} \sin \theta_{ym}}{(1 - \sin^2 \theta_{xm} - \sin^2 \theta_{ym})^{1/2}} \end{bmatrix}$$

$$-\begin{bmatrix} XTWO \\ YTWO \end{bmatrix} + \begin{bmatrix} \cos \psi & \sin \psi \\ -\sin \psi & \cos \psi \end{bmatrix} \begin{bmatrix} XTWAN \\ YTWAN \end{bmatrix} \qquad (3.7)$$

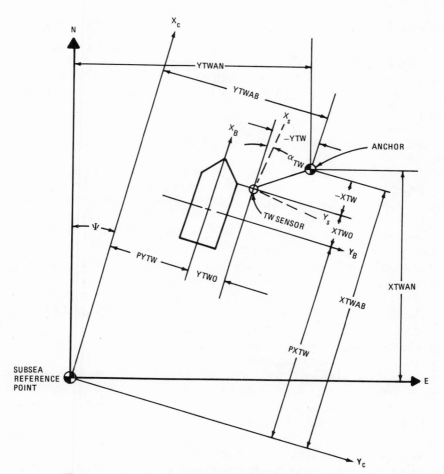

Figure 3-4. Taut Wire Horizontal Plane Geometry

For the case when the taut wire is within a few degrees of vertical, $\sin \theta_{xm}$ and $\sin \theta_{ym}$ are approximately equal to θ_{xm} and θ_{ym}, respectively, and $\sin^2 \theta_{xm}$ and $\sin^2 \theta_{ym}$ are approximately equal to zero. Likewise, if the misalignment of the angle sensor to vessel coordinates is within a few degrees, then $\sin \alpha_{TWS}$ is approximately α_{TWS} which results in an equally valid approximation of $\alpha_{TWS} \theta_{xm}$ and $\alpha_{TWS} \theta_{ym}$ equal to zero. With these approximations Equation 3.7 becomes:

$$\begin{bmatrix} x_{TW} \\ y_{TW} \end{bmatrix} = \begin{bmatrix} \theta_{xm} \\ \theta_{ym} \end{bmatrix} - \frac{1}{h_{TW}} \begin{bmatrix} XTWO \\ YTWO \end{bmatrix} + \frac{1}{h_{TW}} \begin{bmatrix} \cos \psi & \sin \psi \\ -\sin \psi & \cos \psi \end{bmatrix} \begin{bmatrix} XTWAN \\ YTWAN \end{bmatrix} \quad (3.8)$$

Equation 3.8 is derived without any vertical or angular motions of the vessel except the angular motion in the horizontal plane. Even with perfect compensation by the follower/gimbal system there is some variations in the indicated position because of vessel roll, pitch, and heave.

However, the error is small and, generally, does not warrant the effort for compensation, especially when the taut wire is used in a nearly vertical position.

In the actual taut wire system the taut wire is not straight and the fairlead follower/gimbal mechanism and sensor do not reproduce the angle of the taut wire to the vertical perfectly. Each element has as a static behavior as well as a dynamic behavior which affects the manner in which the taut wire system performs as the position reference sensor for a dynamic positioning system.

The static portion of the taut wire system behavior affects only the long-term accuracy of the dynamic positioning system. The dynamic behavior of the system affects not only the dynamic accuracy of the dynamic positioning system but also the stability of the closed-loop control of the dynamic positioning system.

When static, the major element of the taut wire system is the taut wire itself. The other elements will only contribute to the static behavior of the system if they are not operating properly. The static behavior of the taut wire is determined by the distributed forces of water current drag and wire weight (catenary) acting along the length of wire. Fig. 3–5 depicts the two effects acting on the wire and the resulting difference in the actual offset angle and the measured offset angle. To actually compute the difference between the actual offset angle and the measured angle, the classical partial differential equa-

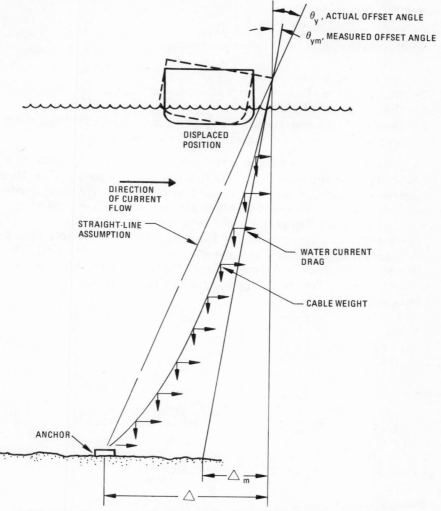

Figure 3-5. The Influence of Current and Wire Weight on the Static Behavior of the Taut Wire System

tion for a stretched string under an axial and lateral load must be solved. In one dimension this equation is given as follows:

$$\frac{\partial}{\partial z}\left(T\,\frac{\partial y}{\partial z}\right) + W\,\frac{\partial y}{\partial z} = \rho A\,\frac{\partial^2 y}{\partial t^2} - F(z,t) \qquad (3.9)$$

where:

  y = the lateral deflection of the string
  z = the vertical distance from the bottom
  t = time
  T = the tension in the string
  W = the weight of the string in water per unit length
  ρA = the mass per unit length of the wire
  F = the lateral loading per unit length (from the hydrodynamic drag
       and the inertia of the water moving with the string).

To solve Equation 3.9 for the static case, all time-related terms are ignored. Then, if the lateral loading function has the proper form, the equation can be solved to obtain a closed form. However, in many cases the loading function does not result in a closed form. Then a computer program can be used to perform the required integrations.

An example of a lateral loading function which results in a closed form solution to Equation 3.9 is for a water current acting along the entire length of the wire whose velocity decreases with the square of the water depth. The resulting solution to Equation 3.9 is given as follows:

$$\frac{dy}{dz}\bigg|_{\substack{\text{upper}\\\text{end}}} = \frac{\Delta}{h_{TW}}\left[\frac{\frac{W}{T}h_{TW}}{\epsilon^{\frac{W}{T}h_{TW}}-1}\right]$$

$$+ \frac{q_0}{W}\left[\frac{\frac{W}{T}h_{TW}}{\epsilon^{\frac{W}{T}h_{TW}}-1}\right]\left[\frac{1}{3}-\left(\frac{T}{Wh_{TW}}\right)+2\left(\frac{T}{Wh_{TW}}\right)^2\right] \qquad (3.10)$$

$$-\frac{q_0}{W}\left[1-2\left(\frac{T}{Wh_{TW}}\right)+2\left(\frac{T}{Wh_{TW}}\right)^2\right]$$

where:

  Δ = the actual offset of the vessel
  q_0 = the surface current loading
  h_{TW} = the length of the taut wire.

The first term of Equation 3.10 represents the catenary effect of the weight of the wire, the second and third terms describe the effect of the water current acting on the wire.

In Fig. 3–6 apparent offset versus actual offset as given by Equation 3.10 is plotted to show the effect of tension, water depth, and surface current. When there are no static errors, the apparent offset equals the actual offset as indicated by the dashed lines in Fig. 3–6. As Fig. 3–6 illustrates, there is no static error when there is no current and infinite tension, zero water depth, or zero actual offset (a perfectly vertical wire).

The conditions of zero current and zero actual offset may occur from time to time but the conditions of infinite tension or zero water depth are, obviously, unreliable. Thus, with a taut wire system there is normally some static error which increases with decreasing tension, increasing water depth, increasing actual offset, and increasing surface current velocity. For more details on the static errors in taut wire systems, see Reference 1.

The predictability of static errors in the taut wire is very good if the current velocity and direction are known as a function of water depth. However, to continually measure current speed and direction along the length of the taut wire is not easy to do. A more convenient way to predict the static error or calibrate the taut wire is to use an acoustic position reference system which is not affected to any degree by water current. Since a taut wire system is very commonly used as a back-up sensor to an acoustic position reference system, the combination of an acoustic position reference system and a taut wire system is very common.

The acoustic calibration of the taut wire system can be implemented either automatically or manually. In either case the static difference between the acoustic and taut wire systems is determined and is used as a bias term to the equation used to compute the indicated taut wire position. In equation form, the new equation for indicated taut wire position is:

$$
\begin{bmatrix} x_{TW} \\ y_{TW} \end{bmatrix} = \begin{bmatrix} \theta_{xm} \\ \theta_{ym} \end{bmatrix} - \frac{1}{h_{TW}} \begin{bmatrix} XTWO \\ YTWO \end{bmatrix} + \frac{1}{h_{TW}} \begin{bmatrix} \cos\psi & \sin\psi \\ -\sin\psi & \cos\psi \end{bmatrix} \begin{bmatrix} XTWAN \\ YTWAN \end{bmatrix}
$$

$$
- \begin{bmatrix} \cos\psi & \sin\psi \\ -\sin\psi & \cos\psi \end{bmatrix} \begin{bmatrix} X_A - X_{TW} \\ Y_A - Y_{TW} \end{bmatrix}
$$

(3.11)

where:

$(X_A - X_{TW})$ = the x-axis static difference between the acoustic and the taut wire position in vessel coordinates in per unit water depth

$(Y_A - Y_{TW})$ = the y-axis static difference between the acoustic and taut wire position in vessel coordinates in per unit water depth.

To determine the static difference between the acoustic and taut wire systems, the uncalibrated signals from the sensors must be filtered to obtain their static component. Also, the difference, which is most conveniently determined in vessel coordinates, should be transformed into earth coordinates because the current, which is acting on the wire, is an earth-referenced phenomenon.

The static accuracy of the taut wire is as important to the accuracy of the system as the static part of the acoustic position reference system discussed in Section 2.6. Likewise the dynamic accuracy of the taut wire system is important to the system accuracy. Even more important, the dynamic accuracy characteristics can affect the stability of the dynamic positioning control system.

At this point a discussion of stability of the dynamic positioning control system is premature. However, before leaving the taut wire system, a simplified model of the dynamics of the taut wire itself is developed. The dynamics of the other parts of the systems, such as the fairlead follower and gimbal structure and the angle sensor, are neglected at this time.

As previously given in Equation 3.9, the motion of a vibrating string in a viscous fluid is described by a partial differential equation involving spatial as well as time-varying terms. As a result, the derivation of a transfer function for a taut wire is not a straight-forward task. In fact, since the final term in the partial differential equation is a nonlinear term, the solution to the equation is nonlinear and changes not only as the boundary conditions change but also as the lateral loading function changes. Thus, no general solution to the motion of the taut wire is possible. As a result, the following brief heuristic argument gives the general nature of the dynamic response of the taut wire.

The dynamic characteristics which are developed in the following presentation are in terms of phase and gain as a function of frequency. Phase and gain characteristics are used because the discussion of stability of a control system in Chapter 7 is in terms of phase and gain in the frequency domain.

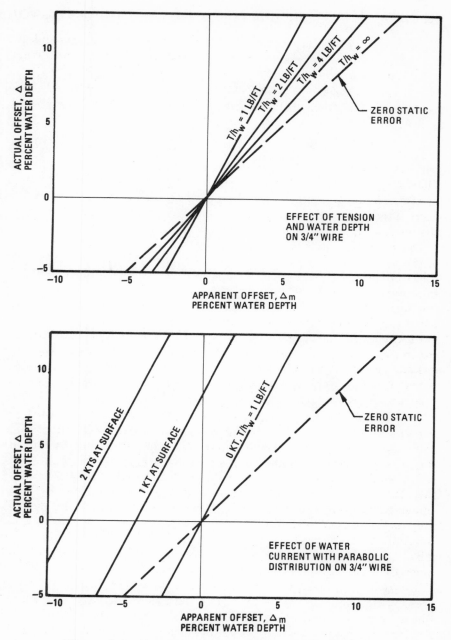

Figure 3-6. Static Accuracy of Taut Wire System

If the taut wire is not in the fluid and the upper end is moved in an oscillatory manner defined:

$$y\,(L,\,t) = y_L \sin \omega t \qquad (3.12)$$

then the general form of the solution is given by the following infinite series:

$$y\,(z,\,t) = \sum_{i=1}^{\infty} Y_i(z)\,[A_i \cos \lambda_i t + B_i \sin \lambda_i t] \qquad (3.13)$$

where:

$y(z,t)$ = the lateral deflection of the string
$Y_i(z)$ = the ith mode pattern of the string
$A_i,\,B_i$ = the constants determined from the boundary conditions
λ_i = the ith natural frequency of the string.

The natural frequencies of the wire are functionally dependent on the following system parameters:

$$\lambda_i \sim f\left[\frac{1}{h_{TW}},\ \sqrt{\frac{T}{\rho A}}\right] \qquad (3.14)$$

When the effects of hydrodynamics are included, the time response of the wire to an oscillatory end point motion is modified considerably. In Fig. 3–7 the gain and phase characteristics as a function of frequency are shown for a 1000-ft taut wire.[2]

None of the resonant response characteristics at the natural frequencies of the wire are visible in the response curves given in Fig. 3–7. The absence of the resonant characteristics of the wire is caused by the hydrodynamic damping of the wire as the upper end is oscillated.

The important dynamic characteristics of the taut wire that Fig. 3–7 clearly shows is the phase lead and amplification that the taut wire introduces in the region of the frequency domain given by the following empirical equation:

$$\omega_{TW},\ \text{rad/sec} = \frac{\pi}{10h_{TW}} \sqrt{\frac{T}{\rho A}} \qquad (3.15)$$

which for steel wire equals

$$\omega_{TW},\ \text{rad/sec} = \frac{1.1}{h_{TW}d_w}\ \sqrt{T} \qquad (3.16)$$

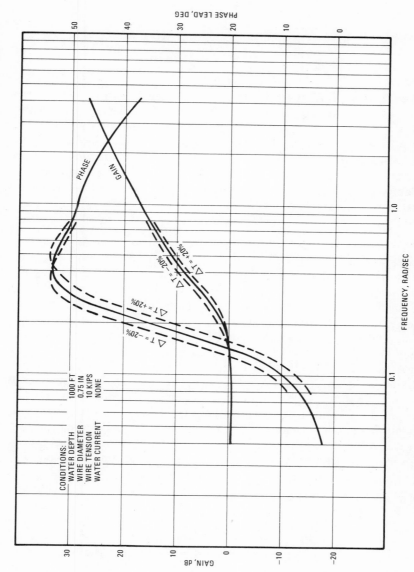

Figure 3-7. Effect of Tension Variation on the Taut Wire Dynamic Model

where:

h_{TW} = the length of the taut wire, ft
T = the tension in the taut wire, lb
d_W = the diameter of the wire, in.

Amplification and phase lead can destabilize the automatic control system of the dynamic positioning system if they are not taken into account in the design of the control system. Likewise, the amplification will exaggerate the indicated motions of the vessel at higher frequencies. This exaggeration is especially true of the wave-induced motions of the vessel. As a result, the display of indicated position from the taut wire in heavy wave action will appear "noisy" compared to an acoustic position reference system.

3.3   Marine Riser System

Between the drillship and the BOP stack on the wellhead there is a large diameter conductor which contains the drill string and a drilling fluid used to clean out the cuttings of the drill, lubricate the drilling process, and control the pressure of the well. The conductor is called the marine riser and is composed of individual sections or joints as shown in Fig. 3–8. The riser is connected to the subsea BOP stack with a flexible joint called the ball joint.

The vertical angle between the ball joint and the marine riser is critical to drilling personnel because if the ball joint and the marine riser are not aligned with a small angular difference, the drill string will rub against the inside of the upper part of the BOP stack. In fact, if the angular misalignment exceeds more than 5 or 6 degrees, the drill string may hang up in the BOP stack and it may be impossible to extract the drill string from the riser.[3]

Because of the importance of the ball joint angle an angle sensor is placed on the riser just above the ball joint.[3] In some cases a second angle sensor is placed on the BOP stack to determine its attitude with respect to vertical. In either case the sensor that is used to measure the angle is a two-channel inclinometer system like the one used for the taut wire mounted in a pressure case. The angular information from the two inclinometers is transmitted to the surface either on a hardwire, as part of subsea BOP stack multiplex system, or acoustically.

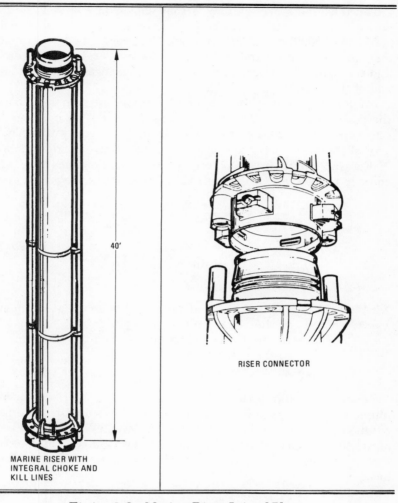

40'

MARINE RISER WITH
INTEGRAL CHOKE AND
KILL LINES

RISER CONNECTOR

Figure 3-8. Marine Riser Joint [7]

To maintain the alignment of the marine riser with the ball joint
and BOP stack, the riser is under tension and the drillship is positioned in a small diameter watch circle over the wellhead. The tension
in the riser is produced by multiple tensioners (6 to 8 units) mounted
around the periphery of the drilling platform with lines feeding under
the drilling floor through sheaves to a collar at the top of the riser.

The tensioners are similar to that shown in Fig. 3–2 for the taut wire. However, they have a much larger capacity and capabilities of 50,000–100,000 lb of tension each. More complete descriptions of the marine riser system and the characteristics as related to their design and the drilling operation are given in References 3, 4, 5, and 6 at the end of this chapter.

Angles measured at the bottom of the riser are also used to compute the position of the drillship, thus providing another back-up position reference system to the acoustic position reference system. The geometry of the riser angle sensor system under ideal conditions is shown in Fig. 3–9. With this geometry the x-y positions of the attachment point to the vessel with respect to the BOP stack are as follows:

$$\frac{XRA^*}{h_{RA}} = \frac{-\sin\theta_{xm}}{\sqrt{1 - \sin^2\theta_{xm} - \sin^2\theta_{ym}}} = -\tan\theta_x \qquad (3.17)$$

$$\frac{YRA^*}{h_{RA}} = \frac{\sin\theta_{ym}}{\sqrt{1 - \sin^2\theta_{xm} - \sin^2\theta_{ym}}} = \tan\theta_y \qquad (3.18)$$

where:

XRA^*, YRA^* = the horizontal distances between the attachment point to the vessel and the ball joint

h_{RA} = the vertical separation between the attachment point to the vessel and the ball joint

θ_{xm}, θ_{ym} = the measured angles.

Generally speaking the measurement axes of the riser sensor is not aligned with geographical earth reference coordinate systems. As a result, the measured offsets from Equations 3.7 and 3.8 must be co-ordinate-transformed by the angular offset between the x-axis of the riser angle sensor and north. This transformation takes the following form:

$$\begin{bmatrix} XRE \\ YRE \end{bmatrix} = \begin{bmatrix} \cos\psi_{RA} & -\sin\psi_{RA} \\ \sin\psi_{RA} & \cos\psi_{RA} \end{bmatrix} \begin{bmatrix} XRA^* \\ YRA^* \end{bmatrix} \qquad (3.19)$$

where:

XRE, YRE = earth axes coordinates of the attachment point to the vessel.

The attachment point of the riser to the drillship is generally not at the position reference point on the vessel. Thus, further transforma-

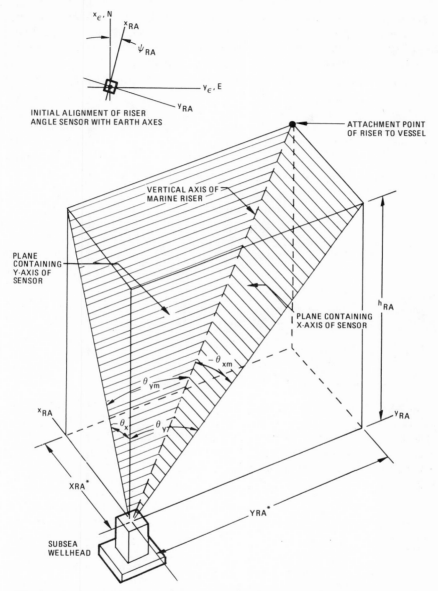

Figure 3-9. Measurement Geometry of the Riser Angle Sensor [7]

tions of the measured riser angle data is required to convert it into comparable data with other position reference systems in use on the drillship. The geometry of the attachment point offsets with respect to Fig. 3–10 and is expressed as follows:

$$\begin{bmatrix} PXRA \\ PYRA \end{bmatrix} = \begin{bmatrix} XRA \\ YRA \end{bmatrix} + \begin{bmatrix} XRAAB \\ YRAAB \end{bmatrix} - \begin{bmatrix} XRAO \\ YRAO \end{bmatrix} \qquad (3.20)$$

where:

XRA, YRA = the distances between the riser attachment point to the vessel and the wellhead in vessel coordinates

XRAO, YRAO = the offsets between the riser attachment point to the vessel and the position reference point on the vessel in vessel coordinates

XRAAB, YRAAB = the distances between the subsea reference point and the subsea wellhead in vessel coordinates

PXRA, PYRA = the distances between the subsea reference point and the position reference point on the vessel in vessel coordinates.

So that the wellhead remains fixed as the heading of the vessel changes, the location of the wellhead is expressed in north-east earth coordinates which transforms into vessel coordinates as follows:

$$\begin{bmatrix} XRAAB \\ YRAAB \end{bmatrix} = \begin{bmatrix} \cos \psi & \sin \psi \\ -\sin \psi & \cos \psi \end{bmatrix} \begin{bmatrix} XRAA \\ YRAA \end{bmatrix} \qquad (3.21)$$

where:

XRAA, YRAA = the location of the wellhead with respect to the subsea reference point in N-E coordinates

ψ = the heading of the vessel.

Likewise, the measurements made by the riser angle sensor in north-east earth coordinates, as expressed in Equation 3.19, must be transformed by the heading of the vessel as given in the following equation:

$$\begin{bmatrix} XRA \\ YRA \end{bmatrix} = \begin{bmatrix} \cos \psi \, \sin \psi \\ -\sin \psi \, \cos \psi \end{bmatrix} \begin{bmatrix} XRE \\ YRE \end{bmatrix} \qquad (3.22)$$

Combining Equations 3.19, 3.21, and 3.22 with Equation 3.20 and redefining PXRA and PYRA in terms of per unit water depth results in

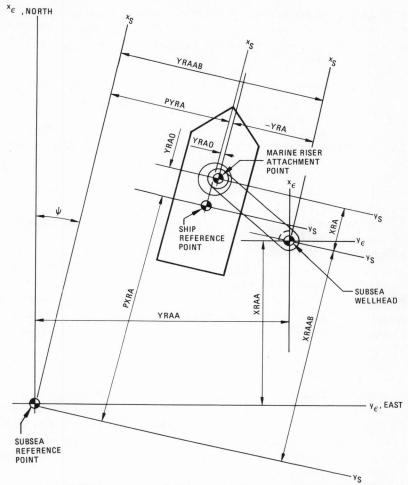

Figure 3-10. Horizontal Plane Geometry [7]

the following equation for the indicated vessel position from riser angle measurements in per unit water depth:

$$\begin{bmatrix} x_{RA} \\ y_{RA} \end{bmatrix} = \frac{1}{h_{RA}} \begin{bmatrix} \cos(\psi - \psi_{RA}) & \sin(\psi - \psi_{RA}) \\ -\sin(\psi - \psi_{RA}) & \cos(\psi - \psi_{RA}) \end{bmatrix} \begin{bmatrix} XRA^* \\ YRA^* \end{bmatrix}$$
$$+ \frac{1}{h_{RA}} \begin{bmatrix} \cos\psi & \sin\psi \\ -\sin\psi & \cos\psi \end{bmatrix} \begin{bmatrix} XRAA \\ YRAA \end{bmatrix} - \frac{1}{h_{RA}} \begin{bmatrix} XRAO \\ YRAO \end{bmatrix} \quad (3.23)$$

Since the tilt of the marine riser is limited to ± 10 degrees, Equations 3.17 and 3.18 can be accurately expressed by small angle approximations which results in the following expression for Equation 3.23:

$$\begin{bmatrix} x_{RA} \\ y_{RA} \end{bmatrix} = \begin{bmatrix} \cos(\psi - \psi_{RA}) & \sin(\psi - \psi_{RA}) \\ -\sin(\psi - \psi_{RA}) & \cos(\psi - \psi_{RA}) \end{bmatrix} \begin{bmatrix} -\theta_{xm} \\ \theta_{ym} \end{bmatrix}$$
$$+ \frac{1}{h_{RA}} \begin{bmatrix} \cos\psi & \sin\psi \\ -\sin\psi & \cos\psi \end{bmatrix} \begin{bmatrix} XRAA \\ YRAA \end{bmatrix} - \frac{1}{h_{RA}} \begin{bmatrix} XRAO \\ YRAO \end{bmatrix} \qquad (3.24)$$

The indicated position of the vessel in terms of riser angle measurements given by Equation 3.24 is derived without any vertical or angular motions of the vessel except in vessel heading. However, the errors caused by vertical motion (heave), roll, and pitch are small because of the size of the vertical motion relative to the length of the riser and the near verticality of the riser. As a result the errors do not generally warrant compensation.

Like the taut wire system, the marine riser system does not perform ideally. It exhibits static and dynamic behaviors which are in variance with the ideal given by Equation 3.24. The static behavior affects only the long-term accuracy of the DP system and can be compensated for, using the acoustic system position measurements as described in the previous section for the taut wire.

The effect of the dynamic behavior on the DP system is twofold: (1) it affects the dynamic accuracy, and (2) it affects the stability of the closed-loop controller of the DP system.

In a manner similar to the taut wire system, the primary element of the marine riser system which determines its static behavior is the riser and its contents. The other elements of the system, such as the sensors or the tensioners, affect the static behavior only if they are not operating properly. The combination of distributed forces caused by water current drag and gravity forces acting along its length determine the actual shape of the riser between the vessel and the ball joint for a given top tension as illustrated in Fig. 3–11.

Fig. 3–11 also depicts the difference in measured riser angle and the ideal or actual offset angle. To actually compute the difference between the measured offset angle and actual offset angle, a complex partial differential equation must be solved. This equation in one dimension is as follows:[7]

$$\frac{\partial^2}{\partial^2 z}\left(EI\,\frac{\partial^2 y}{\partial z^2}\right) - \frac{\partial}{\partial z}\left(T\frac{\partial y}{\partial z}\right) - W\,\frac{\partial y}{\partial z} = F\,(z,t) - \rho A\,\frac{\partial^2 y}{\partial t^2} \qquad (3.25)$$

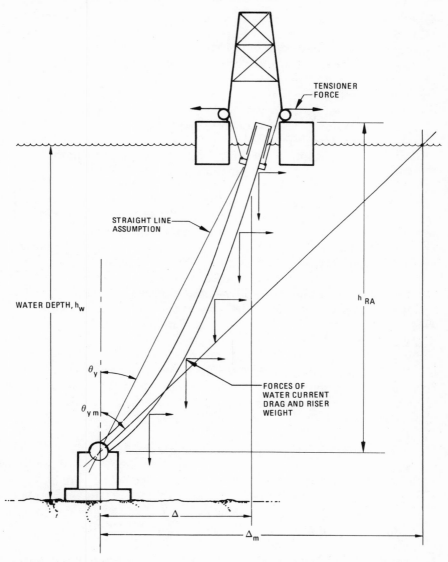

Figure 3-11. Probable Error in the Indicated Riser Angle Position [7]

where:

y = the lateral deflection of the riser
z = the vertical distance from the ball joint
t = time
EI = the bending elasticity of the beam
T = the tension in the beam
W = the weight (in water) per unit length
ρA = the mass per unit length
F = the lateral loading per unit length (from hydrodynamic drag and inertia of the water moving with the riser).

To solve Equation 3.25 in the static case, all time related functions are set equal to zero. Then, for some cases of lateral loading, a closed-form solution can be determined for the equation. More generally it is easier to use a standard computer program to solve such an equation.

Like the taut wire behavior shown in Fig. 3–6 the static behavior of the marine riser is a function of upper tension to the weight per unit foot ratio and the magnitude of the water current. A significant part of the weight per unit foot of the riser is the mud weight. Since mud weight varies depending on drilling parameters, the static behavior of the riser varies unless the top tension is adjusted accordingly.

Yet another variable in the static behavior is the make-up of the riser. To reduce the tensioning requirements, buoyancy modules are attached to the riser. These modules decrease the weight per unit foot of riser, but they also increase the diameter of the riser which results in increased hydrodynamic drag and increased sensitivity to water current.

Determination of the dynamic characteristics of the riser as they relate to the measured angle is even more complex than for the taut wire system. Computer programs have been developed which can be used to determine the dynamic characteristics of the riser angle system. One such program was developed by CONOCO and is available on a commercial time-share computer system. The CONOCO program is unfortunately limited to zero water current conditions which is a serious limitation because of the sensitivity of the dynamic characteristics of the riser to water current. However, the CONOCO program is much more economical to use than time-domain programs which use large quantities of computer time to determine the transfer function characteristics of the riser.

The results of using the CONOCO program to derive the gain and phase characteristics of an 18⅝-in. marine riser are shown in Fig. 3–12 for the indicated parameters. As might be expected, the angle measured at the ball joint of the riser exhibits phase lag (time delay) and amplification (the measured angle is larger than the actual offset angle) as the excitation frequency increases. These dynamic characteristics of the riser angle may affect the dynamic performance of the DP control system if the riser data is used in the same manner as the acoustic data.

Whether the dynamic characteristics of the riser position data affect the DP control system depends on where the phase lag and amplification introduced by the riser occurs relative to the critical frequency range of the DP control system. Because the riser characteristics are a function of so many parameters, some of which vary as a function of the drilling operation and others because of the change in water current, the problem of properly compensating for the riser dynamics is not trivial if compensation is required. The problem may require either an adaptive compensation or supplementation of the riser data with another sensor output such as the taut wire or another riser angle sensor near the top of riser. The two riser angle sensor approach is being proposed by at least one offshore supplier at the present time.

3.4   Other position reference sensors

So far two position reference sensors have been considered. The position of offshore vessels can be measured by several other techniques, including radio, satellite navigation, and inertial navigation. To date, these position reference sensors have been used almost exclusively for navigation and survey purposes and not in dynamic positioning. The USS *Naubuc* successfully used HIFIX as its position reference system as shown in Table 1–1.

The basic reason for this pattern of application of these position reference sensors is that their position accuracy is not suitable for dynamic positioning. Additionally, geographic position is of only secondary importance to most applications to which dynamic positioning has been applied so far.

In the future, however, this situation will change because of applications such as mining and pipelaying and the projected improved accuracy of these position reference sensors.

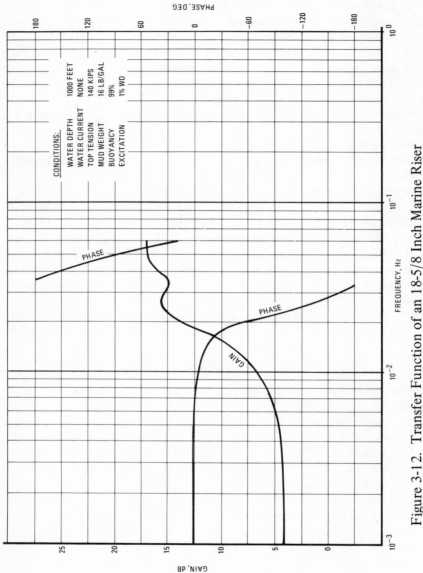

Figure 3-12. Transfer Function of an 18-5/8 Inch Marine Riser (Lower Inclination Angle Versus System Offset Angle) [7]

The discussion in the following paragraphs is very brief because there are many excellent papers and books on these systems.

There are many radio position reference systems.[8] The most commonly used radio position reference system uses three transmitting stations at known geographical positions. One of the stations is designated the master and the other two are slaved to it. By transmitting either a continuous radio signal or a pulse, a hyperbolic grid pattern is established between the master and one slave. Thus, with two master-slave pairs, two overlapping hyperbolic patterns are generated to form a radio grid coordinate system of the area of coverage of the stations as shown in Fig. 3-13. With the knowledge of the geographic location of the three transmitters the hyperbolic grid that they form can be overlaid on maps of the area.

The mobile user who is within range of the transmitting stations and who is equipped with the proper receiver equipment and hyperbolic map, can conveniently locate himself within the grid and, therefore, geographically. The accuracy with which he can locate himself is a function of many variables such as the radio transmission characteristics which are affected by the ionosphere, the time-of-day, local area electrical storms, etc.

The location of the user within the hyperbolic grid is also important as well as the relative location and distance between the two master-slave pairs. For the best accuracy, the baselines connecting the two master-slave pairs should be at right angles to one another and the user should be located in the area where the hyperbolic lines generated by the separate master-slave pairs are more or less orthogonal to each other.

Radio positioning or navigation systems can be categorized by their maximum operating range. The first category, the most accurate, has a maximum range of up to 50 km which is line-of-sight operation. The other three categories are medium range, long range, and global. The maximum range of the medium range system is up to 750 km and long range is up to 2,000 km. Accuracy of the various categories of radio systems generally decreases with range as shown in Table 3-1.

Position-measuring accuracy for radio systems is a function of radio transmission conditions (daytime versus nighttime, electrical storms, etc.) and geometrical considerations. As a result, a general statement of the accuracy of the radio positioning system even from one time of day until another is impossible. However, of the various radio systems, the short-range systems have the greatest potential for use as a position reference sensor for a dynamic positioning system.

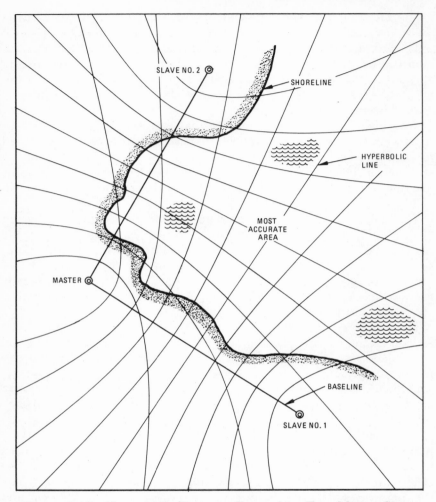

Figure 3-13. Hyperbolic Patterns Created by Two Master-Slave
Combinations

In fact, at least one supplier of dynamic positioning control systems
is configuring his system with a short-range radio system.

The short-range configuration being used as a position reference
sensor consists of two transmit/receive transponders mounted on the
vessel separated by a known baseline distance, d. On a platform in the

TABLE 3-1. RADIO POSITION REFERENCE SYSTEMS

CLASSIFICATION	RANGE	ACCURACY	FREQUENCY
SHORT RANGE	50 km	2 - 20 m	3000 MHz
MEDIUM RANGE	500 km	10 - 100 m	2 MHz
LONG RANGE	2000 km	15 - 400 m	100 kHz
GLOBAL	7000 km	1000 - 5000 m	10 kHz
SATELLITE	Global	40 - 100 m	150 MHz and 400 MHz

vicinity of the operation to be performed, an active transponder is located as shown in Fig. 3–14. Then, by direct ranging from the two vessel-mounted transponders to the platform-mounted transponder, the x-y position of the vessel relative to the transponder can be computed as follows:

$$x = \frac{R_1^2 - R_2^2}{2d} \tag{3.26}$$

$$y = \left[R_1^2 - \left(\frac{R_1^2 - R_2^2}{2d} + \frac{d}{2} \right)^2 \right]^{1/2} \tag{3.27}$$

where the quantities are defined in Fig. 3–14.

The achievable accuracy with such a system as shown in Fig. 3–14 is in the range of a few meters when the range intersection angle is between the recommended limits of 30 and 150 degrees. As a result of this constraint, the accuracy of the system becomes a function of vessel offset from the platform geometry and heading of the vessel as described by the following expression for the intersection angle:[9]

$$\tan \alpha = \frac{4 \, \Delta \, d}{4\Delta^2 - d^2} \cos \beta \tag{3.28}$$

where the quantities are defined in Fig. 3–14.

The significance of this operational limit is illustrated in Fig. 3–15 where Equation 3.28 is plotted for a minimum recommended intersection angle of 30 degrees and various fixed-baseline lengths. As Fig. 3–15 shows, to maintain the high accuracy the vessel offset must not exceed less than twice the separation of the two transponders on the vessel being dynamically positioned.

Since the baseline length on most vessels will be limited to less than 100 m, this range-range radio position reference system is limited

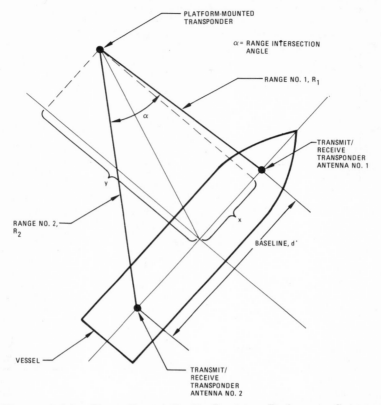

Figure 3-14. Short-Range Radio Position Reference System

to operating close to the platform transponder. Of course, the higher degree of accuracy is generally only necessary near the platform. Thus, when the vessel moves to transponder offsets which no longer give the highest accuracy, but still give line-of-sight measurements, the radio system may still operate with sufficient accuracy to be usable.

Satellite navigation also uses the propagation of radio frequencies to permit users to compute their geographical location. The present satellite navigation system uses satellites in circular polar orbits at an approximate altitude of 600 nautical miles. The system is operated by the U.S. Navy and includes six satellites, each circling the earth in about 107 minutes.

As a satellite circles the earth, it transmits continuously messages

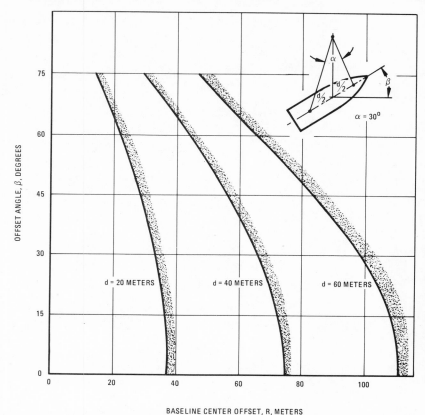

Figure 3-15. Offset Angle as a Functicn of Offset Distance for the
Minimum Intersection Angle

of two minute duration on 150 and 400 mhz. Each message contains
the two-minute message synchronization mark, a 400-hertz reference
signal, and fixed and variable parameters describing the satellites
orbit. The fixed and variable orbit parameters are stored in a memory
in the satellite and are updated periodically from ground tracking
stations operated by the U.S. Navy (every 12 to 16 hours).

Computation of the user's position is similar to the other hyperbolic
radio position reference systems. The obvious difference is that the
satellite acts as the master transmitter and the slave transmitter as it
moves along its orbit as illustrated in Fig. 3–16. Thus, with three

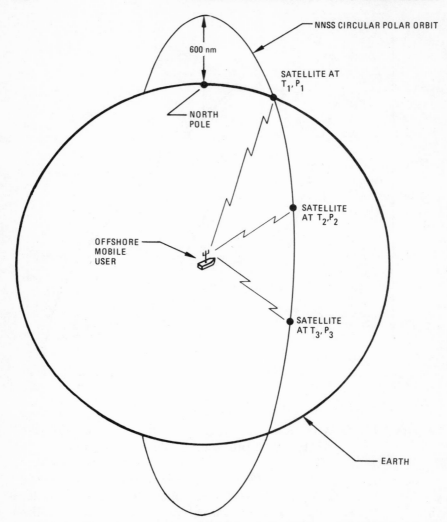

Figure 3-16. Geometry of a Satellite Navigation Position Fix
Calculation

"good" message receptions from the satellite, the user can compute
his location (position fix) from the two intersecting hyperbolic surfaces.
The accuracy of the position fix can be improved by more "good"
message receptions on the same satellite orbit pass. The maximum
number of messages under the most favorable conditions is 9. How-

ever, many factors such as unfavorable location with respect to the orbiting satellite and radio interference can invalidate many of the satellite messages.

To determine his position, the user requires an antenna, a receiving unit, and a computer. The receiving unit provides a measure of the slant range between the user and the satellite and the satellite-orbit parameters for use in the computer. The computer, using preestablished mathematical equations, computes the position of the user in geographic coordinates (latitude and longitude).

The computer can also keep track of when the next satellite will pass, which occurs on the average of every 90 minutes. For more complete descriptions of the entire satellite navigation method see References 11, 12, and 13.

The present satellite navigation gives the offshore user accuracy in the 40–100 m range, depending on conditions. However, it suffers from the fact that users cannot obtain continuous position measurement. To overcome this problem and to improve the position-fixing accuracy, integrated navigation systems have evolved. These systems combine the satellite navigation receiver and its computer with sensors to measure the user's velocity over the surface of the earth, the heading of the user's vessel, and the roll and pitch of the user's vessel.

Then, the satellite navigation system computer can combine the data from all the sensors in an optimum manner to obtain the "most accurate position measurement" of the user's vessel during a satellite pass and during the time between satellite passes. The computational methods being applied in these integrated systems use the Kalman filtering technique.[12]

Typically, a Doppler sonar, a gyrocompass, and inclinometers are used to measure vessel velocity, heading, and roll/pitch, respectively. More advanced (and more expensive) systems add radio position reference system inputs and inertial platform inputs. Then, the software programs are prepared which combine the various sensor data to achieve an optimum position estimate.

With these wide arrays of sensor inputs the totally integrated navigation system can operate in every part of the world, using only the sensors suited for the particular area of interest. Where more than one system can be used to create the same data, the computer of these systems can combine the data to resolve ambiguities and furnish uninterrupted position information as the outputs of the sensors fail or are interrupted.

However, the accuracy of these systems is still beyond the required range for dynamic positioning. A new satellite system referred to as the Global Positioning System (GPS) promises increases in position-fixing accuracy to the range of 10 m and 24-hour continuous position fixing throughout the world.[14] This system, if it becomes a reality, may be accurate enough for certain dynamic positioning applications.

The final position reference system to be discussed is a combination of several already mentioned position reference sensors. The combination involves an inertial platform and at least one other local position reference system as shown in Fig. 3–17. Other local position reference systems which have been proposed are the acoustic position reference system, taut wire, and riser angle sensor.[15]

By itself, an inertial position reference system, composed of an inertial reference unit (IRU) and a digital computer, is a dead-reckoning navigation system. The basic measurement performed by the inertial system is acceleration. Then, by integrating acceleration, velocity is determined, and by integrating velocity position is determined. Initialization of the integrations must be supplied to the system.

Additionally, the platform must be aligned with respect to the earth so that a measurement and computational reference system can be established.

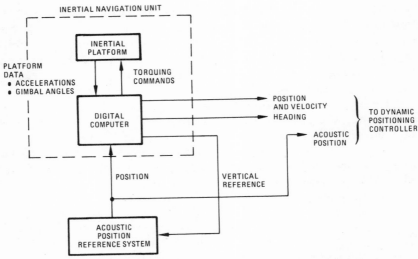

Figure 3-17. An Inertial Position System Block Diagram with Acoustic Position Reference System Augmentation

The acoustic, taut wire, marine riser, or any combination of the three sensors supply the needed measurements for platform alignment and initialization. Also, these support sensors provide measurements necessary to bound the drift in the computed velocity and position caused by various errors in platform measurements and computations.

All this information and the platform data are combined in the digital computer using Kalman filter techniques to obtain the best estimate of the vessel's position and velocity for use in the dynamic positioning controller. Reference 15 gives a complete description of this system.

The inertial platform system provides secondary benefits to the dynamic positioning system. In addition to position and velocity, the platform provides good vertical reference data for use with the short-baseline acoustic position reference system and precise heading information referenced to north. As a result the platform can replace the vertical reference sensor and gyrocompass (the heading sensor) normally used in a dynamic positioning system.

When this book was written, the inertial position reference system had not been applied to a dynamic positioning system. Some of the reasons why this system has not been used, even though a backup system to the acoustic position reference system is needed, are:

1. The commercially-available inertial platforms are capable of backup for only a limited number of minutes because of their drift rate limitations.

2. Commercially-available inertial platforms are sophisticated pieces of equipment which have published MTBF (Mean Time Between Failures) of 3,000 to 4,000 hours.

3. The equipment is quite expensive and requires maintenance at manufacturer's depots, increasing the logistics problems of operating such a system.

4. Even with the drawbacks to the taut wire and marine riser backup position reference systems, the combined position-reference systems perform satisfactorily and are easier to understand and maintain by shipboard personnel.

However, in the future the inertial position reference system will be used in a dynamic positioning system. This will be especially true if the military-grade electrically-suspended-gyro platform is made available for commercial uses because of its very low drift characteristics.

3.5   Heading sensor

In addition to controlling the position of the vessel with respect to some reference point, the dynamic positioning system controls the heading of the vessel. The purpose of heading control depends to some degree on the task of the vessel.

If the vessel is a track-following vessel, then heading control is important in achieving the desired track. Otherwise, control of the vessel's heading is more often used to maintain the orientation of the vessel at a heading which is either favorable to the operation or to the weather or both. In the case of heading control because of the weather, the heading control can maintain the bow or stern of the vessel into the weather to reduce thruster loading and vessel motion and to maintain a near-constant heading for equipment handling.

Control of heading requires the measurement of the vessel's heading. Measurement of the vessel's heading is also necessary to perform any coordinate transformations from vessel coordinates to another coordinate system in the horizontal plane. For example, the location of the acoustic beacon must be coordinate-transformed by the vessel heading if the beacon is not located at the subsea reference point.

The heading of a vessel has historically been measured with reference to north. The first heading sensor of any consequence was the magnetic compass which operates by the alignment of a free-rotating magnet with the earth's magnetic field. The performance of the magnetic compass depends on the earth's magnetic field which is highly variable from one point on the earth's surface to another. Any local magnetic field or magnetic material such as the hull of a steel ship can influence the reading of the compass.

In addition, the mechanism which allows the magnetic needle of the compass to rotate into alignment with the earth's magnetic field must ideally be frictionless, which is impossible. As a result, the accuracy of the magnetic compass is in terms of degrees, not in terms of fractions of a degree.

A more modern heading sensor than the magnetic compass is the gyrocompass. Although the principle by which the gyrocompass works was discovered by Foucault in 1851, the practical application of the principle had to wait until the advent of electric motors to drive the gyros, and Sperry and Anschutz to invent separate versions of the gyrocompass.

A gyrocompass is basically a pendulous suspended gyro as shown in

Fig. 3–18. The sensed component of earth rotation causes the gyro to precess so as to align its spin axis in a northerly direction. Then by measuring the angle between the reference on the case containing the gyro and the gyro-spin axis, the heading referenced to north is determined.

A gyrocompass is basically a two-axis gyro (Fig. 3–18). By producing a torque about the tilt axis proportional to the tilt angle, θ, the gyro can be caused to precess toward the meridian from whichever side of north the spin axis is initially oriented. The torque about the tilt axis can be produced by adding a pendulous mass below the center of gravity of the rotor or by an electric torque motor driven by a signal which is proportional to the tilt angle as shown in Fig. 3–18.

The pendulous-mass technique was used in the early gyrocompasses

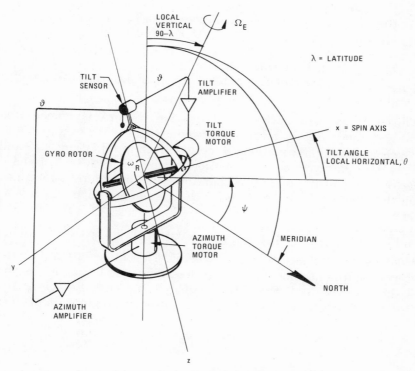

Figure 3-18. A Schematic Diagram of a Gyrocompass

but the torque motor has found more application in modern gyrocompasses because of the advantages of electric control.

Once the spin axis is aligned with the meridian, the heading referenced to north is determined by measuring the angle between the reference on the case holding the outer gimbal and the gyro spin axis. When the spin axis of the gyro is not aligned to the meridian, the gyrocompass can be modeled for small angles of tilt and azimuth errors from north by the following differential equations:[17]

$$I_z \frac{d^2\psi}{dt^2} + H \frac{d\theta}{dt} + \psi H \Omega_E \cos \lambda = 0 \qquad (3.29)$$

$$I_y \frac{d^2\theta}{dt^2} - H \frac{d\psi}{dt} + K_T \theta - H \theta_E \sin \lambda = 0 \qquad (3.30)$$

where:

I = the moment of inertia
H = the angular momentum of the rotor = $\omega_R T_R$
Ω_E = the earth's angular velocity = 72.9211×10^{-6} r/s
λ = the latitude
θ = the tilt angle of the spin axis
ψ = the angle between the spin axis and north, i.e., the heading error
K_T = the tilt or pendulous gain.

The earth rate term in the second equation, $H_E \sin \lambda$, is compensated for in gyrocompasses by torquing the gyro with the tilt-torque motor.

The solution to the equations of motion given in Equations 3.29 and 3.30 is the undamped harmonic motion for any initial offset in ψ and θ. The frequency of the undamped harmonic motion can be determined from the characteristic equation for the system equation which is as follows:

$$I_z I_y s^4 + (H^2 + K_T I_z + H I_y \Omega_E \cos \lambda) s^2 + K_T H \Omega_E \cos \lambda = 0 \quad (3.31)$$

where s is the Laplace transform operator. From Equation 3.31 the frequency of the harmonic motion is determined by letting $s = j\omega$ and solving Equation 3.31. Since H is much greater than K_T, I_y, I_z, and $H\Omega_E \cos \lambda$, the solution to Equation 3.31 is approximately as follows:

$$\omega_{n1,2} = \sqrt{\frac{H^2}{2I_y I_z} \left[1 \pm \sqrt{1 - \frac{4K_T I_y I_z H \Omega_e \cos \lambda}{H^4}} \right]} \qquad (3.32)$$

where $\omega_{n1,2}$ = the two primary harmonic frequencies at which the gyrocompass oscillates when given an initial offset angle. Equation 3.32 can be further approximated because of the dominant size of H with respect to the other quantities. The resulting harmonic frequencies are:

$$\omega_{n1} = \sqrt{\frac{H^2}{I_y I_z}} \tag{3.33}$$

$$\omega_{n2} = \sqrt{\frac{K_T \, \Omega_E \, \cos\lambda}{H^2}} \tag{3.34}$$

or in terms of natural periods:

$$T_1 = 2\,\pi\sqrt{\frac{I_y I_z}{H^2}} \tag{3.35}$$

$$T_2 = 2\,\pi\sqrt{\frac{H^2}{K_T \, \Omega_E \, \cos\lambda}} \tag{3.36}$$

The first natural period, T_1, is very short (a fraction of a second) and represents the transient or short-term behavior of the gyrocompass. The second natural period, T_2, is many minutes, is a function of the latitude, and greatly affects the initial alignment time of the gyrocompass.

The periods determined for the gyrocompass are the same in both the tilt and azimuth axes. Likewise, the motion in both axes is oscillatory with a dominate period equal to the longer period. The oscillations in the two axis will be approximately 90 degrees out-of-phase because of smallness of the first natural period. As a result, the end of the spin axis will trace out an elliptical path if the spin axis is given a small initial azimuth error.

The amplitude of the oscillations in each axis can be determined by simplifying Equations 3.29 and 3.30, since the acceleration terms have little affect on the response of the gyrocompass.

The simplified equations are:

$$\frac{d\theta}{dt} = -\psi\Omega_E \cos \lambda \tag{3.37}$$

$$\frac{d\psi}{dt} = \frac{K_T}{H}\,\theta \tag{3.38}$$

Since the resulting motion to an initial angular offset in either axis is sinusoidal (oscillatory), the tilt and azimuth angles can be expressed approximately as follows:

$$\psi = \psi_o \cos \omega_{n2} t \qquad (3.39)$$

$$\theta = \theta_o \sin \omega_{n2} t \qquad (3.40)$$

where:

ψ_o, θ_o = the peak amplitude of the azimuth and tilt angles, respectively

t = time.

Then, by substituting Equations 3.39 and 3.40 into Equations 3.37 and 3.38 and solving for the ratio of peak amplitudes, the result is

$$\frac{\psi_o}{\theta_o} = \sqrt{\frac{K_T}{H\Omega_E \cos \lambda}} = \frac{\omega_{n2}}{\Omega_E \cos \lambda} \qquad (3.41)$$

The relative size of the parameters in Equation 3.41 for a gyrocompass makes the ratio significantly greater than one. Thus, for all the previous equations to be valid, the tilt angle of the gyrocompass must be considerably less than one degree.

The gyrocompass as modeled by Equations 3.29 and 3.30 is useless because of its oscillatory nature. By adding damping to a gyrocompass, it is made a useful navigation device. Damping can be added either around the horizontal axis or around the vertical axis. In this book only the latter will be discussed. Reference 16 gives a discussion of the horizontal damping technique.

Damping is added to the gyrocompass around the vertical axis through the azimuth torque motor. The amount of torque applied around the vertical axis is proportional to the tilt angle which is measured by the tilt sensor. Fig. 3–18 illustrates schematically the arrangement of the tilt sensor and the azimuth torque motor.

The effect of applying a torque around the vertical axis on the model equations for the gyrocompass can be seen in the following equations:

$$\frac{d\theta}{dt} = -\psi\Omega_E \cos \lambda - \frac{K_D \theta}{H} \qquad (3.42)$$

$$\frac{d\psi}{dt} = \frac{K_T \theta}{H} \qquad (3.43)$$

where K_D = the damping gain. When these equations are solved for the heading error, ψ, a damped second-order equation results; i.e.:

$$\frac{d^2\psi}{dt^2} + \frac{K_D}{H}\frac{d\psi}{dt} + \psi\,\frac{K_T}{H}\,\Omega_E\cos\lambda = 0 \qquad (3.44)$$

which has a natural frequency equal to that of the undamped gyro-compass and an equivalent damping factor equal to:

$$\zeta_G = \sqrt{\frac{K_D^2}{4HK_T\,\Omega_E\cos\lambda}} \qquad (3.45)$$

Thus, with the addition of vertical damping the response of the gyro-compass to an initial offset such as during initial alignment will be a damped sinusoid, causing the path of the end of the spin axis to be a convergent spiral. The time required for the offset to decay to a small fraction of its initial value will depend on the damping factor and the longer natural period of the gyrocompass, i.e., T_2. However, the time will be no less than one period, which is generally over an hour.

Once the spin axis of the gyrocompass is aligned with the local meridian, the angle between the spin axis and the reference line of the gyrocompass case is the indicated heading of the gyrocompass with respect to geographic north. Thus, if the reference line of the gyro-compass case is aligned to the centerline of the vessel on which the gyrocompass is mounted, the heading of the vessel with respect to geographic north can be read directly by an angular sensor mounted between the gimbaling system and the case as shown in Fig. 3–18.

The spin axis of the gyrocompass remains aligned to the local meridian unless it is subjected to external disturbance torques about its horizontal and vertical axes. Since the bearing between the gimbaling system and the gyrocompass case is designed to have minimum friction, the response of the gyrocompass to changes in vessel heading is instantaneous and has a negligible effect on the alignment of the spin axis of the gyrocompass to the local meridian. As a result, the gyrocompass exhibits negligible dynamics in the measurement of the heading of the vessel with respect to geographic north.

Because of the response of the gyrocompass to other external disturbance torques, the measurement of the heading of the vessel contains errors from various sources. Each gyrocompass includes compensating elements to minimize the effects of the sources of error. The errors included in the gyrocompass measurement are as follows:[16]

1. Latitude
2. North speed
3. East speed
4. North acceleration
5. East acceleration
6. Gimbal
7. Random

Many of these errors are very small, especially for vessels which are not moving fast, e.g., station-keeping vessels. Additionally, the manufacturers of marine gyrocompasses have been ingeniously successful in minimizing these errors.[16] As a result, the gyrocompass has an accuracy of better than one degree even in heavy seas.

To electrically interface the heading measured by the gyrocompass into the dynamic positioning system, two devices are integrated into the gyrocompass hardware. The most common is a synchro transmitter which outputs three ac voltages whose phase differences are functionally related to the measured angle.[18]

Another interfacing device is a sine/cosine potentiometer which converts the measured angle into two analog voltages which are directly proportional to the sine and cosine of the measured angle. Then, the heading angle can be computed using an arc tangent function of the ratio of the sine voltage to the cosine voltage.

3.6   Vertical reference sensors

There are many devices on a dynamically positioned vessel which require the determination of the angles between the local vertical and a reference plane on the vessel. The device which is used to measure these angles is called a vertical reference sensor.

There are many types of vertical reference sensors. Their performance can generally be summarized by the phrase, "you get what you pay for." In this section three types of vertical reference sensors are discussed to show the manner in which their response characteristics affect the devices that they are a part of, or that they support.

The vertical reference sensor is used on a dynamically positioned vessel in at least four ways, two of which have already been mentioned. The first is the compensation for hydrophone plane variations in the short baseline acoustic position reference system. In this application the vertical reference sensor operates in parallel with the acoustic system, i.e., if the vertical reference or the acoustic system fails, the other continues to perform.

In the second and third applications the vertical reference sensor acts as an inclinometer to measure the slope of the taut wire and the marine riser, respectively. As such, the vertical reference sensor is a primary in-line device for which the taut wire and marine riser system will not operate if the vertical reference sensor is not operating.

The vertical reference sensor is also used to sense the attitude of the vessel for display to shipboard personnel or for use in ballast control systems.

The first vertical reference sensor to be considered is the pendulous mass type sensor. It is the simplest, least expensive, and poorest performer of the various types. The principle by which the pendulum mass operates is simple: a pendulous mass tends to seek alignment on the average along the local vertical. Thus, if the mass is free to rotate with respect to its reference frame, e.g., the horizontal plane of the vessel, and if a means of measuring the angle between the pendulous mass and the reference frame is provided, then the pendulous mass, as it seeks the local vertical, forms an angle to the reference frame which is the desired measured angle. An example of the pendulous mass sensor is shown in Fig. 3–19.

The basic problem with the pendulous mass sensor is not the measurement of the angle between the reference frame and the pendulous

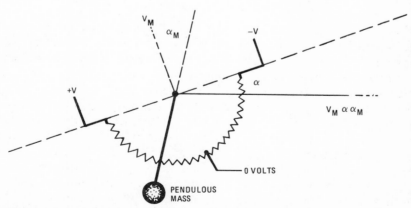

Figure 3-19. Schematic Representation of the Pendulous Mass Vertical Reference Sensor

mass, but that the angle between the mass and the reference frame is not, except on the average, the correct angle. The primary reason the angle is not correct is that the pendulous mass reacts to lateral accelerations. This effect is illustrated in Fig. 3–20. In terms of a mathematical expression the effect of lateral acceleration is:

$$\alpha_M = \alpha + \frac{\ddot{x}}{g} \qquad (3.46)$$

where the quantities are defined in Fig. 3–20 and g is the acceleration caused by gravity with a commonly used value of 32.2 ft/sec/sec.

Sources of lateral acceleration include the lateral acceleration of the vessel and angular accelerations acting through lever arms when the sensor is not mounted at the center of rotation of the vessel. The length of the lever arm is equal to the distance of the sensor from the center of rotation. For this reason these sensors should be mounted as near the center of rotation as possible.

Another problem which can affect the performance of the pendulous mass is the means of measuring the angle between the pendulous mass and the sensor reference frame. One common means is to use a resistive potentiometer in which the wiper of the potentiometer is coupled to the pendulous mass. Then, the voltage at the wiper (Fig. 3–19) is

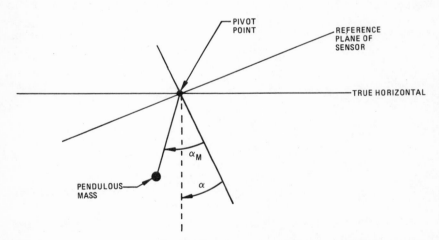

Figure 3-20. Diagram Showing the Effect of Lateral Acceleration
on the Measurement of the Vertical Angle

proportional to the angle to be measured. The mechanism is inexpensive and simple, but the wiper, rubbing on the resistive element of the potentiometer, exhibits a frictional drag and has a life limit of only several million cycles. For waves with a 10-second period, one million cycles is only 2,800 hours.

To illustrate the frictional and acceleration effect on the angular measurement by the pendulous mass sensor, two comparative time histories are shown in Fig. 3–21. The upper time history is from a pendulous mass sensor and the lower time history is from a vertical gyro. The pendulous mass sensor used to record the time history given in Fig. 3–21 was mounted near the center of rotation of the vessel.

The frictional effects in the pendulous mass can be eliminated by using a nonmechanical coupling between the pendulous mass and the reference frame of the sensor. These couplings include variable reluctance or optical devices. However, they involve more electronics generally and are, therefore, more complicated and expensive. Their quoted life times are as much as 10 billion cycles.

A significant improvement in the measurement of roll and pitch can be achieved with a vertical gyro. A vertical gyro consists of a rotor, spinning at very high speed, i.e., 24,000 rpm. The rotor is maintained vertically by a sensitive torquing circuit which senses local vertical. A schematic diagram of a vertical gyro is shown in Fig. 3–22.

The vertical gyro is also sensitive to lateral accelerations, but in quite a different manner than the pendulous mass sensor. Fig. 3–23 illustrates the sensitivity of the vertical gyro to lateral acceleration. The overall transfer function for the block diagram shown in Fig. 3–23 is:[19]

$$\frac{\theta}{A}(s) = \frac{1}{g} \frac{1}{\dfrac{J}{gK_1K_2} s^3 + \left(\dfrac{D + JK_2}{gK_1K_2}\right) s^2 + \dfrac{s}{K_2} + 1} = \frac{1}{g} F_A(s) \quad (3.47)$$

where:

J = moment of inertia of pendulum about its pivot, 15.7 dyne · cm · sec^2

D = viscous damping of pendulum, 1.7×10^5 dyne · cm · sec

K_1 = pendulum constant, 5.52 gm · cm

K_2 = transfer function of the erection loop, 0.0714 rad/sec/rad

g = earth's gravity, 980 cm/sec^2.

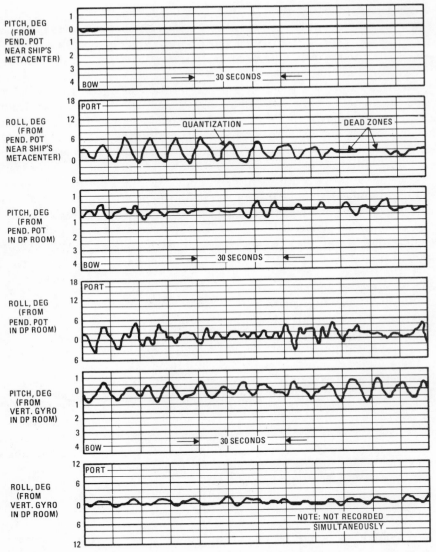

Figure 3-21. Roll/Pitch Measurement for Pendulum Pot and Vertical
Gyro Sensors [2]

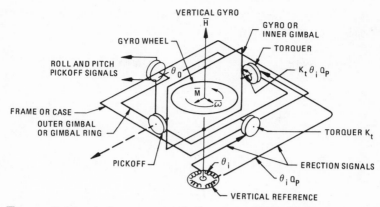

Figure 3-22. A Schematic Diagram of a Vertical Gyro [15]

Since the moment of inertia is so large, Equation 3.47 can be approximated as follows:

$$\frac{\theta}{A}(s) = \frac{1}{g}\frac{\omega_n^2}{s^2 + 2\zeta\,\omega_n s + \omega_n^2} = \frac{1}{g}\,F_A(s) \tag{3.48}$$

where:

$$\omega_n = 0.0479 \text{ rad/sec}$$
$$\zeta = 0.332$$

A new vertical reference sensor consists of a combination of linear and angular accelerometers. The linear accelerometers are used to measure the low-frequency roll and pitch. The angular accelerometers measure acceleration and upon double integration the high-frequency roll and pitch results. A special blending circuit is used to transition from the low-frequency to the high-frequency measurements.

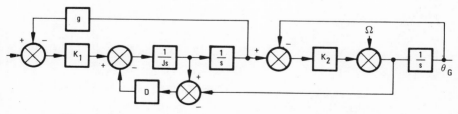

Figure 3-23. A Block Diagram for a Vertical Gyro [7]

The transfer function given by the manufacturer for this accelerometer vertical reference system is as follows:

$$\frac{\theta}{A}(s) = \frac{1}{g}\frac{K_1(1 + a_1s + a_2s^2)}{1 + a_1s + a_2s^2 + a_3s^3 + a_4s^4} = \frac{1}{g}F_A(s) \qquad (3.49)$$

where:

$$K_1 = 57.3 \text{ degrees}$$
$$a_1 = 78$$
$$a_2 = 1{,}140.8$$
$$a_3 = 8{,}483$$
$$a_4 = 25{,}570$$

which factors into the following form:

$$\frac{\theta}{A}(s) = \frac{1}{g}\frac{K(s + 0.0513)(s + 0.0171)}{(s + 0.1645)(s + 0.0162)(s + 0.0755 \pm j\,0.0945)} \qquad (3.50)$$

where $j = \sqrt{-1}$.

As a compensatory device to the acoustic position reference system, the vertical reference system operates in parallel with the acoustic measurement system (Fig. 3–24). The outputs of the vertical reference system correct for the angular variations of the hydrophone plane caused by vessel roll and pitch motions. In equation form the vertical reference sensor correction is given as follows:

$$\theta = \theta_m - \left(\alpha + \frac{1}{g}F_A(s)\,A(s)\right) \qquad (3.51)$$

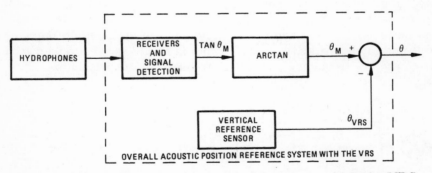

Figure 3-24. A Block Diagram of the Manner in which the VRS Combines with the Short-Baseline Acoustic Position Reference System

where:

θ_m = the acoustical measured offset angle
α = the actual angle of variation of the vessel from the horizontal
$F_A(s)$ = the lateral acceleration transfer function of the vertical reference sensor
$A(s)$ = the lateral acceleration
s = the Laplace transform operator

The manner in which the vertical reference sensor combines with the acoustic system affects the overall transfer function of the acoustic position reference system. This phenomena can be seen by considering Equation 3.51 with only the linear portion of the lateral acceleration term. Starting with the approximation that the acoustically measured offset angle is equal to the following expression for small angles:

$$\theta_m \cong \frac{x}{h_w} + \alpha \qquad (3.52)$$

then Equation 3.51 can be written approximately in transfer function form:

$$\theta_m = \left(1 - s^2 \frac{h_w}{g} F_A(s)\right) \qquad (3.53)$$

where:

h_w = the water depth
x = the position displacement of the vessel.

An important functional relationship that Equation 3.53 demonstrates is the dependence of the transfer function of the overall compensated acoustic position reference system on the water depth. The effect of the depth will be in the design of the control system for the dynamic positioning system.

Unlike the short baseline acoustic position reference system, the vertical reference sensor is used in a direct measurement fashion in the taut wire and riser angle systems. As a result the vertical reference sensor becomes a serial element of the taut and riser angle systems. This change of roles for the vertical reference sensor between the acoustic and the other two systems also results in a change in the effect of the vertical reference sensor on the transfer function of the other two systems. For example, for the taut wire illustrated in Fig. 3–25, the measured angle of the vertical reference sensor, θ_{ms}, is equal to:

$$\theta_{ms} = \theta_m + \frac{1}{g}\, F_A(s)\, A(s) \tag{3.54}$$

where $F_A(s)$ and $A(s)$ are defined previously in Equation 3.51. Thus, because the angle measured by the vertical reference sensor is the angle used to compute the position instead of to compensate for the hydrophone plane angular variation, the second term of Equation 3.54 resulting from lateral acceleration is additive as opposed to subtractive.

The change in sign of the acceleration term from the acoustic system application to the taut wire system changes the entire overall transfer function of the taut wire system.

Another factor which changes the effect of the vertical reference sensor on the taut wire system as opposed to the effect on the acoustic

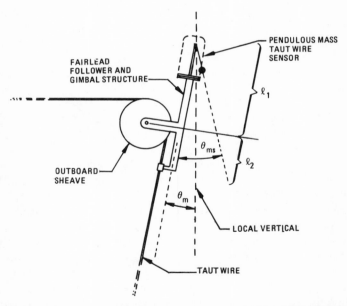

Figure 3-25. A Schematic Representation of the Taut Wire Sensor

system is that the angle of the wire, θ_m, which is a measure of the offset of the vessel, cannot be approximated as a simple ratio of position off-set divided by the water depth as was the case in the derivation of Equation 3.52 for the acoustic system.

The approximation is invalid because of the dynamics of taut wire which create a transfer function between the actual offset and the indicated offset can be approximated for small angles as the simple ratio of position offset divided by water depth, the angle at the top of the wire is something else than the actual angle, or in equation form

$$\theta_m \neq \theta \cong \frac{x}{h_w} \tag{3.55}$$

In functional form the correct relationship is:

$$\theta_m = f(\theta) = f\left(\tan \frac{x}{h_w}\right) \tag{3.56}$$

The most commonly-used vertical reference sensor in taut wire and riser angle systems is the pendulous mass sensor. The pendulous mass sensor enjoys this widespread use not because of its dynamic per-formance but because of its low price, relative ruggedness, and ability to be conveniently packaged in a compact, extremely rugged container for use in an exposed environment.

The vertical gyro cannot be applied to either the taut wire or riser angle applications. The new accelerometer-type vertical reference sensor has the life and the size so that it may be possible to use in the taut wire and riser angle applications if repackaged into a very rugged container as the pendulous mass sensors are. However, the cost of the accelerometer-type sensor may not outweigh the gain in performance over the pendulous mass sensor.

Fig. 3–25 illustrates that the mounting location of the taut wire sensor on the fairlead follower and gimbal structure can influence the performance of the taut wire sensor. As the lever arm to which the sensor is mounted increases in length, the taut wire sensor experi-ences increasing lateral accelerations induced from angular accelera-tions of the wire through the lever arm combination of ℓ_1 divided by the sum of ℓ_1 and ℓ_2. Thus, if ℓ_1 is equal to zero, the sensor will not experience any induced lateral acceleration from the wire. For the pendulous mass sensor, the minimization of sources of lateral accelera-tion will improve the performance of the overall system.

3.7   Wind sensor

Wind speed and direction is generally regarded as an environmental parameter, unrelated to the control of vessel position. However, as analysis and application have demonstrated, position control can be improved by using the wind speed and direction to compute thruster commands. In later chapters the exact manner in which the computations are performed and how they improve position control are discussed. In this section the sensors which are used to measure the wind speed and direction are discussed, along with installation considerations.

The two parameters, wind speed and direction, are measured using different sensors. However, measurement units exist which combine the two sensors, thereby simplifying installation and ensuring that the two parameters are measured at the same location.

The most common wind speed sensor employs a propeller to drive a small dc voltage generator. The voltage generated is nearly directly proportional to the speed of the propeller. Then by using a propeller whose speed is proportional to the speed of air forcing it to move, the output voltage of the generator can be calibrated as a measure of the wind speed. Generally speaking, the propeller must be designed with free-running bearing and not excessive mass so that it can respond reasonably quickly to wind speed variations, i.e., wind gusts. The measurement of wind gusts is important if an improvement in position control is going to be achieved.

Measurement of wind direction is commonly achieved with a vane which rotates to track the direction of flow of the air by the vane. Attached to the vane is an angle-measuring sensor which exerts minimum drag on the vane. Commonly used angle sensors are linear potentiometers and synchro transmitters. Unfortunately there is a discontinuity in the angle measurement using potentiometers and the friction of the potentiometer wiper tends to wear them out. The synchro repeater does not suffer from either of these problems, so it is preferred.

Locating wind sensors on the vessel is an important and often difficult task. In order to achieve the best possible wind compensation for position control, the wind must be accurately measured from all directions of wind relative to the vessel. On vessels with prominent masts the wind sensor can be mounted there to obtain "good" wind data, but on vessels which have other dominating structures, such as derricks and helicopter pads, installing wind sensors becomes a problem.

3.8 Environmental and motion sensors

Although not directly required for information input to the control system portion of the dynamic positioning system, the environment and motion of the vessel is often measured for operator display and data recording. The parameters that are measured include ocean current speed and direction, wave height and direction, wind speed and direction, roll, pitch, and heave.

Roll and pitch can be measured by a vertical reference sensor discussed earlier. If the vertical reference sensor is required for tilt compensation of the hydrophone plane in the short baseline acoustic position reference system, then the roll and pitch information can be provided for display to the operator by the same sensor. Otherwise, a stand-alone vertical reference sensor must be provided.

Wind speed and direction can serve a dual purpose: for use in the control system to give anticipatory commands to the thrusters to counter wind gusts, and for operator display and data recording. The display of wind speed and direction for the operator is as important as position and heading because the wind is the major disturbing environmental element on the vessel.

The DP operator can better understand his thruster loading pattern and determine the most favorable vessel orientation to the weather if he knows the wind speed and direction.

Wave height and direction are also useful to the DP operator in determining preferred vessel orientations to the sea. In the case of wave height there are several choices of sensors, but wave direction is an oceanographic parameter which still defies accurate measurement.

Currently either direct observation of the waves or reliance on the measured vessel motions, or both, are used to orient the vessel at a preferred direction to the waves.

One commonly-used wave height sensor for a ship-shaped vessel uses two pressure transducers, one mounted on each side of the vessel. The pressure transducers measure the local head of water above them. Then, the outputs from the two transducers are averaged and scaled in units of distance. To the averaged output of the transducers, the heave of the vessel is added and the depth of transducers below the still-water line is subtracted. These concepts are illustrated in Fig. 3–26 where the wave height is calculated as follows:

$$\text{Wave height} = \frac{a + c}{2} + \text{heave} - b \qquad (3.57)$$

Another form of vessel-mounted wave height sensor is referred to as a wave staff. The wave staff consists of two parallel conductors. The staff is mounted vertically on the vessel and the height of water level on the staff is determined by measuring the resistance between the two conductors. Like the pressure-measuring wave height sensor, heave is added to the measured water level and the still water level is subtracted from the measured water level.

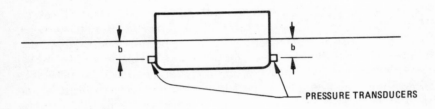

(a) ZERO WAVE HEIGHT CASE

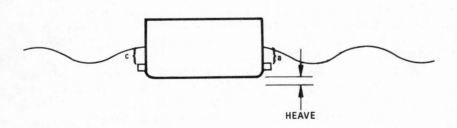

(b) WAVE HEIGHT MEASUREMENT

Figure 3-26. Pressure Transducer Method of Measuring Wave Height

This sensor is well-suited for semisubmersibles where the staff is mounted between the upper deck and lower pontoons away from the columns which definitely affect the measured wave height because of the wave piling up against the columns.

Wave height can, also, be measured with a radar. The radar is mounted on an upper level of the vessel looking down on the free surface of the water as shown in Fig. 3–27. The output of the radar is then compensated for the heave of the vessel. Additional compensation for the roll and pitch of the vessel may be necessary since the radar antenna may be some distance from the center of rotation of the vessel.

The most accurate method of measuring wave height is with a free-floating, instrumented buoy. The buoy measures the wave height using a doubly-integrated accelerometer and transmits the measurement to the vessel over a radio link. The advantage of this technique is that the buoy is free of the vessel. The disadvantage is that the buoy must be deployed away from the vessel and will eventually drift out of radio range with the vessel unless the buoy is tethered. The tethering process can introduce errors into the measurement.

Heave has already been mentioned in regard to the measurement of wave height to compensate for the movement of the platform to which the wave height sensor is mounted. Heave also gives operating personnel a measure of the vertical motion of the vessel for purposes of judging if the limits of the operating equipment are being approached.

To measure heave, the output of a vertically-mounted accelerometer

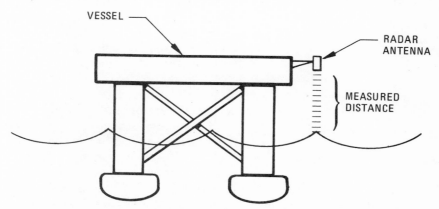

Figure 3-27. Radar Measurement of Wave Height

is "doubly integrated" to obtain vertical displacement. The double integration is only over a certain portion of the frequency domain, making the value of heave accurate only at certain frequencies. Another source of error is that the vertically-mounted accelerometer is not accelerated vertically because of the roll and pitch of the vessel.

Also, if the vertically-mounted accelerometer is not at the center of rotation, then the accelerometer output will contain components of acceleration equal to the angular acceleration times the offset of accelerometer from the center of rotation. Of course, these effects could be compensated for, but generally they are not.

The measurement of current speed and direction has been performed by oceanographers for many years by mounting the sensors on mooring lines on buoys or on taut wires from oceanographic survey vessels. The most common sensor used by the oceanographers has been the rotating transducer. In the rotating transducer the current speed is measured by an impeller rotor and the current direction by a vane.[21] Similar to the wind sensor, the rotating transducer inertia gives the instrument a slow response to dynamic motions.

Slow response for the measurement of current for the oceanographer may be a problem. However, to monitor current for a DP system the slow response should not be a significant problem. For the DP application the mechanical mounting of the sensor package to the vessel is a greater difficulty since the sensor is most conveniently mounted on a cable and not on the hull of a vessel.

A more convenient current sensor from the standpoint of hull mounting is the static current sensor. Three different static current sensors exist. The most common is the electromagnetic sensor. This sensor employs a transducer which generates an electromagnetic field in the region of four orthogonally arranged electrodes. The amount of induced voltage between two electrodes in the same orthogonal axis is proportional to the water flow velocity in that axis.

The total current speed and direction is then determined by the square root of the sum of the squares of the two measured voltages and the arctangent of the ratio of the x-axis velocity to the y-axis velocity, respectively. As can be imagined, the shape and location of the transducer is very important to the accuracy of the electromagnetic current sensor.

The other two types of static current sensors use vortex shedding and acoustic Doppler effects to measure current. More details on the operation of these sensors are given in Reference 21.

REFERENCES

1. Adams, R. B., "Accuracy of the Taut Line Position Indicator for Offshore Drilling Vessels," *J. of Engineering for Industry*, Trans of the ASME, May 1967, pp 1–11.

2. Benz, R. C. and Morgan, M. J., "An Interim Report on the Investigation of Taut Wire Position Reference System," Honeywell Document #2800, 9 January 1976.

3. M. A. Childers, D. Hazlewood, W. T. Ilfrey, "An Effective Tool for Monitoring Marine Risers," *Journal of Petroleum Technology*, March 1972.

4. L. M. Harris, *An Introduction to Deep Water Floating Drilling Operations*, Petroleum Publishing Company, 1972.

5. G. W. Morgan and J. W. Peret, "Applied Mechanics of Marine Riser Systems," *Petroleum Engineer*, Series of Articles beginning in July, 1974 through 1976.

6. B. G. Burke, "An Analysis of Marine Risers for Deep Water," OTC 1771, 1973.

7. R. C. Benz, "An Investigation of a Riser Angle Position Reference System," Honeywell Document 2911, 31 March, 1976.

8. G. E. Beck, Editor, *Navigation Systems*, Van Nostrand Reinhold, New York, 1971.

9. I. G. Raudsep, "Ship Position-Determining Systems," *Oceanology International*, March/April, 1968.

10. C. Powell, "An Informal Review of Shore-Based Radio Position-Fixing System," Joint Meeting of Hydrographic Society and the Society for Underwater Technology, September 6, 1973.

11. T. A. Stansell, Jr., "Using the Navy Navigation Satellite System," Proc. of the ION National Marine Navigation Meeting, Second Symposium on Manned Deep Submergence Vehicles, November 3–4, 1969, San Diego, California.

12. Harry Halamandaris, "An Integrated Marine Navigation System," Proc. of the ION National Marine Navigation Meeting, Second Symposium on Manned Deep Submergence Vehicles, November 3–4, 1969, San Diego, California.

13. A. R. Dennis, "Satellite Positioning and Navigation for Offshore Applications: Past, Present, and Future," OTC 2170, 1975.

14. L. E. Ott, "A Totally Integrated Navigation System," OTC 2461, 1976.

15. Buechler, D. L. and Hanna, J. T., "Inertially-Aided Dynamic Positioning," OTC paper 2417, 1975, Volume III, pp 777–784.

16. Paul H. Savet, *Gyroscopes: Theory and Design*, McGraw-Hill Book, New York, 1961.

17. Richard F. Deimal, *Mechanics of the Gyroscopes*, Dover Publications, New York, 1929.

18. John T. Truxal, Editor, *Control Engineer's Handbook*, McGraw-Hill, New York, 1958.

19. Singer reference, "High Accuracy Displacement Gyros C70 4104 Series," D66–0772, 3/30/66 Revised 7/1/72.

20. Lichtenstein, Bernard, "Gyros, Platforms, Accelerometers," Technical Information for the Engineer, Kearfott Div., General Precision, Inc., Fifth Edition, June 1962.

21. Gerald F. Appell and William E. Woodward, "Review of Current Meter Technology," *Undersea Technology*, June, 1973, pp. 16–18.

4

Thrusters

4.1 Introduction

As has already been discussed, the element of a dynamic positioning system which produces the forces and moment to counter the disturbance forces and moments acting on the vessel is the thrusters. Here, thrusters include any commandable device which produces thrust. In this context, all propellers are thrusters. Likewise, water jets can be classified as thrusters.

In fact, if propellers, jets, or rockets, normally used for aircraft, were used above the water on the vessel, they would be called thrusters.

Propellers have been in use on ships for many years and their design is well known. However, the normal function of these propellers is to move the ship from one point to another. More recently, smaller propellers have been mounted laterally in the bow of ships to aid in maneuvering and docking. In the special application of tugboat operations in a crowded harbor the special steerable thruster has been used.

In the case of the propeller for moving ships, the design concentrated on maximizing thruster output at maximum speed. The propellers used for docking and maneuvering are designed for maximum output at low speed. Low speed design is also important for dynamic positioning.

Many factors go into the design of thrusters for dynamic positioning. For example, speed of response, thruster size, reliability, efficiency, servicability, and quietness all play a part in the design of a good dynamic positioning thruster.

4.2 The thruster system

So far the only member of the thruster system to be mentioned is the thrust producing mechanism, i.e., a propeller or a jet. There are

144

other elements in a complete thruster system as illustrated in Fig. 4–1. Usually, the type of thruster includes the method of changing the thrust output. For example, the thrust output from a controllable pitch propeller is varied by changing the pitch of the propeller.

As illustrated in the block diagram of a single thruster system, in addition to the thrust-producing mechanism, there is a thrust control system. The purpose of the thrust control system is to implement the command variable in the thrust-producing mechanism as quickly and accurately as possible.

For example, for a controllable pitch propeller the control system sets the pitch of the propeller to be equal to the commanded pitch. As a result, the control system is generally a feedback control system.

The thrust control system can be electrical, hydraulic, mechanical, or a combination. The type of actuating mechanism used by the control system will depend on the type of control.

Another major element of the thruster system is the prime mover of the thrust-producing mechanism. The purpose of the prime mover is to furnish power to the thrust-producing mechanism which converts the power to thrust. Generally, the prime mover is an electric motor. However, hydraulic motors can be used also. Electric motors are in general use because of their convenient integration with the power system of the vessel and their availability.

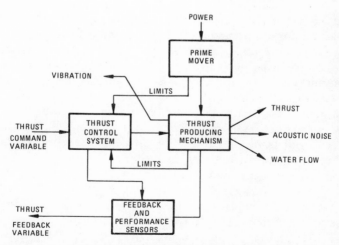

Figure 4-1. A Block Diagram of a Single Thruster System

With certain types of thrusters the control system controls the prime mover directly. For example, in the case of the variable-speed propeller, the control system controls the speed of the dc motor driving the propeller. For the controllable-pitch propeller the control system will control only the load on the prime mover. In either case, the prime mover as well as the thrust-producing mechanism will impose limits on the operation of the control system.

For the prime mover the limits will be caused by overloads in motor output or other operating parameters, i.e., temperature, oil pressure, etc. In the case of the thrust-producing mechanisms, limits are imposed on the control system because of mechanical parameters such as blade loading, bearing temperatures, etc.

Because of limitations of both the prime movers and the thrust-producing mechanisms, the thruster control system has built-in limiting properties. For example, most thrust control systems will change the thrust control variable at no more than a constant rate past a certain value. This feature prevents overloading the prime mover and the mechanical structure of the thruster.

In order to monitor the performance of the thruster system, a sensor system is required to measure the various parameters. Some of the measured parameters are used directly by the dynamic positioning system, such as the thrust command variable. Others are used only to detect faulty performance so that the thruster can be taken out-of-service.

The dynamic positioning controller could monitor all performance parameters from the thruster system and sound the alarms indicating off-nominal performance. However, the monitoring is generally performed at another point on the vessel where the personnel responsible for repairing and servicing the thrusters are located.

From this intermediate point a common alarm is then transmitted to the dynamic positioning controller to alert the controller that the thruster should not be used for dynamic positioning. In addition to the alarm, a signal is needed from the thruster service area telling the controller when the thruster is ready for use. A signal from the controller area requesting that a thruster be made ready for use is also needed. These signals are summarized in Fig. 4–2.

Unfortunately, thrusters produce undesirable outputs in addition to the required thrust. Some of these undesirable outputs are shown in Fig. 4–1, and include acoustic noise which is radiated into the water and into the structure of the vessel. The first is a source of noise to the

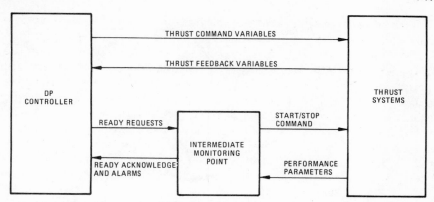

Figure 4-2. Signals Between the Thruster System and the DP
Controller

acoustic position reference system and the second is an annoyance to
the personnel of the vessel. Also, the thruster produces vibration which
is annoying to the vessel personnel and is wear-and-tear on the
thruster mechanism and mounting structure.

Another product of the thrusters which is necessary for their operation, but which can affect the performance of other thrusters and hull-
mounted devices, is the discharge flow of water. If the flow is into
another thruster, the downstream thruster performance will be adversely affected. Likewise, any hull-mounted devices which are
influenced by water motion will be affected, such as current meters.

Each active element of the thruster system has dynamic properties.
These dynamics usually do not affect the performance of the thrusters
in a manually-controlled system, such as when the vessel is underway
from one port to another.

However, in a dynamic positioning system the dynamics of the entire
thruster system are important. In the discussion of each thruster
which follows, the dynamics will be included as related to performance
and the overall dynamic positioning system.

4.3   Thruster types

The most common thruster type is the open propeller, illustrated in
Fig. 4-3. It acts like a screw in the water and, therefore, is generally
called a screw. These propellers are produced in many configurations.

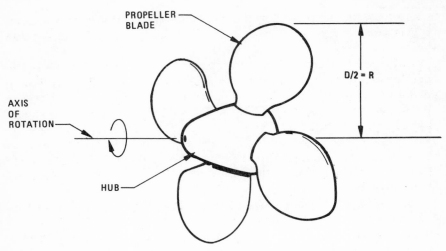

Figure 4-3. Open Propeller-Type Thruster

For example, they can have a minimum of three blades and as many as six blades. They are capable of a few hundred pounds of thrust and as much as hundreds of thousands of pounds of thrust. Some typical parameters for several open propellers are given in Table 4–1.

Open propellers can be either fixed pitch with controllable speed, or fixed speed with controllable pitch. The reasons one type of control

TABLE 4–1
EXAMPLES OF TYPICAL OPEN AND DUCTED PROPELLER
PARAMETERS

Types	Number of blades	Horsepower	Thrust, lb	RPM	Diameter, ft
Open	3	1,500	33,000	150	10
Open	4	2,100	48,500	200	
Open	5	35,000	–	240	15
Open	4	20,000	–	235	13
Ducted	4	500	11,300	400	4.8
Ducted	4	1,000	24,300	260	6.6
Ducted	4	1,500	36,400	240	7.9
Ducted	4	2,200	55,000	200	9.2
Ducted	4	3,000	77,200	161	11.2

is selected over the other can be complicated and will not be discussed at this time. Both types have been used on dynamically positioned vessels.

In all recent cases, the use of the open propeller on DP vessels has been limited to the main propulsion which is used not only for dynamic positioning, but also moving the vessel from one location to another.

A second type of thruster places the open propeller in a specially-shaped housing called a duct. This thruster is illustrated in Fig. 4–4. The primary reason for adding the ducting to the open propeller is to increase its efficiency or thrust output per unit horsepower. The increase in efficiency can be as much as 20% at the low speeds used during dynamic positioning.

The ducting can either be as shown in Fig. 4–4 or as shown in Fig. 4–5 where the open propeller is placed in a tunnel. Commonly, in the latter case the thruster is called a tunnel thruster as opposed to a ducted thruster.

Ducted thrusters are primarily used on dynamically positioned vessels as controllable-pitch propellers with a fixed axis of thrust output at right angles to the centerline of the vessel, i.e., the lateral direction. Less frequently they are either controllable speed or main propulsion, as is shown in Table 4–2

Until recently, ducted propellers have not been produced in large horsepower units. Likewise, they are generally three or four-bladed. In the case where the propeller cannot be driven by a straight shaft from the prime mover or its gear box, the size of the ducted propeller is limited to about 3,000 hp. The reason generally given for the limit in the capacity of the propeller has nothing to do with the duct. Instead the reason given is that for a right-angle drive from the prime mover (see Fig. 4–4), the right-angle gears cannot be manufactured to transmit more than 3,000 hp.

The ducted propeller can be significantly more efficient in one direction than the other. The difference in efficiency is not only a function of the shape of the duct, but also the shape of the blades. In the case of fixed-axis thrusters, the loading on the vessel will be such that the thrust output of the thruster should be as symmetrical as possible.

To achieve symmetrical thrust output, the maximum efficiency of the ducted propeller must be reduced. The reduction will be as much as 10 or 20% of the maximum efficiency when the propeller is designed for thrust in a predominant direction. When a thruster is designed

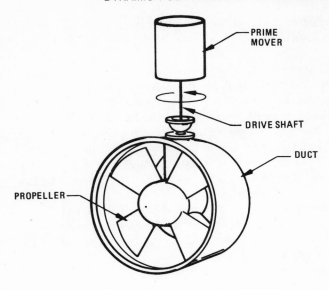

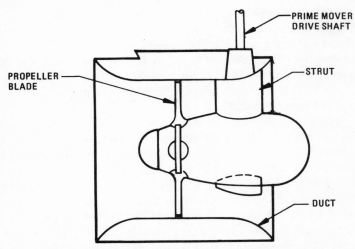

Figure 4-4. A Typical Ducted Propeller

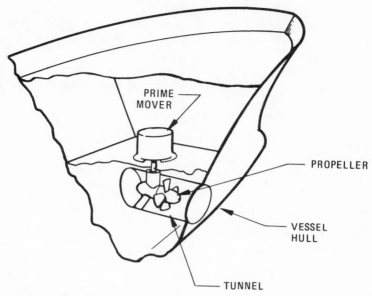

Figure 4-5. A Typical Ducted Propeller in a Tunnel-Type Installation

for one predominant direction, the reduced efficiency for the opposite direction can be more than 50% of the maximum efficiency.

Besides the reduction in efficiency, the symmetrically-ducted propeller suffers from another drawback when used as a fixed-axis propeller. Because the fixed axis propeller can only produce thrust along a single azimuth with relation to the centerline of the vessel, the propeller is less able to counter disturbance forces which are not colinear with its line of thrust.

For this reason, another type of thruster which can change the azimuth of its thrust output with respect to the vessel coordinate system has been produced. The azimuthing thruster can be an open or ducted propeller and can have controllable pitch or controllable speed. However, the duct must be free to rotate and, therefore, cannot be a tunnel in the hull of the vessel.

As Fig. 4-6 shows, a typical azimuthing thruster has not only a prime mover to rotate the propeller to produce thrust, but also a drive mechanism to rotate the azimuth of the thruster. Thus, the rate at which the direction of the thrust can be changed depends on the speed

TABLE 4–2

THRUSTER TYPES AND HORSEPOWER FOR SECOND-GENERATION
DYNAMICALLY POSITIONED VESSELS

	Thruster System					
Vessel	Lateral Type	Nbr	Total hp	Longitudinal Type	Nbr	Total hp
Sedco 445	CS, FA, D*	11	8,800	CS, FA, O	2	4800
Saipem Due	CP, CY, O	3	9,000	**	–	–
	CP, AZ, D	1				
Pelican-Class	CP, FA, T	5	7,300 or greater	CP, FA, O	2	7000
Discoverer 534	CP, FA, D	6	15,000	CS, FA, O	2	8000
Arctic Surveyor	CP, FA, D	2	1,200	CP, FA, O	2	2000
Wimpey Sealab	CP, AZ, D	4	4,000	–	–	–
Kattenturn	AZ	2	1,026	CP, FA, D	2	1930
Seven Seas	CP, FA, D	6	15,000	CS, FA, O	2	8000
Sedco 709	CP, AZ, D	8	24,000	**	–	–
Sedco 470-series	CS, FA, D	12	9,600	**	–	–
Uncle John	CP, FA, T	6	6,000	CP, FA, D	2	6000

* Legend
   CS - Controllable speed
   CP - Controllable pitch
   FA - Fixed axis
   CY - Cycloidal
   AZ - Azimuthing
    O - Open
   D - Ducted
** For cycloidal and azimuthing thrusters the thrusters perform as lateral and longitudinal thrusters, even though they are listed in the lateral thruster column.

of the azimuthing drive system. If the azimuthing drive system is too slow, the advantages gained by being able to steer the thruster in the direction of the forces acting on the vessel and to maximize thruster efficiency will be lost.

Another type of thruster which can steer its thrust direction is the cycloidal propeller shown in Fig. 4–7.[7 8] Not only does the cycloidal propeller steer its thrust output differently than the azimuthing thruster shown in Fig. 4–6, but also it produces its thrust in a different manner.

The rotational axis of the cycloidal propeller is vertical. Through a complicated mechanical coupling the pitch of each vertical blade is changed as the blade makes one revolution. The name of the propeller is derived from the way the pitch of the blade traces out a cycloidal

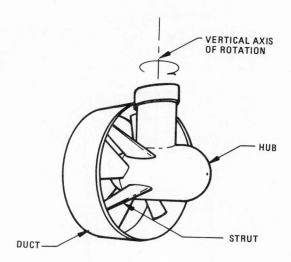

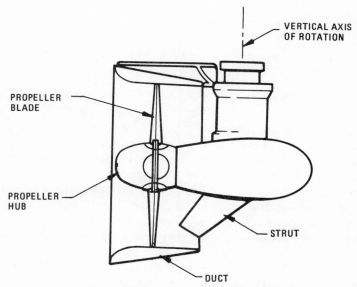

Figure 4-6. A Typical Azimuthing Ducted Propeller

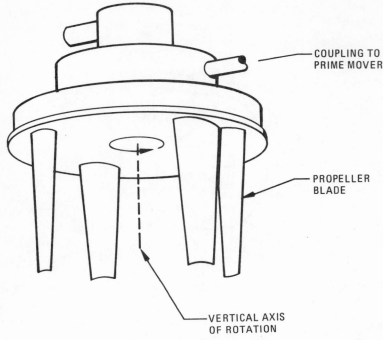

COUPLING TO
PRIME MOVER

PROPELLER
BLADE

VERTICAL AXIS
OF ROTATION

Figure 4-7. A Cycloidal Propeller

pattern as the blade completes a revolution. By properly positioning
the pitch control mechanism, the thrust magnitude and direction can
be controlled. More discussion of the cycloidal propeller is given in the
section on thruster characteristics.

Because of their highly flexible thrust azimuthing capability,
cycloidal propellers have been used on vessels requiring high maneuverability, such as tug boats. They also have been used on the dynamically positioned vessels (*Duplus* and *Saipem Due*).

The reason they have not received wider use on DP vessels is as
complex as the process of choosing any thruster type on any dynamically positioned vessel.

4.4   Thruster Configurations

In Chapter 1 an example of the location of thrusters for several
dynamically positioned vessels is shown. A survey of all dynamically

positioned vessels would not show any greater consistency than shown in Chapter 1. However, there are several general rules regarding thruster configurations on dynamically positioned vessels.

Before these rules can be stated, several basic requirements must be established about the thruster system with respect to holding a vessel against a given environment while performing a given task. First of all, forces acting on the vessel generate a requirement for thrust along the longitudinal and lateral axes of the vessel plus thrust to generate a counter moment.

The most efficient direction to produce moment-generating thrust is at right angles to the line connecting the location of the thruster and the center of rotation of the vessel. If the direction of thrust is not along this line, then some inefficiency must be accepted.

The first requirement for thrust capability establishes the first rule for thruster configurations. The first rule is that at least two thrusters are required. The reason that at least two thrusters are required can be understood by writing the equations which express the distribution of the force and moment requirements among the thrusters.

With a longitudinal force, a lateral force, and a moment requirement, there are three equations to be satisfied. However, with only one thruster there are only two possible unknowns. The unknowns are thrust magnitude and direction. By adding another thruster, there are sufficient unknowns to satisfy the equations as shown in the following force-moment equations:

$$F_x = F_1 \cos \theta_1 + F_2 \cos \theta_2 \tag{4.1}$$
$$F_y = F_1 \sin \theta_1 + F_2 \sin \theta_2 \tag{4.2}$$
$$M_z = \ell_1 F_1 \sin (\phi_1 - \theta_1) + \ell_2 F_2 \sin (\phi_2 - \theta_2) \tag{4.3}$$

where:

F_x, F_y = the forces required to counter the forces acting on the vessel in the x-y direction, respectively

M_z = the moment required to counter the moment caused by the forces acting on the vessel about the vertical axis

F_1, F_2 = the thrusts from the first and second thruster, respectively

θ_1, θ_2 = the directions of the thrust from the first and second thruster, respectively

ℓ_1, ℓ_2 = the known distance from the center of rotation of the vessel to the first and second thruster, respectively

ϕ_1, ϕ_2 = the known angle of the first and second thruster to the line from the center of rotation of the vessel to the first and second thruster, respectively.

The reference direction of the angles to be used in this book are shown in Fig. 4–8.

In the case of fixed-axis thrusters, as they are most frequently arranged with one set longitudinally for transiting propulsion and another set laterally, the angles of thrust directions in Equations 4.1, 4.2, and 4.3 are fixed at either 0° or 90°. Under these conditions the allocation equations reduce to the following form:

$$F_x = F_1 \tag{4.4}$$
$$F_y = F_2 \tag{4.5}$$
$$M_z = \ell_1 F_1 \sin \phi_1 - \ell_2 F_2 \cos \phi_2 \tag{4.6}$$

where $\theta_1 = 0°$ and $\theta_2 = 90°$. As Equations 4.4, 4.5, and 4.6 show, again there are more equations than unknowns. Thus, for fixed-axis propellers, a minimum of three thrusters are required to satisfy the force-moment requirements. With the addition of another thruster, the force-moment equations become:

$$F_x = F_1 \tag{4.7}$$
$$F_y = F_2 + F_3 \tag{4.8}$$
$$M_z = \ell_1 F_1 \sin \phi_1 - \ell_2 F_2 \cos \phi_2 - \ell_3 F_3 \cos \phi_3 \tag{4.9}$$

These equations can be solved for the thruster forces to achieve the following results:

$$F_1 = F_x \tag{4.10}$$

$$F_2 = \frac{M_z - \ell_1 F_x \sin \phi_1 + F_y \ell_3 \cos \phi_3}{\ell_3 \cos \phi_3 - \ell_2 \cos \phi_2} \tag{4.11}$$

$$F_3 = \frac{M_z - \ell_1 F_x \sin \phi_1 + F_y \ell_2 \cos \phi_2}{\ell_2 \cos \phi_2 - \ell_3 \cos \phi_3} \tag{4.12}$$

If the x-axis propeller is located on the centerline of the vessel along with the two lateral propellers, then Equations 4.10, 4.11, and 4.12 reduce to:

$$F_1 = F_x \tag{4.13}$$

$$F_2 = \frac{M_z \pm F_y \ell_3}{\pm \ell_3 \mp \ell_2} \tag{4.14}$$

$$F_3 = \frac{M_z \pm F_y \ell_2}{\pm \ell_2 \mp \ell_3} \tag{4.15}$$

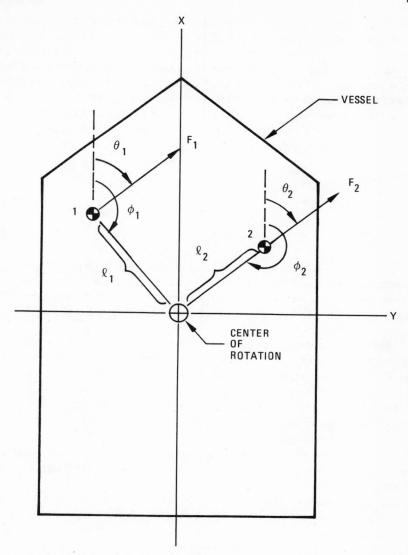

Figure 4-8. The Reference System for the Thruster System of this Book

where the upper sign on a moment arm signifies that the thruster is aft of the center of rotation of the vessel and vice versa for the lower sign.

To minimize F_2 and F_3 (likewise the power required), Equations 4.14 and 4.15 show that the two lateral thrusters should not be on the same side of the center of rotation of the vessel. Thus, by arbitrarily placing the second thruster forward of the center of rotation of the vessel and the other thrusters aft, the simplest fixed-axis thruster force-moment operations become as follows:

$$F_1 = F_x \tag{4.16}$$

$$F_2 = M_z \left(\frac{1}{\ell_2 + \ell_3}\right) + F_y \left(\frac{\ell_3}{\ell_2 + \ell_3}\right) \tag{4.17}$$

$$F_3 = -M_z \left(\frac{1}{\ell_2 + \ell_3}\right) + F_y \left(\frac{\ell_2}{\ell_2 + \ell_3}\right) \tag{4.18}$$

Equations 4.16, 4.17, and 4.18 give a unique allocation of thrust to each thruster for a three-thruster fixed-axis propeller system. For this reason they are referred to in this book as thrust allocation equations which will be needed in the controller of the DP system to be discussed in the next chapter.

Several important features of the allocation of thrust as given by Equation 4.16, 4.17, and 4.18 should be noted. First, the weighting for the distribution of the moment requirement is equal for each thruster, but opposite in direction to create a couple of lengths equal to the sum of the lever arms. In contrast, the weighting for the allocation of the lateral force requirement is proportional to the lever arm of the other lateral thruster and in the same direction so that the lateral allocation does not generate any moment. The effect of this allocation is that one thruster is generally loaded heavier (producing more thrust) than the other.

Another feature of the allocation given by Equations 4.16 through 4.18 which is caused by the unequal loading, is that once one thruster reaches the limit of its output, the other thruster output is limited even though it is below its limit. As a result, full use of installed thrust capability is not generally possible except if the moment requirement is zero and the lever arm lengths are equal.

If the vessel is designed to stay with one end always in the direction of the heaviest load, then advantage can be made of this unbalancing situation between thrusters by making the more lightly loaded thruster smaller. However, generally both thrusters should have the same thrust capability and approximately the same lever arm.

For vessels with significant moment requirements, such as ship shape vessels, the fixed-axis allocation equations show that the lever arms should be as large as possible.

If, instead of only three thrusters, there is more than one thruster at any one of the locations used for Equations 4.16 through 4.18, the allocation at each location can be made with Equations 4.16 through 4.18 and then divided equally among the thrusters at each location.

However, generally thrusters are not placed one above the other. They are usually arranged next to one another in a horizontal plane in the same general area to form a group. In this case Equations 4.16, 4.17, and 4.18 can still be used if the thrust to the longitudinal thrusters is distributed so that no moment is generated which results in the following relationships:

$$F_{port} = \frac{\ell_s}{\ell_p + \ell_s} F_x \tag{4.19}$$

$$F_{stbd} = \frac{\ell_p}{\ell_s + \ell_p} F_x \tag{4.20}$$

where:

F_{port}, F_{stbd} = the thrust for the thrusters to port and starboard of the longitudinal centerline of the vessel, respectively

ℓ_p, ℓ_s = the average lever arm of the thrusters to the port and starboard of the longitudinal centerline of the vessel, respectively

The distribution of the thrust to the individual thrusters on the same side of the centerline of the vessel is made by the following expression:

$$F_{pi} = \frac{F_{port}}{N_{port}} \tag{4.21}$$

$$F_{si} = \frac{F_{stbd}}{N_{stbd}} \tag{4.22}$$

where:

F_{pi}, F_{si} = the thrust allocation for the ith thruster on the port and starboard side, respectively

N_{port}, N_{stbd} = the number of thrusters on the port and starboard sides of the centerline of the vessel, respectively

The average lever arm is computed as follows:

$$\ell_p = \frac{1}{N_{port}} \sum_{i=1}^{N_{port}} \ell_{pi} \qquad (4.23)$$

$$\ell_s = \frac{1}{N_{stbd}} \sum_{i=1}^{N_{stbd}} \ell_{si} \qquad (4.24)$$

where:

ℓ_{pi}, ℓ_{si} = the lever arm of the ith thruster on the port and starboard side of the centerline of the vessel, respectively

TABLE 4–3
LATERAL THRUSTERS FOR THREE DP DRILLSHIPS

	Discoverer 534 [4]	*Havdrill* [5]	*Neddrill* [5]
Horsepower	2,500	1,726	2,071
RPM	198	219	219
D, ft	10.33	7.74	9.05
h_s, ft	27.2	18	35.4
Thrust, lb	67,000	42,500	52,000
Tip speed, $\frac{ft}{sec}$	107	88.6	107.3
σ_R	0.31	0.3	0.38
C	0.974	0.951	0.939
Thrust/horsepower	26.8	24.6	25.5

For the simplest and most common case (Table 4–4) of two equally-spaced main propellers, the total longitudinal force requirement is divided equally between the two propellers.

In the case of the lateral thrusters, the thrusters on the same side of the center of rotation are considered as one equivalent thruster with an equivalent lever arm as follows:

$$\ell_F = \frac{1}{N_F} \sum_{i=1}^{N_F} \ell_{Fi} \qquad (4.25)$$

$$\ell_A = \frac{1}{N_A} \sum_{i=1}^{N_A} \ell_{Ai} \qquad (4.26)$$

TABLE 4-4
THRUSTER CONFIGURATIONS ON DP VESSELS

Vessel	Forward	Aft	Main Screws
Sedco 445	Two groups (2,4)	Two groups (4,1)	2
Pelican-Class*	3	2	2
Saipem Due	2	2	
Discoverer 534	3	3	2
Glomar Explorer	3	2	2
Arctic Surveyor			
(Connector)	1	1	2
Seaway Falcon	2	2	2
Kattenturm (Hansa)	1 (Steerable)	1 (Steerable)	2
Arctic Seal	2	2	2
Sedco 470 Series	Two groups (2,4)	Two groups (4,2)	2
Sedco 709	4 (Steerable)	4 (Steerable)	

* There are six or seven vessels of this design. The Pelican was the first and the *Havdrill* the second. The design is a product of IHC Holland, Gusto.

where:

ℓ_F, ℓ_A = the equivalent forward and aft lever arms, respectively
N_F, N_A = the number of thrusters forward and aft of the center of rotation, respectively
ℓ_{Fi}, ℓ_{Ai} = the lever arm of the ith thruster forward and aft of the center of rotation, respectively

The allocation of thrust to each thruster is, then, performed as follows:

$$F_{Fi} = \frac{F_F}{N_F} \qquad (4.27)$$

$$F_{Ai} = \frac{F_A}{N_A} \qquad (4.28)$$

where:

F_{Fi}, F_{Ai} = the thrust allocated to the ith forward and aft thruster, respectively
F_F, F_A = the thrust given by Equation 4.17 and 4.18, respectively

Commonly, as Table 4-4 illustrates, many DP vessels have thruster configurations which conform to the requirements for using the above given fixed-axis thrust allocation equations with multiple thrusters in each group. The notable exceptions are the *Saipem Due* and those vessels with steerable or azimuthing thrusters.

The *Saipem Due* is an exception because its stern propellers act as both longitudinal and lateral thrusters and are actually steerable thrusters. When the *Saipem Due* is underway, the stern propellers act as main screws and the bow thrusters are retracted into the hull.

The equal allocation of thrust to each thruster on the same side of the axis of rotation as just described should only be done if the thrusters have approximately the same thrust capability. Otherwise, full advantage of the larger units cannot be made. The proper distribution for thrusters of unequal maximum thrust capability will be completely developed in Chapter 5 on thrust allocation logic.

With azimuthing propellers where the direction of the thrust can be varied as well as the magnitude, there are more unknowns than there are allocation equations. As a result, without at least another equation to establish a definite rule regarding the distribution of thrust, a nonunique answer for the allocation of thrust results. A complete discussion of the additional allocation rules will be given in Chapter 5 on thrust allocation logic.

For now, the main points to be established are that two steerable thrusters, preferrably with one forward and one aft of the center of rotation, are required for dynamic positioning. In addition, the thrusters should be located some distance from the center of rotation of the vessel to obtain sufficient moment capability to properly maintain the heading of the vessel in the design environment.

The locations available to mount thrusters on a vessel influence, to a great extent, thruster configuration. Also, the requirement that the dynamically positioned vessel must not only hold station, but also must sail to the work site, has a definite influence on the thruster configuration for a dynamically positioned vessel.

In the case of satisfying the requirement for propulsion, most ship shape vessels use a conventional sailing propulsion configuration. This consists of two propellers mounted equidistant from the centerline of the vessel at the stern of the vessel. Generally, open propellers are used with either controllable pitch or speed.

Improved thrust efficiency could be achieved with nozzled propellers. However, since these propellers are generally sized to move the vessel at a transit speed in excess of 10 knots, they are generally capable of supplying more than enough thrust for dynamic positioning. This is not always true in the case when the main screws are used to counter environmental forces on the stern of the vessel because the propellers are normally designed with their predominant thrust direction forward. As a result, when thrusting aft, the maximum thrust output can be reduced over 40%.

The question of location for the lateral thrusters is considerably more complicated than the main screws. The solutions to the tradeoffs for the various dynamically positioned ship shape vessels have resulted in two primary configurations for fixed-axis thrusters. The first configuration places the propeller in a lateral tunnel through the hull of the vessel. The second places the propeller in a ducted structure below the keel of the vessel.

In either configuration the design of the tunnel or the duct, and the placement of the propeller in the duct or the tunnel, is performed so that the thrust in one direction is as nearly equal to the thrust in the other direction for the same command.

Hull shape has a strong influence on the location of the tunnel thrusters. The length of the tunnel influences the efficiency of the thruster. The longer the tunnel is, the lower the efficiency. As a result, the tendency is to locate the tunnels in the narrower portions of the hull. This means that they are located as far forward and aft as possible for ship shape vessels.

Naturally there are limits. For instance, in the case of the bow thruster, there are always limits with regard to equipment areas and regulations which require that the thrusters be located behind certain restraining bulkheads for safety in the case of collisions.

In the case of the aft tunnel thrusters, the question of location is complicated by the presence of the longitudinal propellers (the main screws). However, generally the skeg of a vessel can be used as a place to locate the lateral tunnel thrusters. A further complication to locating the tunnel thrusters near the main propellers of vessel is that the output thrust of the tunnel thrusters as well as the output thrust of the main propellers can affect one another. If this cross-axis coupling is severe, the design of the control system will be more complicated. For the present dynamically positioned vessels using stern tunnel thrusters, this problem has apparently been solved.

The same problem of tunnel length does not exist for the ducted lateral propeller suspended below the keel of the vessel. However, the ducted propeller presents a different problem which can be more serious. The problem is that they should be retractable to reduce drag when the ship is underway and to protect the thruster when the vessel enters a harbor.

If, in making a thruster retractable, access to the thruster can be provided so that the thruster can be lifted free of the vessel, then the entire thruster can be replaced or repaired without the vessel going into dry dock. The ability to repair or replace thrusters without a dry dock visit can be an advantage for the entire vessel in terms of maintaining the highest possible use factor.

For this reason the ducted thruster can supply a superior installation to the tunnel thruster because the tunnel thruster must be replaced or repaired in dry dock.

As a result of the requirement to retract the ducted thruster into the hull, finding space for the thruster and its retracting mechanism may place serious constraints on the areas which are available to the thruster. Also, the retracting mechanism will represent a significant added cost to the ducted propeller thruster system. Not only must some means be provided to lift the thruster, the duct, and the prime mover of the thruster, but also the entire unit must be isolated from the vessel to ensure a water-tight security. To achieve this security, the thruster system can be placed in a water-tight capsule with the ducted propeller attached to the bottom of the capsule as illustrated in Fig. 4–9. For azimuthing propellers the capsule system is required because the propeller is generally in a duct and, more importantly, is free to rotate.

Another factor which can completely eliminate the choice of the tunnel thruster over the ducted thruster is the amount of required thrust. If the required thrust is high, then either a large number of small units must be used or vice versa. Both alternatives do not favor the tunnel thruster.

For the first alternative – a large number of units – the hull cannot support all the penetrations because of space and strength. For the second alternative, the diameter of the thruster becomes so large that the thruster is not sufficiently submerged to prevent cavitation or the suction of air from the surface. Both cavitation and air suction reduce the thrust output of the thruster and can degrade the performance of an acoustic position reference system. Thus, the ducted propeller must be used because either larger diameters or more smaller units can be tolerated in the design of the vessel.

The *Discoverer 534* is an example of the use of larger ducted propeller units[4] and the *Sedco 470* series vessels are examples of the use of many smaller ducted propeller units.

The situation on semisubmersibles is somewhat different than for the ship shape vessel because the flotation for the vessel is provided by multiple submerged hulls. During transiting these hulls are generally partially exposed or at a shallow draft. However, during opera-

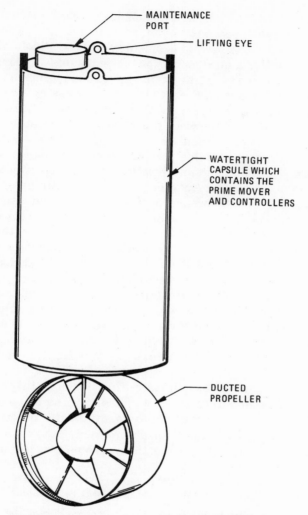

Figure 4-9. A Capsule for a Ducted Propeller

tion the hulls are at depths greater than those for a ship shape vessel with the same displacement.

In addition, the hull shapes and dimensions are such that tunnel thrusters are not conveniently installed. As a result, the ducted azimuthing thruster has a definite application for semisubmersible thruster installations which it does not have for ship shape vessels.

Examples of drafts and hull dimensions for several semisubmersibles are given in Table 4–5.

Another practical aspect of semisubmersibles which makes the azimuthing thruster a good choice for the semisubmersible is that the hulls provide increased separation among thrusters. Increased separation is important because if the wake from one thruster is directed into another thruster, the downstream thruster can suffer a considerable loss of thrust output. Likewise, the increased separation gives the thruster system better moment-generating capacity.

The azimuthing or steerable thruster appears to be well-suited for operation on a semisubmersible because the thruster should perform at its maximum efficiency. Another positive aspect in using steerable thrusters on a semisubmersible is that less installed horsepower is required than with a fixed-axis thruster installation. This is true because with most semisubmersibles the loading caused by the environment and work task does not vary nearly as much as for a ship shape vessel as a function of angle of attack. As a result, the thruster installation in the longitudinal and lateral axis are nearly equal.

With a fixed-axis thruster installation, the total installed horsepower is equal to the sum of the required horsepower in the lateral and longitudinal axes. Thus, for a semisubmersible with nearly equal horsepower requirements in the lateral and longitudinal axes, the total installed horsepower is approximately twice the single-axis requirement.

With the steerable thruster the same thruster capability can be used in whatever direction is necessary. Thus, for the semisubmersible with nearly equal horsepower requirements in the two vessel axes, the steerable thruster installation is no greater than 1.4 times the single-axis requirement. Therefore, the steerable thruster installation

TABLE 4–5
SEMISUBMERSIBLE OPERATING DRAFTS AND HULL DIMENSIONS

	Operating* Draft, ft	Pontoon dimensions, ft			
		Length	Depth	Width	Separation
Sedco 700 Series	75	295	21	50	145
Aker H-3	70	355	22	36	149
Aker H-4	82	370	26	49	143
Aker H-5	72	354	23	45	142
Penrod 70 Series	70	288	30	70	76

* The actual draft varies by a few feet from the given value depending on conditions.

is smaller than the fixed-axis installation by a significant margin.
More discussion of this topic appears in Appendix A.

For the semisubmersible the question of thruster repair and service
is different than for the ship shape vessel in several respects. First,
the draft of the semisubmersible can be changed from the 50- to 80-foot
operating depth to a value where the hulls containing the thrusters
are sufficiently exposed to give access to the upper portions of the hulls.

At the shallower draft, the possibility exists for the thruster capsules
to be lifted by a crane on the upper deck and transferred to a support
vessel for either on-the-spot servicing or transport to a service port
onshore.

If the thruster is built into the lower hull or if the capsule is not
accessible from the upper deck or at shallow draft, the thruster cannot
be replaced at sea. Several of the thruster manufacturers are developing schemes so that the ducted propeller can be mounted and dis-
mounted while the vessel is at sea.[9] [10]

In making the thruster retractable for a semisubmersible, the advantages gained for the ship shape vessel do not provide sufficient
weight to make them practical. First of all, the semisubmersibles are
not designed for as high a transit speed as a ship shape vessel and yet
they require some form of transit propulsion. Consequently, the same
thrusters used for dynamic positioning provide a convenient transit
propulsion source. A notable exception to this is the Aker semisubmersible design which uses a ducted controllable pitch propeller on
each pontoon for transiting propulsion.

Secondly, semisubmersibles do not normally enter the same harbors
as ship shape vessels. This fact is accepted when a semisubmersible
design is selected. Thus, being able to retract the thrusters for easier
harbor entry is not as important as for a ship shape vessel.

However, if a semisubmersible design was being made which gives
the features of high transit speeds and easy harbor entry, then either
tunnel thrusters or retractable thrusters would be required. Such an
application is the fire-fighting vessel which must reach a fire in minimum time and operate in very significant sea states. As a result, a
"high" speed semisubmersible is a logical design.

4.5   General thruster properties

All thrusters operate by producing a flow of water. The flow may be
created by a rotating propeller which has a fixed pitch and a variable

speed or vice versa.* The rotation may be about the major thrust axis or normal to it. In the case of a jet, a pump pulls the water in an forces it out in a jet. Thus, the properties of a thruster are determined by hydrodynamic flow which is beyond the scope of this book.[1 2] However, the general results of this flow are contained in this section as they relate to dynamic positioning.

The most common thrusters in use today for dynamic positioning consist of a propeller which is either controllable speed or controllable pitch. In either case, the propeller rotates around the predominant thrust axis as illustrated in Fig. 4–3. The shape of the blade and the direction of rotation will determine the direction of the thrust. The magnitude of the thrust produced by the propeller depends on several parameters expressed by the following equation:[1]

$$\text{Thrust, } T = \rho n^2 D^4 K_T \qquad\qquad (4.29)$$

where:

ρ = density of water
n = rotational speed of the propeller, revolutions/sec
D = diameter of propeller
K_T = thrust coefficient

To produce the thrust, a torque must be applied to the propeller to make it rotate and force water to move through the blades. The amount of torque can be expressed functionally as:[1]

$$\text{Torque, } Q = \rho n^2 D^5 K_Q \qquad\qquad (4.30)$$

where K_Q is the torque coefficient.

The values of the thrust and torque coefficients are functions of the physical characteristics of the propeller. These characteristics include the number of blades, the pitch of the blades, and the shape of the blades. The pitch of propeller is defined by the following relationship:[1]

$$\text{Pitch} = 2\pi r \tan \phi \qquad\qquad (4.31)$$

where:

r = the radius at which the pitch is measured.
ϕ = the pitch angle shown in Fig. 4–10.

---

* The pitch of a propeller varies as a function of the radius. Thus, all propellers have variable pitch. However, in the context used here the variability of pitch refers to changes in pitch caused by some other parameter than radius.

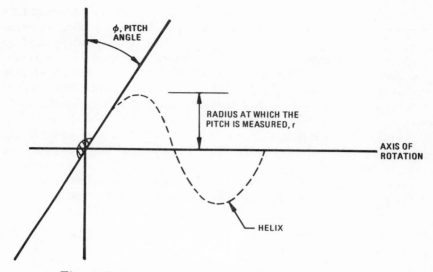

Figure 4-10. The Definition of Propeller Pitch [1]

The helix shown in Fig. 4–10 is the path of a point on the propeller at radius r as the propeller advances through one complete revolution without slipping. Normally, the pitch of the propeller blades are not constant as a function of radius. In these cases, the pitch at the point of approximately maximum lift is used as the defining pitch for the propeller. Typically, the maximum lift occurs at approximately 0.7R.[1]

Commonly propeller performance is given as a function of the P/D ratio for a propeller which is rotating at a constant speed. As Fig. 4–11 shows, the propeller requires torque or input power ($K_Q \neq 0$) for zero output thrust ($K_T = 0$) at zero pitch.

From the thrust and torque expressions, an efficiency of the propeller can be defined as follows:[1]

$$\eta_0 = \frac{TV_A}{2\,nQ} \tag{4.32}$$

where η_0 is the efficiency of the propeller in the open water, free of the vessel, and V_A the relative speed of the propeller to the water at a distance sufficiently removed from the propeller not to be influenced by a propeller. Equation 4.32 can be expressed in terms of the thrust and torque coefficients as:

$$\eta_0 = \frac{V_A K_T}{nDK_Q} = J\frac{K_T}{K_Q} \qquad (4.33)$$

where J is referred to as the advance ratio.

In Figure 4-12, a typical plot of K_T, K_Q, and η_0 as a function of J is shown for one value of pitch-diameter ratio. The region of curves shown in Fig. 4-11 which are pertinent to dynamic positioning are near zero, i.e., low relative speeds. As Fig. 4-11 shows, the propeller is not efficient according to the definition given in Equation 4.33.

However, Fig. 4-11 shows that the thrust output near zero advance is at its maximum. As a result, a more meaningful measure of performance for the dynamic positioning propeller is a static figure of merit. Two commonly used static figure of merits are given as follows:

$$C = \frac{T}{550SHP}\sqrt{\frac{T}{\rho\pi\,\dfrac{D^2}{4}}} = \frac{K_T^{3/2}}{\pi^{3/2}K_Q} \qquad (4.34)$$

and:

$$\eta_s = \frac{T}{550SHP}\left(\frac{550SHP}{\rho\,\pi\,\dfrac{D^2}{4}}\right)^{1/3}\frac{1}{2^{1/3}} = \frac{K_T}{K_Q^{2/3}}\frac{1}{2^{1/3}\,\pi} \qquad (4.35)$$

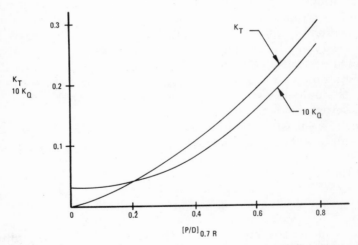

Figure 4-11. Thrust and Torque Coefficients as a Function of P/D Ratio for Zero Speed of Response

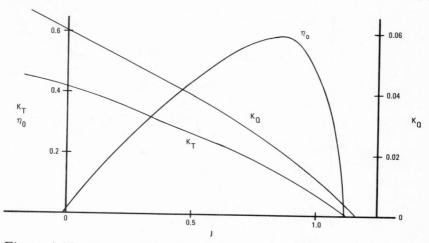

Figure 4-12. Typical Propeller Thrust and Torque Coefficients and Efficiency Curves in Open Water for a Single Value of Pitch

where SHP is the shaft horsepower and η_s is the Bendemann static thrust factor.

A typical plot of C and η_s as a function of P/D ratio is shown in Fig. 4–13. In Table 4–3, values for the figure of merit C are given for the lateral thrusters of three dynamically positioned vessels. The thrusters on the *Havdrill* are tunnel-type thrusters whereas the thrusters for the *Discoverer 534* and *Neddrill* are symmetric ducted propellers.

There are many types of thrusters. The figures of merit are one method of comparing the thrust producing capabilities of each type of thruster. In Table 4–6, the figures of merit, C, for five types of thrusters are given. Significantly, for any propeller a higher figure of merit can be achieved in one direction than the other. Thus, to achieve a propeller with nearly equal thrust capabilities in both directions, the figure of merit must be less than the maximum that can be achieved in one direction.

As Table 4–6 shows, the difference between the dominant and reverse direction for a nonsymmetric thruster is greater for a ducted propeller than a nonducted propeller.

Another parameter which is important to propeller design and, especially, important to a dynamically positioned vessel is the cavitation number. Propeller cavitation not only can result in a reduction in thrust, vibration, and erosion to the propeller blades and ducts,[1] it

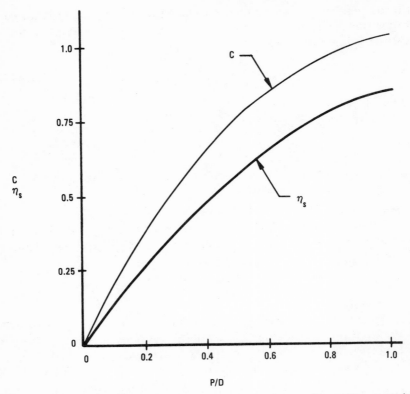

Figure 4-13. The Figures of Merit, C and η_s, as a Function of P/D
Ratio

TABLE 4-6
FIGURES OF MERIT FOR SEVERAL TYPES OF THRUSTERS

Thruster Type	Typical Value of the Figure of Merit
Ducted propeller	
Predominant direction	1.25
Reverse direction	0.4
Symmetric ducted propeller	1.01
Nonducted propeller in the	
predominant direction	0.88
Nonducted propeller with flat blades	0.71

also can generate noise which can reduce the effectiveness of the acoustic position reference system.[45] The local cavitation number can be expressed as a function of radius as follows:[1]

$$\sigma(r) = \frac{P_A + \rho gh_s - \rho gr - P_v}{1/2\rho \ (V_A{}^2 + 4 \ \pi^2 n^2 r^2)} \qquad (4.36)$$

where:

P_A = atmospheric pressure
r = radius from the center
g = acceleration of gravity
h_s = submergence of the centerline of the propeller
P_v = vapor pressure of water
ρ = density of water

If the atmospheric and vapor pressures of water are replaced by equivalent heads of water, then Equation 4.36 can be rewritten as follows:

$$\sigma(r) = \frac{2g(h_A + h_s - r - h_v)}{(V_A{}^2 + 4 \ \pi^2 n^2 r^2)} \qquad (4.37)$$

At the tip of the propeller where cavitation is very likely to occur for a dynamic positioning thruster, the local cavitation number is:

$$\sigma_R = 6.525 \left[\frac{32.5 + h_s - D/2}{n^2 D^2} \right] \qquad (4.38)$$

where σ_R is the local cavitation number at the maximum radius of the propeller R, h_A is approximately equal to 33 ft, h_v is approximately 0.5 ft, and the speed of advance, V_A, is zero for the dynamic positioning station keeping case.

Normally, in order to design a propeller so that it will not cavitate, a lower bound is established on the local cavitation number. The bound will generally be selected on experience or model basin data and very likely varies from one designer to the next.

As a result of bounding the cavitation number, a more useful form of Equation 4.38 becomes a solution for the rotational frequency of the propeller as a function of cavitation number. This solution is:

$$n, \ rps = \frac{2.55}{D} \sqrt{\frac{32.5 + h_s - D/2}{\sigma_R}} \qquad (4.39)$$

where all quantities are in feet.

So far, expressions involving thrust, rpm, diameter, efficiency, and local cavitation numbers have been given. There are other parameters which influence the overall design of the thrusters on a dynamically positioned vessel.

For example, the shape of the fairing on the inlet and outlet of the thruster, the length of the duct or tunnel which contains the propeller, the mechanical support of the thruster with the duct or tunnel, minimum submergence, and minimum separation of adjacent thrusters are all parameters which must be accounted for in a thruster installation design. For the interested reader, there are many publications which consider in great detail the design of propellers.[1 2 3 4 5 6 11]

The interaction of the various parameters is illustrated by the following simplified example. The parameters of the example are taken from no known vessel, but they are representative of a dynamically positioned drillship. The design considers the lateral thrusters only.

Table 4–7 contains the assumed dimensions of the vessel and the constraints imposed on the design by the vessel layout. Furthermore, it is assumed that the areas allocated for the thrusters in the hull take into account sufficient space so that the tunnel length and fairing shape for tunnel thrusters are proper for an efficient thruster design, or that there is space to retract ducted propellers into the hull.

TABLE 4–7
ASSUMED PARAMETERS FOR THE EXAMPLE THRUSTER DESIGN

Vessel parameters		
Vessel length between perpendiculars		500 ft
Operating draft		25 ft
Areas available to thrusters	Forward area:	40 ft
	Aft area:	40 ft
Maximum required thrust	Forward:	200,000 lb
	Aft:	200,000 lb
Minimum number of thrusters	Forward:	2
	Aft:	2
Minimum distance from the keel	Forward:	0.5 diameters
if mounted in the hull	Aft:	0.5 diameters
Propeller parameters		
Minimum centerline submergence		1.0 diameters
Minimum horizontal separation		0.5 diameters
Minimum local cavitation number		0.355
Maximum available horsepower per thruster		3,000 hp

Similarly, the thrusters are assumed to be designed and mounted so that they give equal thrust to port and to starboard for an equal command. Also, the owner of the ship has stated that all the thrusters should be exactly alike for purposes of maintenance and spare parts.

The questions the designer must answer in this hypothetical design are:

1. Should a tunnel or a retractable ducted propeller be used?
2. How many thrusters are needed?
3. What is the required diameter of the thrusters?
4. What is the required rpm of the thrusters?
5. What is the required horsepower?

As will become apparent, the answer to each of these questions is interrelated with other questions.

The problem has been designed so that the forward and aft requirements are identical. As a result, by solving the design for either set of thrusters, the overall design is complete. Obviously, this is not the usual case.

The first question is complex, involving not only hull and structural design tradeoffs, but also machinery, and competition among suppliers. For this example, the first question is simplified by considering only a thruster in the hull or suspended under the ship. As a result, one of the constraints on the design may make one type of thruster or the other impractical.

In determining the number of required thrusters an inequality can be written based on maximum available horsepower per thruster, i.e., 3,000 hp, and the maximum required horsepower. The result is:

$$\frac{200,000 \text{ lb}}{N_T \, \eta_T} \le 3,000 \text{ hp} \qquad (4.40)$$

where N_T is the number of thrusters and η_T is the static thrust per horsepower ratio. If Equation 4.40 is plotted and the fact is recognized that the number of thrusters can only be integer values, the result is shown in Fig. 4–14.

A realistic range of values for η_T for tunnel and ducted propellers is 24 to 28 lb/hp (Table 4–3). With the range of η_T the minimum number of thrusters is three (Fig. 4–14). Naturally, more thrusters can be used, but more space will be required.

There are two constraints on space: one horizontal and one vertical.

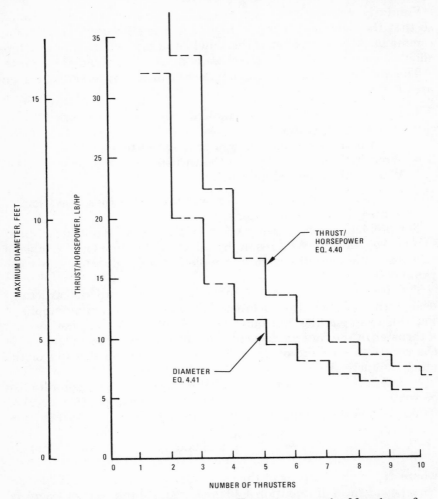

Figure 4-14. Plots of Constraint Equations on the Number of Thrusters

In inequality form, these constraints can be expressed as follows:
Horizontal constraint:

$$H_T \geq N_T D + (N_T - 1)\frac{D}{2} \qquad (4.41)$$

Vertical constraint:

Tunnel thrusters:
$$h_d \geq 3/2D + D + 1/2D = 3D \qquad (4.42)$$
Ducted thrusters:
$$h_d \geq 3/2D - L_s \qquad (4.43)$$

where:

H_T = horizontal space limit, which is 40 ft in the example
h_d = draft of the vessel, which is 25 ft in the example
L_s = separation of the ducted thruster from the keel of the vessel which is generally no more than a fraction of a propeller diameter. For this example L_s will be assumed to be 0.1 D.

Obviously the larger the separation from the keel, the larger the diameter of the propeller can be. However, there are several natural limits on the separation distance. One is the mechanical difficulties in extending the thruster at greater distances from the hull. Another is that the greater the separation distance from the hull, or more importantly, the roll center, the greater the roll moment and roll angle at maximum lateral thrust output. This roll-coupling with lateral thrust output can also affect the dynamic positioning control system if not properly compensated for.

The inequalities above assume that the thrusters are arranged in the same horizontal plane. Similarly, the spacing between adjacent thrusters reflects that only a fixed-axis thruster installation is being considered. If azimuthing thrusters are being considered, more spacing is required to reduce thrust spoiling.

The vertical constraint inequalities place maximum limits on the propeller diameter as soon as the operating draft of the vessel is specified. In the example, with a vessel draft of 25 ft, the maximum diameter for a tunnel thruster is 8.3 ft and for a ducted thruster is 17.9 ft.

Obviously, the tunnel thruster is much more severely limited in diameter than the ducted propeller which will ultimately result in a thrust capability limit also. In fact the difference in thrust limits for the two types of propellers will be even more pronounced because thrust varies as the fourth power of the diameter (see Equation 4.29).

In the case of the horizontal constraint inequality, the limits of propeller diameter are a function of the number of thrusters as well as the horizontal space limit. For this example, the functional relationship between diameter and number of thrusters is illustrated in Fig. 4-14. As Fig. 4-14 shows, if the thrust requirement can be satisfied with three thrusters, the vertical constraint is the limiting factor on propeller diameter for the tunnel thruster and vice versa for the ducted thruster.

With these limits established on the diameter by space considerations the next question is the choice of the diameter of the propeller to furnish the necessary thrust. One way of determining the necessary thrust is to combine the thrust relationship given in Equation 4.29 and the rotational rate relationship given in Equation 4.39 into one equation. The resulting equation is:

$$T = \frac{6.525 \, \rho}{\sigma_R} (37.5 + h_s - D/2)D^2 \, K_T \tag{4.44}$$

In the case of the tunnel thruster with the constraints given in the example problem, the submergence depth, h_s, can be expressed as the draft of the vessel minus the diameter of the propeller. Thus, the thrust for a tunnel thruster can be expressed as:

$$T = \frac{6.525\rho}{\sigma_R} (57.5 - 1.5D) \, D^2 \, K_T \tag{4.45}$$

For a ducted thruster the submergence depth, h_s, is equal to:

$$h_s = \text{draft} + L_s + D/2 \tag{4.46}$$

When Equation 4.46 is substituted into Equation 4.44, the thrust for the ducted propeller becomes:

$$T = \frac{6.525\rho}{R} (57.5 + 0.1D) \, D^2 \, K_T \tag{4.47}$$

As previously mentioned, the local cavitation number, σ_R should be larger than some minimum value. Similarly, the thrust coefficient, K_T, for readily available thrusters will span some range. Thus, by selecting the minimum value of the local cavitation number and a range of thrust coefficients, thrust can be expressed as a third-order function of the propeller diameter.

Although the third-order function expressing the relationship between thrust and propeller diameter can be solved, plotting thrust as a function of propeller diameter is a simple straightforward way to determine the propeller diameter for a given thrust. Fig. 4-15 is such

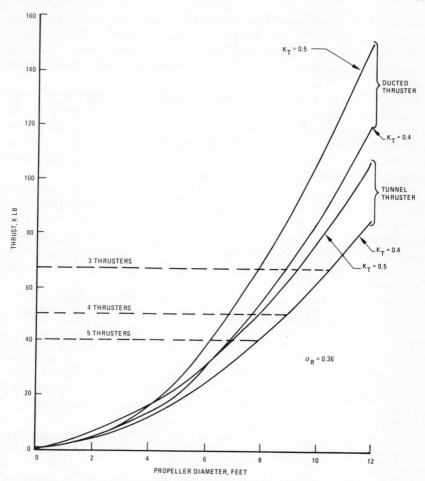

Figure 4-15. Thrust as a Function of Propeller Diameter for Ducted
and Tunnel Thrusters

a plot for the tunnel thruster (Equation 4.45) for $K_T = 0.4$ and 0.5 and
$\sigma_R = 0.36$.

Recalling that the maximum allowable propeller diameter for a
tunnel thruster in the example is 8.33 ft for three thrusters or 67,000 lb
of thrust per thruster. Fig. 4-15 shows immediately that for a tunnel
thruster with a maximum propeller diameter of 8.33 ft, the thruster

must have a thrust coefficient greater than 0.5 (nearly 0.6 according to Fig. 4–15) which is most likely not achievable with a commercially available propeller. Furthermore, for four and five tunnel thruster combinations, thrusters with thrust coefficients greater than 0.5 are required.

Another option to achieving the necessary thrust with the smaller propeller is to accept a smaller local cavitation number (larger rotational rate). However, the potential penalty is higher levels of cavitation which are mechanically undesirable for the thruster and may generate disruptive acoustic noise which may adversely affect the dynamic positioning system.

For the ducted propeller the thrust is higher for the same cavitation number and thrust coefficient because the propeller is deeper which allows the propeller tip velocity or rpm to be higher for the same cavitation number. As Fig. 4–15 shows, the increased thrust output of the ducted propeller for the same diameter is sufficient to satisfy the constraints on the diameter resulting from space considerations for the three- and four-thruster case.

However, for five thrusters the maximum propeller diameter of 5.75 ft cannot produce sufficient thrust without an increase in thrust coefficient or a decrease in the local cavitation number.

Without further discussion of the effect of adjustments to the cavitation number and the thrust coefficient a diameter for the ducted propeller can be selected from Fig. 4–15 at the midpoint between the two curves. For the three-thruster case the diameter is 8.4 ft and 7.3 ft for the four-thruster case. Then, the rotational rate or rpm of the propeller can be calculated from Equation 4.39.

For the three-thruster case the propeller rpm is 232 and for the four-thruster case the propeller rpm is 267. Finally, the power for the two cases can be calculated from Equation 4.34 using a value of $C = 1.0$. The resulting horsepowers for the two cases are 2,995 hp for the three-thruster case and 2,222 hp for the four-thruster case.

Using the smaller propeller size results in a slightly smaller total installed horsepower (8,888 versus 8,985) and gives more redundancy in the thruster system. However, the added complexity to the overall thruster system and perhaps cost may favor the three-thruster case.

For either propeller diameter the thrust-per-horsepower ratio is easily achievable (less than 23 lb/hp). In fact, a higher ratio is desirable to achieve more thrust per horsepower or more efficiency or fuel economy. The easy way to increase the thrust-to-horsepower ratio is to increase the diameter of the propeller to reduce the loading per unit area on the propeller.

Furthermore, by increasing the diameter of the propeller the local cavitation number (propeller tip speed) can be increased and still maintain the thrust coefficient in the 0.4 to 0.5 range. With reference to Fig. 4–15, the increase in the local cavitation number can be as great as from 0.36 to 0.5 by increasing the propeller diameter to the maximum allowable diameter of 10 ft for three thrusters while maintaining the thrust coefficient at 0.45. This is a 40% increase in the cavitation number which should result in a quiet propeller.

The corresponding rotational speed and horsepower for a 10-ft propeller are 165 rpm and 2,520 horsepower, respectively. Thus, the total installed power in the bow group of three thrusters is 7,560 hp compared to 8,985 hp with the small propeller. This increase in efficiency or economy is reflected in the thrust-to-horsepower ratio of 26.6 lb/hp for the larger diameter propeller.

The same approach could be applied to the four-thruster configuration. However, the maximum allowable diameter for four thrusters is 7.4 ft. Thus, only a very small increase in thrust-to-horsepower ratio can be achieved. So the increase is not reasonable, especially within the accuracy of the equations used in this analysis to predict real-world results.

All the previously given questions for the sample design are now answered. In summary form the answers are:

1. A ducted propeller should be used to achieve an efficient propeller with sufficient thrust and a low local cavitation number and with the proper diameter to fit within the space allocated for the thrusters.
2. Three thrusters are required.
3. The diameter of each propeller should be 10 ft.
4. The rotational rate of each thruster should be 165 rpm.
5. The required horsepower for each unit is 2,520 hp.

Naturally, there may be other factors which negate the above given answers. However, the example does illustrate the interaction between propeller parameters and some of the practical aspects of selecting a propeller for a dynamically positioned vessel.

So far in this section, the emphasis has been on the effect of propeller parameters, i.e., pitch, rpm, and diameter, on thrust output and power consumption under static or bollard conditions.

Other parameters which affect thrust include hull shape, inlet and outlet shape and radius, upstream and downstream objects which modify the flow of water into the propeller. These effects are discussed in the references[12] and are strongly dependent on the physical geometry of the entire system. For this reason model basin testing has become the accepted method of evaluating the influence on the thruster performance of any of these effects.

As a result, further discussion of the prediction of these effects on the thruster performance is unwarranted, with the exception of the simple inflow of water.

Inflow of water into a thruster occurs constantly because of the motion of the vessel caused by wind, waves, and other disturbances. Also, any surface current acts as an inflow to the thruster. Thus, the net inflow of water is a function of the vector combination of surface current and the current caused by relative motion of the vessel with respect to the water.

This factor of relative current inflow velocity in combination with the nonlinearities associated with the angle of attack of the relative current inflow velocity to the direction of static thrust makes the inflow question very complex. Adding the further complications of hull effects and interactions with other thrusters makes the problem a candidate for the model basin and beyond the scope of this book.

However, the effect of inflow current on thruster output can be successfully approximated on a first order level for many thruster configurations.

For the case when the inflow current is along the thrust axis of the thruster, the effect of the inflow on the thrust output can be modeled approximately as a linear decrease in thrust output. The validity of this model, at least as a first-order approximation, is apparent from Fig. 4–12 where for increased speed of advance (or increasing inflow velocity) the thrust coefficient, K_T, decreases. Thus, in the equation for thrust, Equation 4.29, for constant propeller rotational speed and diameter, the thrust will decrease if the thrust coefficient decreases with increasing speed of advance. The effect of inflow current can be expressed approximately as:

$$F_T, \text{lb} = T_o \left(1 - K_{IT}V_C\right) \qquad (4.48)$$

where:

F_T = output thrust of the thruster with current inflow

T_o = output thrust of the thruster without current inflow and constant rpms and pitch ratio

V_C = inflow current velocity relative to the thruster

K_{IT} = inflow current thrust coefficient in per unit thrust per corresponding units of current with V_C

Inflow current velocity is the resultant velocity of the water current flow at a distance from the thruster not to be affected by the vessel or the thruster and the motions of the thruster. Thus, even if there is no surface water current, the motions of the vessel can reduce the thrust output of the thruster. The amount of thrust change or the value of the inflow coefficient is typically 10% per knot of current or 0.1 per knot, respectively.

The direction of the inflow current in Equation 4.48 is in the direction of flow of water through the propeller without current inflow or in the opposite direction to the thrust. If the inflow is in the same direction as the thrust, then Equation 4.48 predicts that the thrust will increase because of the sign reversal in the current velocity.

As Fig. 4.12 indicates, this increase in thrust does occur because the thrust coefficient does increase over its bollard value. However, the linearity of the effect of inflow current in the same direction as the thrust is restricted over a smaller range of current speeds than in the other direction. In either case, the linear relationship of the effect of inflow current in the same direction as the thrust is restricted over a smaller range of current speeds than in the other direction.

In either case, the linear relationship of the effect of inflow on thrust output cannot be carried more than two or three knots before the degradation in thrust output becomes notably nonlinear, generally a second- or third-order reduction. In the reverse direction the nonlinear effects are even greater with a tendency toward a reduction in thrust from the predicted linear increase.

If a more precise model is required and if the functional relationships are available from a model test, then the relationships can be stored in a computer memory. Naturally, when seeking such modeling precision, a functional relationship for each thruster may be necessary. If a vessel has many thrusters, for example the *Sedco 400* series with 14 thrusters, the computer memory requirements can become excessive.

At first glance the effect of inflow current may appear to be a villain in that it reduces thrust output when thrusting into the current. When thrusting in the same direction as the current, the thruster output increases which appears to be an unexpected windfall, i.e., increased thrust for the same propeller diameter, rotational speed, and P/D ratio.

However, inflow current affects the required input power to the thruster. The effect of inflow on required power is much the same as for thrust (Fig. 4-12) with increasing speed of advance, J (analogous to increasing current inflow velocity). Thus, as the thrust output changes with inflow current, the required input power changes in the same relative manner, which means that even though thrust is lost or gained with inflow, the required power is correspondingly reduced or increased.

If the change in required input power caused by inflow current behaves with the same functional relationship as the change in thrust, then the thrust per horsepower for the thruster will remain constant in the presence of current inflow. However, even though Fig. 4-12 implies that a linear power relationship is equally valid over the same ranges of current variations as used for the thrust, the current inflow coefficient is slightly larger. Thus, the thrust-per-horsepower ratio for a thruster with inflow will increase slightly for increasing inflow current velocity.

In terms of performance in a dynamic positioning control system the question arises regarding the effect of thrust changes resulting from inflow current. If the control system of the thruster is open-loop, as in the case of standard helm control from the bridge, less thrust will be produced as inflow increases for the same rotational speed or pitch command. With less thrust the vessel will lose position if the commands to the thrusters balance the forces acting on the vessel prior to the increase in the inflow current.

Additionally, an increase in inflow current, if it is caused by surface current, will result in an increased drag force on the vessel, causing an even further loss of vessel position. To counter these effects, the helmsman will necessarily increase the command to the thruster.

In a like manner, the control system of a dynamic positioning system will automatically increase the command to the thruster in response to the measured loss of position from the position reference sensor system.

As a result, the change in thrust will be automatically compensated for by an appropriate change in command to the thruster within the limits of the command variable. Once one of the limits in the command variable is reached, no further compensation is possible, and the loss of position can no longer be prevented. In the next section the limiting factors on the thrusters are discussed in more detail.

Before the thrust change compensation is performed in the method just given, the vessel must drift off station. For changes in ocean surface currents caused by wind or tides, this method is generally fast enough to prevent perceptible excursions in vessel position. If the compensation for inflow was desired on a more rapid basis, either measurement of the inflow current or power might be used. In the case of inflow current measurement, the greatest problem is obtaining an accurate, reliable current measurement.

For the power method the problem is simpler because the power to the thruster can be measured to the prime mover of the propeller. Thus, a control system on the thruster which maintains a constant power for a particular thrust command should regulate the thrust output of the thruster resulting from inflow current.

However, as mentioned before, the thrust and power changes caused by inflow are not equal so the regulation of power will not compensate thrust output perfectly for inflow effects. Additionally, greater expense and complexity in the thruster controller are required which may not be justified except in unusual current conditions.

The previous discussion assumes that the inflow current is aligned with output thrust vector of the thruster. Naturally, the most common case is when the inflow current is from some other angle of attack to the thruster.

The previously-used linear model of the effect of inflow degrades as a model of the real world as the angle of attack of the inflow current to the thruster increases. The degree of degradation depends on the type of thruster and its installation configuration in the hull and installation with respect to other thrusters. A complete discussion of this topic is not possible here. In fact, model basin testing is required generally to make a reasonable estimate of this effect and unless the thrusters are tested in the proper hull and thruster grouping configuration, the tests may misrepresent the true effects of inflow current.

Without model basin information or expert predictions, the system designer can use the obvious approximation for the effect of the angle of attack of the current inflow on the thrust output of a thruster. This approximation is simply the resolution of the inflow current into a component in the direction of thrust under zero inflow conditions and a component at right angles to the zero inflow thrust direction. Then, the coincident component is used in the model given by Equation 4.48, or in equation form:

$$F_T, \text{lb} = T_o (1 - K_{IT} V_C \cos \alpha_C) \tag{4.49}$$

where α_C is the angle of attack between the zero inflow thrust vector and the inflow current, remembering that the inflow current is the composite of surface current and vessel motion.

Thus, with the model expressed by Equation 4.49, when inflow current is at right angles to the direction of thrust with zero inflow, the output thrust is not influenced by the inflow current which is generally not true. However, the effect of inflow is generally small when the inflow is at right angles to the zero inflow thrust direction.

A more precise model of the effect of current inflow shows that not only the magnitude of the thrust is changed by the inflow current, but the effective direction of the output thrust is also changed. This results in a component of thrust being produced at right angles to the commanded direction. The quadrature component acts as a cross-coupling force into another control axis. Since the magnitude of the cross-coupling force is a function of the control commands in another control axis and of inflow current magnitude and direction, the cross-coupling forces acts as a disturbance force in the control axis to which it couples.

Thus, the disturbed control axis must produce a thrust to compensate for the current inflow effect in the other control axis. If the current inflow is a slowly changing phenomenon, then the compensation caused by cross-coupling will only be a change in the average load on the distributed axis thrusters. In some cases, the change will aid (decrease the average load) and in other cases the change will be a hindrance (increase the average load).

4.6   Thruster drive and control system

As shown in Fig. 4–1, there are several other elements to the thruster system than the thrust-producing mechanism. First, there is a prime mover which furnishes power to the thrust-producing mechanism. Second, there is the thrust control system and, third, feedback and performance sensors. Coupling in the form of electrical and mechanical limits exists among the various thruster system elements. Therefore, these remaining elements of the thruster system are closely related and require consideration as a unit.

The discussion here is limited to two types of systems which represent a very high percentage of the thruster systems now in use or planned for the future. The two system types are designated controllable pitch (CP) and controllable speed (CS) and refer to the type of control used for the propeller.

There is more to a thruster system than just a propeller, and dynamic positioning places certain unique requirements on this portion of the thruster system which are not required for standard marine propulsion applications.

In the case of the controllable pitch system, the prime mover rotates the propeller at a nearly constant rotational speed and the thrust control system sets the pitch of the propeller blades based on a commanded pitch from the dynamic positioning control system. Since the propeller is rotating at a nearly constant rotational rate, the prime mover can conveniently be an ac induction motor, especially if the power system is planned to be a diesel-electric generator system, driving a common ac bus.

The thrust control system for the controllable pitch propeller, as the name implies, controls the pitch of the propeller. When speaking of controlling pitch of the propeller, pitch must be properly defined because the pitch of any propeller is variable as a function of radius. Thus, the normal meaning of controlling the pitch of the propeller is with reference to the pitch at a given radius. Usually the reference radius is 70% of the total radius as mentioned in the previous section.

Varying the pitch of the propeller for control purposes is achieved by rotating each propeller blade about the radial centerline of the blade. The angular rotation is achieved through a mechanical coupling which is driven by a hydraulic piston acting along the rotational centerline of propeller. This mechanism is illustrated diagrammatically in Fig. 4–16.

If the mechanical coupling between the hydraulic piston and propeller blade is strictly translational motion to rotational motion as illustrated in Fig. 4–17, then the propeller blade angle as a function of hydraulic piston displacement is:

$$\theta_p, \text{ degrees}, = 57.3 \arcsin \left[\frac{x_p}{R_p} \right] \qquad (4.50)$$

where:

θ_p = propeller blade angle which is zero for zero pitch
x_p = displacement of the hydraulic piston which is zero for zero pitch
R_p = radius of rotation of the coupling point on the propeller blade to the mechanical coupling to the hydraulic piston

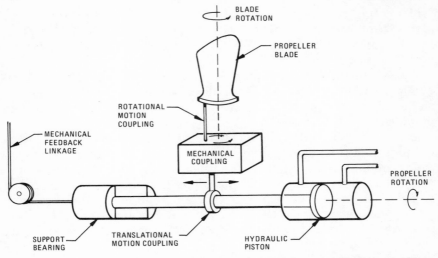

Figure 4-16. A Sketch of the Propeller Pitch Setting Mechanism for a CP Propeller

The hydraulic piston and the mechanical coupling mechanism shown in Fig. 4–16 form the hub of the propeller shown in Fig. 4–3. To limit the size of the hub, the piston displacement and the radius of rotation of the coupling point should be as small as possible. As is noted in the previous section of this chapter, the P/D ratio of the propeller should be equal to approximately one to achieve the best performance from the thruster at maximum thrust output.

A P/D ratio of unity translates into a blade angle of about 25° according to Equation 4.31. With a blade angle of less than 30°, the ratio of piston displacement to rotational radius is less than 0.5. Thus, the total piston displacement to rotational radius is less than 0.5. The total piston displacement for bidirectional thrust is equal to the rotational radius, R_p, on the base of the propeller blade.

The hydraulic piston, in addition to being sized to furnish sufficient stroke to vary the pitch angle over a ±25° range, must be sized to furnish the force required to hold the propeller blade at maximum pitch angle under maximum load. The parameters of the piston which determines the force which it exerts on the rotating mechanism of the propeller blade are its area or diameter and the hydraulic system operating pressure.

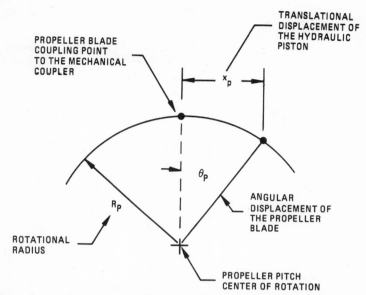

Figure 4-17. Geometrical Relationship Between Translation and a CP Propeller Rotation

Since the piston is a part of the propeller hub, the diameter of the piston is very critical to sizing the propeller hub. As a result, a tradeoff is made between hub diameter, piston diameter, and hydraulic system pressure. The tradeoff is necessary because although the hub size can be decreased by higher hydraulic pressure, a lower pressure hydraulic system is desirable to reduce the problems associated with high-pressure hydraulic systems such as with seals.

Another part of the mechanism which sets the blade angle of the propeller is a mechanical linkage for the measurement of the position of the hydraulic piston. With the measurement of the hydraulic piston position a feedback control system can be used to control the position of the piston either with a closed-loop servo system or an open-loop operator control on the bridge. As Fig. 4-16 indicates, the feedback linkage attaches to the piston shaft and is routed mechanically away from the hub.

If a closed-loop servo system is used to regulate the position of the hydraulic piston and, therefore, the propeller blade angle, the feedback linkage is used in the servo system. In Fig. 4-18 an electro-

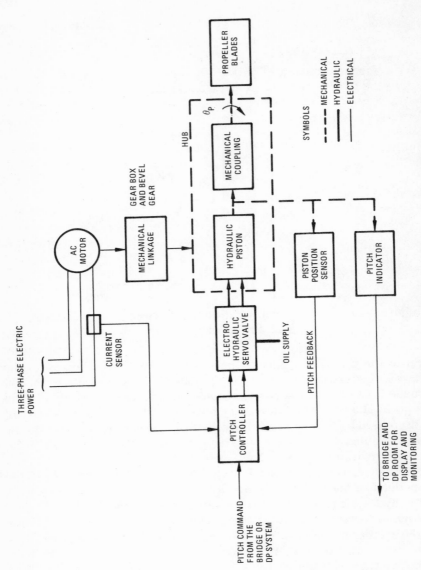

Figure 4-18. The Pitch Control Servo System for a CP Propeller

hydraulic system is illustrated. The key element of the system is the pitch controller which gives the servo valve the commands to move the hydraulic piston in the propeller hub.

To perform this task, the pitch controller compares the pitch feedback to the pitch command and commands the servo valve to move the hydraulic piston according to the magnitude and sign of the difference. For differences larger than a preset threshold, the command will result in the hydraulic piston moving at a constant rate. Also, the amount of amplification that the pitch controller applies to the difference in the command and feedback is selected so that the feedback control loop is stable and has a good time response to small changes in the input command.

Besides selecting the amplification or gain of the pitch control loop for stable operation, the designer of the pitch control system selects the constant rate that the piston moves with significant changes in command. For typical dynamic positioning applications, an adequate length of time to move the hydraulic piston so that the blade angle (or pitch) goes from one limit in one direction to the other limit in the other direction is 10 to 15 seconds.

As indicated in Fig. 4–18, the electric current in the motor driving the propeller at a constant rotational rate is measured. If the measured current is greater than a preset threshold, the pitch command will be reduced by a preprogrammed amount. This is a safety feature to prevent overloading the motor at large pitch commands.

Usually large changes in pitch commands will not overload the motor because of the command rate limit of the movement of the hydraulic piston. However, if there was an overload caused by a command change, the motor current feedback circuit would protect the motor.

As shown in Equation 4.50, the propeller blade angle or pitch is a nonlinear function of the hydraulic piston displacement. Thus, if the command from the bridge or dynamic positioning system is for propeller pitch, then at maximum pitch commands there will be an error between the commanded pitch and the actual propeller pitch. The error can be compensated for in the pitch controller by operating on either the command, the feedback, or the difference signal. If propeller pitch was measured instead of hydraulic piston position, then the error would be corrected by the closed-loop feedback pitch control servo.

In the case of feedback compensation where the command signal is the P/D ratio for 0.7 times the radius of the propeller, measured feedback P/D ratio of the propeller is given by the following function of the measured hydraulic piston displacement, x_p:

$$P/D_{0.7R(measured)} = 0.7 \, \pi \, \frac{\dfrac{x_p}{R_p}}{\left(1 - \dfrac{x_p^2}{R_p^2}\right)^{1/2}} \qquad (4.51)$$

For small piston displacement, e.g., $x_p \leqq 0.1 \, R_p$ the P/D ratio is a linear function of the piston displacement. Even at a P/D ratio of unity, the correction is only 10 to 15% of the linear value. Thus, the compensation may not be warranted because the output thrust of the thruster may not be predictable for a particular inflow condition to any greater accuracy than 10 to 20%.

Other compensation factors can be included in the pitch controller if a deterministic relationship is known which relates the effect to a measurable parameter. For instance, if the thruster is known to produce more thrust in one direction than the other because of the geometry of hub or the propeller support struts in the tunnel or entrance to the duct, then, a function generator can be included in the pitch controller to weight the commands in one direction differently than in the other direction.

Again, the increased complexity of the compensation may not be warranted because the effect is less than the accuracy with which thrust output can be predicted.

For the controllable speed propeller, the speed of the propeller is variable and the pitch as a function of propeller radius remains constant. To vary the speed of the propeller, the speed of the prime mover driving the propeller is varied. If the prime mover is a diesel engine, a throttle system can be used to control the rotational speed of the engine. If the prime mover is an electric motor, the motor is a dc motor whose speed can be controlled very accurately using commonly available electrical speed control techniques. One such technique is illustrated in Fig. 4–19.

If the input command to the thruster control system in Fig. 4–19 is the rotational speed of the propeller, then the command variable controller is a speed controller. The speed controller compares the commanded speed with the speed feedback and adjusts the command to the armature circuit or field circuit controllers.

Usually the armature circuit controller is used to change the speed of the propeller and the field is held constant. Only when the limit of

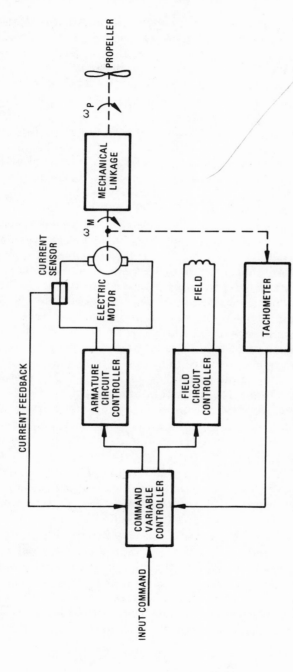

Figure 4-19. A Controllable Speed Propeller Control System

the armature circuit control is reached and additional thrust is required is the field changed to increase the speed of the motor.

To reverse the direction of rotation of the propeller and, therefore, the thrust, the propeller is stopped, the field is collapsed, the polarity of the field is reversed, and the field circuit energized to its normal level. As a result, a time delay occurs when the thrust command reverses. With proper design, the delay can be minimized to a few seconds which will not significantly affect the performance of the dynamic positioning system.

The delay generally does not even affect performance at all because the thrusters are producing a nonzero average thrust to counter the average environment acting on the vessel. As a result, the thrust output never reverses and the delay does not occur.

There is no time delay for pitch reversals in a controllable pitch propeller because the propeller continues to rotate at a constant speed even at zero thrust.

Like the controllable pitch propeller, the rate at which the commanded variable can react to changes in the input command is limited to prevent overloads to the motor. Thus, the propeller can go from forward to reverse in a time no less than determined by the command rate limit. For a thruster on a dynamically positioned vessel a full forward-to-full reverse time of 10 to 15 seconds is quite acceptable.

The thrust output of a controllable speed propeller varies approximately as the square of the propeller rotational speed (Equation 4.29). In contrast, the thrust output of the same propeller when the input command variable is the armature current varies approximately linearly as a function of the armature current. From the standpoint of control, there is an advantage to a linear relationship between the thrust and command variable. Therefore, a current control system may be preferred for the controllable speed propeller.

However, if there is any stiction in the mechanical coupling between the motor and the propeller, a current control system has the disadvantage that the current is properly controlled, but the propeller is not turning because of the stiction. This condition will continue until the current command is large enough to provide the break out torque to overcome the stiction. Then, the propeller will perform properly until it goes through another thrust reversal.

The manner in which this affects the performance of the dynamic positioning system is to create a slow periodic cycling in position holding. The cycling occurs because of a combination of the feedback in the dynamic positioning control system and dead zone and delay in the thrust output caused by the reversal process and stiction. However, the effect is only prevalent when there are frequent thrust reversals, such as during periods of slack environment.

Depending on the amount of stiction in the mechanical coupling, the amplitude of the periodic cycling is small and the period is very long and as soon as average thrust output increases so that thrust reversals no longer happen, the limit cycling stops. An increase in the average environment will stop the thrust reversals. Also, creating a pseudo-environment with another thruster in a manual control mode will prevent the cycling.

Where the input command variable is rotational speed, the problem with stiction is not present because the controller automatically increases the command to the armature circuit controller, if the propeller rotational speed is not equal to the input command.

If the thruster is an azimuthing type, the magnitude of the thrust can be controlled by either a controllable pitch or control speed system. Usually for either type of control system, the thrust is reversed by turning the propeller through 180°. As a result, the thrust magnitude control system can be designed for unidirectional operation which is a definite simplification for a controllable speed system. Also, the thruster can be designed to give the maximum achievable thrust as was discussed earlier in this chapter.

With the azimuthing thruster a second control system is required to control the direction in which the thruster is pointing, i.e., the thruster azimuth. Fig. 4–20 illustrates one method of controlling the thruster azimuth. The azimuth motor can be either hydraulic or electric.

In either case the system will respond to large changes in command with a constant rate. A rate of 12°/second or faster is adequate for most dynamic positioning applications.

The azimuth controller in its simplest form compares the command to the measured azimuth of the thruster and, if the differences are larger than a present dead zone, it commands the azimuth motor to turn at its maximum rate in the proper direction to reduce the difference. In this form the controller acts as a logic switch and is generally referred to as a bang-bang controller.

A more sophisticated controller is a variable command system whose command is proportional to the difference between the command and the measured azimuth for differences less than a preset value. For

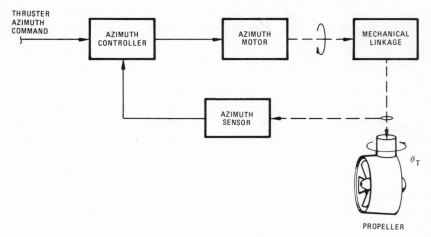

Figure 4-20. A Thruster Azimuth Control System

differences greater than the preset value the command rate to the azimuth motor is the motor's maximum design azimuth rate.

The proportional controller gives more precise pointing control with less wear-and-tear on the motor and the mechanical linkage than the bang-bang system. These advantages may be sufficient to make the proportional system worth the added cost both in terms of performance and maintenance and repair costs over the life of the system.

REFERENCES

1. John P. Comstock, Editor, *Principles of Naval Architecture*, Soc. of Naval Architects and Marine Engineers, New York, 1967.

2. Harold E. Saunders, *Hydrodynamics of Ship Design, Volume I and II*, Soc. of Naval Architects and Marine Engineers, New York, 1957.

3. John L. Beveridge, "Design and Performance of Bow Thrusters," Marine Technology, Volume 9, Nbr 4, October 1972, pp. 439–453.

4. Neal A. Brown and John A. Norton, "Thruster Design for Acoustic Positioning Systems," Marine Technology, Volume 12, Number 2, April 1975, pp. 122–137.

5. C. C. Schneiders and C. Pronk, "Performance of Thrusters," OTC 2230, 1975.

6. George R. Stuntz, Jr., and Robert J. Taylor, "Some Aspects of Bow-Thruster Design," Trans SNAME, Volume 72, 1974.

7. Dr. Hans F. Mueller, "Recent Developments in the Design and Application of the Vertical Axis Propeller," SNAME, Spring Mtg., Philadelphia, May 19 and 20, 1975.

8. Michael A. Rickards, "Cycloidal Propulsion of Submersibles," AIAA 2nd Advanced Marine Vehicles and Propulsion Meeting, Seattle, Washington, May 21–23, 1969.

9. C. A. Lindahl, "Consideration on Thruster Design and Installation," DNV Symposium Proceedings on Safety and Reliability of Dynamic Positioning Systems, Nbr 1. November 17–18, 1975.

10. Anders M. Liaaen, "Discussion of Liaaen's Drop-Down type thruster design," DNV Symposium Proceedings on Safety and Reliability of Dynamic Positioning Systems, Nbr. 1, November 17–18, 1975, pp. 314–316.

11. T. E. Hannan, *Strategy of Propeller Design,* Thomas Reed Publications Ltd. 1971.

12. J. W. English, "Thruster Hull Interaction," DNV Symposium Proceedings on Safety and Reliability of Dynamic Positioning Systems, Nbr. 1, November 17–28, 1975, pp. 294–306.

5

Controllers

5.1 Background

The oldest controller in the marine world is the helmsman. He controls the course, position, and speed of the vessel by observing his instruments (sensor outputs) and commanding the steering and throttle control system of the vessel to achieve the desired result. Normally, the helmsman is used on a vessel moving from one location to another. However, in the first dynamic positioning test in the open sea on the *Cuss I* [12] a helmsman was used to keep the vessel within a suitable watch circle on a given heading for the drilling experiment.

In performing the control function the helmsman reads the sensor outputs, compares them with the desired results, and formulates steering and throttle commands to place the vessel as close to the desired position as possible. As such, the helmsman must perform mental calculations rapidly enough to maintain the vessel "steady as she goes." A natural progression as control system technology developed was to replace the helmsman by an automatic control system, keeping the helmsman strictly as a monitor and supervisor for the automatic system.

For the applications requiring dynamic positioning, the replacement of the helmsman by an automatic system became more of a requirement than just a natural progression of technology. The helmsman task in a dynamic positioning system was far too demanding for most operators, especially in heavy sea states over extended periods of time. In addition, the automatic control system gives much more consistent performance.

The first controllers for dynamic positioning systems followed the lead of other automatic control systems in that they were analog systems. Then, as digital technology and minicomputers reached maturity, digital systems began appearing in dynamic positioning systems. Today, digital controllers dominate the dynamic positioning market. The future should be no different, except the digital technology used to implement the controllers will change.

In order to better understand the dynamic positioning controller this chapter will describe the basic control problem and a solution which can be implemented conveniently in an automatic control system. Additional functions that can be performed to greatly improve the performance and usefulness of the controller to the overall vessel system are described.

5.2   The basic control problem

In Fig. 5-1, the geometry of a vessel in the horizontal plane is illustrated with respect to a reference point and a reference heading. The obvious control problem is to position the vessel so that it is at the reference point with a heading equal to the reference heading. To solve the control problem, the position and the heading of the vessel are measured with respect to the reference quantities. Then, from the measured differences, thrust commands are calculated using a control law, which will reduce the differences. The thrust commands as a function of measured differences can be expressed as follows:

$$T_x = f(\epsilon_x) \tag{5.1}$$
$$T_y = f(\epsilon_y) \tag{5.2}$$
$$M_z = f(\epsilon_\psi) \tag{5.3}$$

where:

$T_x, T_y =$ the x-axis and y-axis thruster commands, respectively

$M_z =$ the yaw moment commands

$\epsilon =$ the difference between the measured value and the reference value of the respective variable

$f(\ ) =$ a functional relation between the argument of the function and commanded variable

There are any number of control laws that might be devised to convert the measured variables to thrust commands. A very common control law for a positional control system is given as follows:

$$f(\epsilon) = K_P\epsilon + K_I \int \epsilon \, dt + K_D \frac{d\epsilon}{dt} \tag{5.4}$$

where:

K_P = the proportional multiplier or gain factor
K_I = the integral gain factor
K_D = the derivative gain factor

A common name for the control law given in Equation 5.4 is Proportional plus Integral plus Derivative (PID) control. The units of the gain factors convert difference units into thruster commands.

Each term given in Equation 5.4 has a specific purpose. The proportional term gives the control system a restoring thruster command proportional to the difference in the reference and measured variables. Thus, as the vessel moves further from the reference value, the thruster commands become even greater. The rate or derivative term

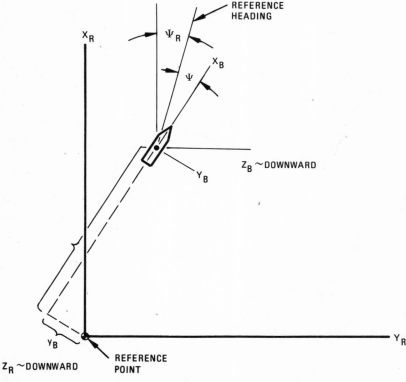

Figure 5-1. The Control Problem Geometry

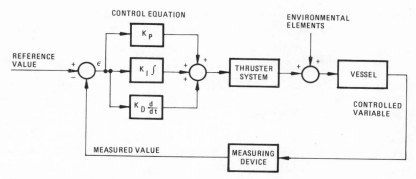

Figure 5-2. A Block Diagram of a Dynamic Positioning System

gives the necessary anticipation to the thruster commands to give the dynamic positioning system the desired dynamic response to disturbances.

The rate term can either be derived rate from the measured error, actual rate from measuring the rate of change of the corresponding control variable, or a blend of derived rate and measured rate.

The final term in the control law is proportional to the integral of the difference between the reference and the measured variable. The purpose of the integral compensation term is to counter the slowly varying or steady-state forces acting on the vessel. If the integral term is not included, a difference in the measured and reference variables must be tolerated so that the control law through the proportional term can generate the necessary steady-state counterforce.

The dynamic positioning system in one control axis using Equation 5.4 as a control law appears as shown in Fig. 5–2. The other control axes are similar. An obvious characteristic of the dynamic positioning system shown in Fig. 5–2 is the closed-loop feedback structure.

Although the control law given in Equation 5.4 solves the basic dynamic positioning control problem, another fundamental feature which can be added to the basic control law which greatly increases system performance is active wind compensation. With active wind compensation the control law given in Equation 5.4 becomes:

$$f(\epsilon, \alpha_A, v_A) = K_P\epsilon + K_I \int \epsilon \, dt + K_D \frac{d\epsilon}{dt} + F_A(\alpha_A, v_A) \qquad (5.5)$$

where:

$F_A(\alpha_A, V_A)$ = a counterforce or moment which is functionally related to the wind speed and direction relative to the vessel.

v_A = wind speed

α_A = wind direction

One convenient way to calculate the wind term of the control law is to measure the wind speed and direction and use the following standard aerodynamic drag equations:

$$F_{XA}(\alpha_A, v_A) = -C_{XA}(\alpha_A)v_A{}^2 \qquad (5.6)$$
$$F_{YA}(\alpha_A, v_A) = -C_{YA}(\alpha_A)v_A{}^2 \qquad (5.7)$$
$$M_{ZA}(\alpha_A, v_A) = -C_{ZA}(\alpha_A)v_A{}^2 \qquad (5.8)$$

where $C(\alpha_A)$ is the counter drag coefficient with units of force or moment per wind speed units squared. The block diagram of the dynamic positioning system with active wind compensation added is shown in Fig. 5.3. The drag coefficients used in Equations 5.6 through 5.8 can be determined either from model tests or from accepted techniques of calculation such as:[1]

$$C(\alpha_A), \text{ lb/knot}^2 = 0.00338 \; C_sC_hA(\alpha_A) \qquad (5.9)$$

where:

C_s = the shape coefficient from Table 5.1.
C_h = the height coefficient from Table 5.2.

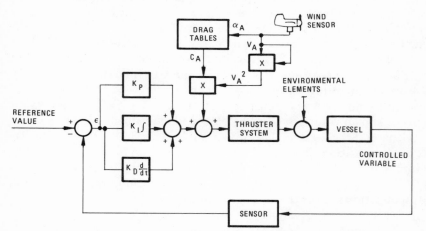

Figure 5-3. A Block Diagram of a Single-Axis DP System with Active Wind Compensation

TABLE 5-1
VALUES FOR SHAPE COEFFICIENTS[1]

Shape	C_s
Cylindrical shapes	0.5
Hull (surface type)	1.0
Deck house	1.0
Isolated structural shapes	
(cranes, angles, channels, beams, etc.)	1.5
Under deck areas (smooth surfaces)	1.0
Under deck areas (exposed beams and girders)	1.3
Rig derrick (each force)	1.25

$A(\alpha_A) =$ the projected area in square feet of all exposed surfaces in either the upright or heeled condition as a function of the angle of attack of the wind.

For a ship-shaped vessel the drag coefficients have the general shape shown in Fig. 5–4. In many cases a first-order approximation for the functional relationship of the drag coefficient to the angle of attack of the wind are given as follows:

$$C_{XA}(\alpha_A) = C_{XA}\cos\alpha_A \qquad (5.10)$$
$$C_{YA}(\alpha_A) = C_{YA}\sin\alpha_A \qquad (5.11)$$
$$C_{ZA}(\alpha_A) = C_{ZA}\sin2\alpha_A \qquad (5.12)$$

where C_{XA}, C_{YA}, and C_{ZA} are computed for an angle of attack of the wind on the bow, on the beam, and on the quarter, respectively.

The most accurate method of determining the wind drag coefficients

TABLE 5-2
VALUES OF HEIGHT
COEFFICIENTS[1]

Height range, ft	C_h
0–50	1.0
50–100	1.1
100–150	1.2
150–200	1.3
200–250	1.37
250–300	1.43
300–350	1.48
350–400	1.52

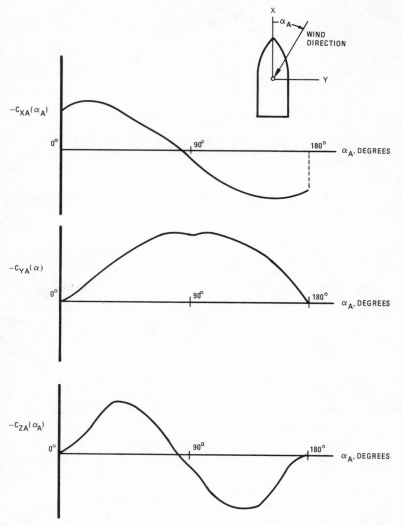

Figure 5-4. Aerodynamic Drag Coefficient as a Function of Angle
of Attack for a Ship-shaped Vessel

is the model tests. Two different model tests are used. The first, and perhaps more standard, is performed with a model of the vessel in a wind tunnel. The second is performed in a hydrodynamic tank by placing a model of the vessel upside down in a tank. The results of the second tests have to be scaled not only for size but also for the differences between the densities of water and air. In either case the model tests are more expensive than the calculation technique.

Even though the drag coefficients are imperfect regardless of the method used to determine them, active wind compensation has been shown by simulation and actual practice to be beneficial in improving position holding accuracy.[2][3] There are two reasons for this success: first, active wind compensation can create a counter thrust caused by a change in the wind without the vessel moving and, second, the wind is a major disturbance element acting on the vessel.

Thus, active wind compensation gives the control system additional anticipation of a major disturbance element, the wind. As Fig. 5–3 shows, the anticipation is achieved by feeding the estimated counter wind force or moment forward of the PID control equations output. For this reason this form of control is referred to as feed forward control.

Once the thrust commands are computed by the control equations, the commands must be transmitted to the thruster and its control system, which are described in Chapter 4. The units of the thrust commands from the control equations are either units of force or moment, e.g., pounds or foot-pounds. The control variables for the thrusters are normally either propeller pitch or propeller rotational speed, rpm. As a result, the thrust command from the control equation must be converted to a thruster control variable.

Most often the relationship between thrust and the thruster control variable is nonlinear. Thus, the thrust-to-thruster control variable converter illustrated in Fig. 5–5 is a function generator which has a limited output which corresponds to the maximum allowable value of the thruster control variable.

Another aspect of the basic control problem, which is common to any control system — especially automatic feedback control systems — is the degree of stability or response characteristics of the system. First, the control system should be stable, which means that the time response of the control system to a disturbance converges as time increases as opposed to increasing without bound (diverging). Technically a control system is considered "stable" if its response to a dis-

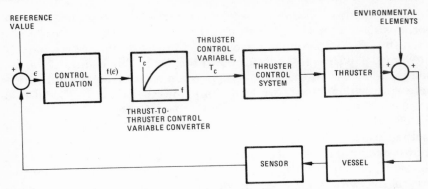

Figure 5-5. A Block Diagram of the Conversion of the Thrust
Commands to Thruster Control Variables

turbance is purely oscillatory. However, from a practical standpoint
a system which oscillates is of little use.

Second, the rate of convergence should meet certain standards which
are determined by the performance requirements of the overall control system. The question of control system characteristics is discussed
in detail in the chapter on Control System Analysis, Chapter 7.

In addition to response characteristics, the basic control problem includes the effect of sources of random variations on certain system
parameters. One system parameter of concern in the basic control
problem is the thruster command variable. The thruster command
variable is of concern because random variations in this variable
cause wear-and-tear on the thruster system and unnecessary consumption of power or fuel. The random variations in the thruster command variable are commonly referred to as thruster modulation.

The most common sources of thruster modulation are noise sources
in the sensors, the wind, and the waves. Clearly, sensor noise creates
nonproductive thruster commands. However, earlier the statement
was made that a component of the control equation output was to
counter the effect of the wind. So why is wind listed as a contributor to
thruster modulation? The reason is simply that only a portion of the
thruster command caused by the wind has any effect on countering
wind. The remaining part is wasted and is, therefore, a component of
the thruster modulation.

Similarly, the only portion of wave-induced motion of the vessel
can be used productively to control the position or heading of the vessel.

The remainder of the motion which is measured by the sensor and operated on by the control equation does not contribute to the control of the vessel, but it does cause variations in the thruster command variable and is, therefore, regarded as thruster modulation.

As mentioned earlier, the basic dynamic positioning problem involves the simultaneous control of X-Y position and heading. Thus, the basic dynamic positioning controller must include three feedback control loops: a X-position, a Y-position, and a heading loop. Each loop should include a control equation with proper coefficients and wind drag model for that vessel axis. Also, a thrust command-to-thruster control variable converter is required for each thruster. If each thruster in the three control axes are the same, the controller can be simplified using three identical converters.

In a digital computer which is discussed later in this chapter the same converter can be used for all thrusters if they are equivalent.

To achieve x-axis control, the thruster is installed with its thrust-producing axis aligned with the x-axis of the vessel. Similarly, the y-axis control is achieved with a thruster aligned with the y-axis of the vessel. If the x-axis and y-axis thrusters are not acting through the center of rotation of the vessel, they produce a moment.

To control the heading axis, the moment command must be generated by a thruster producing a force through a lever arm. The moment-producing thruster can be mounted so that its line of action is aligned with the x-axis or the y-axis or both. The larger the lever arm of the moment-generating thruster, the smaller the thrust that is required to generate a given moment command.

As a result, the moment-generating thruster is aligned in the port-starboard axis (the y-axis) for a ship-shaped vessel because the vessel is much longer than it is wide. For a squarer vessel, such as a semisubmersible, both the x-axis or y-axis can be used to generate the moment command.

If only one thruster is used to generate moment at a given distance from the center of rotation, the same thruster produces a y-axis thrust which generally will not be equal to the required y-axis thrust command. The result is that in satisfying the moment command the y-axis command is no longer satisfied. To resolve this problem of cross-coupling between control axes, the force and moment commands from the controller must be allocated among the thrusters so that all three axes commands are satisfied simultaneously.

The allocation equation is given for the simplest three thruster configurations in Equations 4.16 through 4.18. More details regarding more complex thruster configurations are given in Chapter 4 and later in this chapter.

When the force or moment commands exceed the maximum output from the thrusters, the thrust allocation equations can no longer be satisfied. If the thruster is one that is producing moment and x-axis or y-axis force, then priority can be given to fulfilling either the moment or the other control axis command. For a ship-shaped vessel which will tend to broadside to the weather, the priority is given to fulfilling the moment command. If there is any thrust capability left after fulfilling the moment command, it is allocated to the y-axis command in such a way as to produce zero moment.

In summary, the basic dynamic positioning controller includes three elements, some of which are identical in form. The elements are:

1. Control equations, one in each of the three control axes.
2. Thrust allocation equations, one for each thruster.
3. Thrust command-to-thruster control variable conversion functional relationships.

5.3　Sensor data processing

The sensors shown as a part of the basic dynamic positioning system in Fig. 5–3 measure the various input data which the controller uses to compute the thruster commands. The various sensors measure their respective parameters with varying degrees of accuracy. In addition, the sensor measurements are noisy and sensors are subject to failures which may result in an erroneous input to the controller.

These practical aspects of actual sensor measurements affect the dynamic positioning system in different ways. This section briefly discusses these effects of real-world sensors on the DP system and the features that can be added to the controller to minimize or eliminate the bad effects. A more complete discussion in terms of control system analysis appears in Chapter 7.

Even when the sensors are operating normally, they cannot measure their respective variables precisely. A certain portion of the imprecision is caused by the response characteristics of the sensor which are reflected in the dynamic model or transfer function of the sensor. The transfer functions of the three basic dynamic positioning system sensors are discussed in Chapters 2 and 3.

The effect of the dynamic response characteristics of a sensor on the measurement of the variable is to change the amplitude and phase of the variable so that the output of the sensor is not equal to the input. The change is a function of the frequency and amplitude of the variable being measured and the function may be linear or nonlinear. A linear function is defined as one for which the law of superposition is valid.

A typical example of the linear and nonlinear effects of dynamic response characteristics of a sensor is given by the model of an inclinometer which uses a fluid damped pendulum mass to measure the angle of the reference frame to the vertical. Because of its inertia and damping, the pendulum mass can only respond to angular motions of a certain frequency. Similarly, the pendulum mass can only rotate through a limited number of degrees. The result is a linear response characteristic for angular motions less than the angular limit of travel and a constant output equal to the angular limit for motions greater than the limit. In block diagram form, the model of an inclinometer is given in Fig. 5–6 showing the linear and nonlinear portions of the model.

Another portion of the imprecision of the sensor output is the result of calibration errors and slowly-varying errors generated by the devices which make up the sensor. For example, the electric circuit from the wind speed sensor to the controller is actually a part of the wind sensor. If its series resistance changes, the measured wind speed at the input to the controller is in error. The effect of this type of error on the output of the sensor is to change the amplitude of the variable being measured by a proportional amount. A proportional change is equivalent to changing the gain of the linear portion of the sensor transfer function.

There also may be component devices in the sensor whose outputs

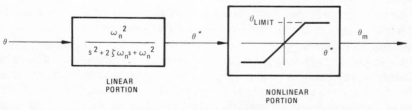

Figure 5-6. A Simplified Block Diagram of an Inclinometer Sensor

are summed with the input to the sensor to compensate for a measurable known error in the sensor input. For example, the output of the vertical reference sensor used to compensate the roll/pitch error in the short baseline acoustic position reference system measurement sums directly with the measured angular displacement. If there is a constant error in the output of such a device, then the same constant error is in the output of the sensor with, at most, a different sign. A constant error in the output of a sensor acts as an offset or bias in the measured variable.

The last portion of the sensor imprecision under normal operation is time-varying uncertainty or noise in the output of the sensor. The source, or sources, of noise depends on the sensor and its method of measuring the given variable. For example, with an acoustic position reference sensor there is a measurement uncertainty in the detection of the arrival of an acoustic pulse from the subsea acoustic source.

The uncertainty is caused by acoustic noise picked by the hydrophone, signal path attenuation, and electronic noise in the hydrophone and detection electronics. As a result, the parameter, either range or bearing, which is computed directly from the interval of time is in error. The timing error is random and is generally modeled as white noise with zero mean and a variance which is dependent on the hydrophone and detection electronics and a minimum signal-to-noise ratio at the hydrophone.

The degree of effect of the noise and bias errors on the output of the sensor depends on which point in the sensor the error originates. As Fig. 5–7 illustrates, noise and bias errors can originate at the input, the output or some intermediate point, internal to the sensor. As shown more completely in Chapter 7, the response characteristics of the sensor through which noise and bias errors pass in reaching the output of the sensor influence the characteristics of the noise and bias errors in the output of the sensor.

Thus, the response characteristics of the sensor enter into the effect of errors caused by the sensor in another way than the direct way mentioned earlier.

In summary, the sensors under normal operation measure their assigned variable imprecisely in both amplitude and time (phase). They add offsets or biases to the measured variable and corrupt it with noise. When the sensor malfunctions or fails, the output of the sensor is unpredictable. The sensor output may also contain measured data which is accurate but unwanted for control purposes. Signal process-

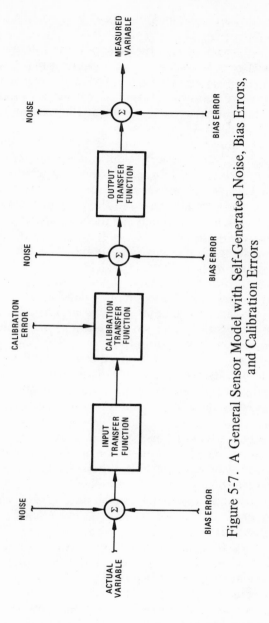

Figure 5-7. A General Sensor Model with Self-Generated Noise, Bias Errors, and Calibration Errors

ing can be included in the controller to at least reduce the effects of some of the sensor problems.

The classical approach to eliminating unwanted data is to pass the data through a filter which attenuates the unwanted data and passes the other data. For dynamic positioning, the unwanted data is the high-frequency wave motions and wind variations. Since the basic frequencies of the control system are lower than these unwanted frequencies, the filtering task seems simple.

However, as the chapter on control system analysis shows, a trade-off develops. At any rate, the sensor data processing of the controller must include filters for high-frequency wave and wind data.

As shown in Fig. 1–17, the spectrum of the high-frequency wave data increases as the seas increase and the peak of the wave spectrum shifts to lower frequency. Similarly, as the seas decay, the peak of the wave spectrum moves to higher frequency. Thus, the filters in the sensor data processing should ideally be adaptive to track the spectrum of the wave-induced motion of the vessel.

One adaptive technique developed by Harthill and Van Calcar is shown in Fig. 5–8 and is explained in more detail in Reference 11. This filtering technique is used in many of the presently operational dynamic positioning systems and is quite effective in reducing thruster modulation caused by waves and other sources in the system. Reference 4 considers an even more advanced concept of filtering of sensor data based on the system shown in Fig. 5–8 and Reference 11.

These same filters will also attenuate the noise picked up or introduced by the sensors, and any other source between the sensors and the controller.

There are other sources of noise in the path between the sensor and the thruster system which can affect the performance of the DP system and the thruster modulation. Likewise, there are sources of calibration error and bias error. These sources are a function of the electronics used in the path, the implementation of the electronics, and the type of controller.

Filters may not provide sufficient attenuation over the proper frequency range. As a result, additional noise filtering may be required. The design of the noise filtering will be involved in trade-off in the same way as the other filter is involved.

When the output of a sensor contains a bias or an offset error, there is no effective way to determine, let alone correct, the error. As the bias error develops, the feedback action of the control system moves the

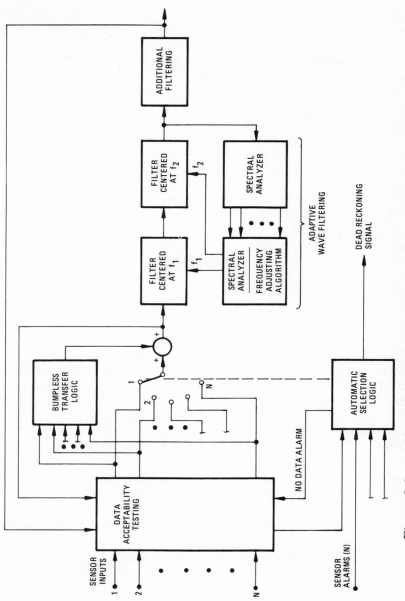

Figure 5-8. A Block Diagram of a Single Axis of Sensor Data Processing

vessel until the measured variable equals the commanded variable. If the development of the bias error is sufficiently slow, the movement of the vessel will be undetected by the operator.

If there is another measurement of the same variable, then the two measurements can be compared. The question then arises as to which sensor measurement is correct. In some cases a third measurement is required.

The controller can contain processing which compares any number of the same variables from different sensors. If the comparison fails by a certain amount, the controller can alert the operator and under certain situations, change from one sensor to another. This processing is only applicable when multiple sensors measuring the same variable are available and active in the DP system.

Another monitoring function which should be included in the sensor data processing is for sensor failures. For some sensors, they contain fault monitors themselves that can be used to alert the controller of the sensor's failure. Likewise, the sensor input to the controller can be monitored and if it does not change for some time, the sensor should become suspect of a problem.

In any case, if a sensor failure is detected in the controller, either by a fault monitor or test, the controller should change to another sensor measuring the same variable and alert the operator to the problem.

There is another item of sensor data processing which relates to filtering or screening noise from the input to the controller. Generally speaking, the noise described previously is of such an amplitude that there is no logical way to separate it from actual data. As a result, a filter is included to attenuate at least a part of the noise.

Under certain conditions, there will be changes in the output of a sensor which are not physically possible. The undetected failure of the sensor or malfunction somewhere in the input is large enough and long enough, the controller will calculate a large erroneous change in the thruster commands. The large erroneous change in the thruster commands will drive the vessel away from the desired position and the highly undesirable condition of a drive-off will occur. The time required for a drive-off to develop depends on many factors, but in the worst case, even for a large drillship, a significant drive-off can occur in one minute or less.

The drive-off will continue until the error in the output of the sensor is corrected or the controller transfers to another sensor measuring the desired variable correctly. In fact, the controller is most likely better off with no data than highly erroneous data.

To prevent such an occurrence caused by data which is obviously erroneous, sensor data processing can be included to monitor the quality of the incoming data. If the data is obviously physically impossible, then the controller can be transferred to a back-up sensor or perhaps no sensor at all.

The task of monitoring the incoming data for erroneous values is not as easy as it might sound. In the first place, the rate of change of the data and the amplitude of the variations in the data increases naturally as the wind and the waves increase. Thus, setting the threshold between "good" data and "bad" data changes with the weather, which implies that some form of adaptive data processing is needed. More is said about the requirements for this very important sensor data processing function in the control system analysis in Chapter 7.

Other than the filtering of the sensor data, the sensor data processing described to this point monitors sensor data and, if a problem is detected, the operator is alerted. In some cases, the controller uses data from another sensor if one is available.

Two questions arise at this point. First, what action should the controller take if there is not another sensor available? Second, what happens if there is a significant steady-state difference or offset between the outputs of the old and new sensor? Switching between sensors automatically creates the problem of which sensor does the controller use.

If another sensor is not available when the output of the operational sensor is declared unacceptable, then the controller must take some action. The choices are to continue to use the output of the operational sensor, to stop operating, or to enter a special mode which is referred to in this book as dead reckoning.

In the dead reckoning mode, the controller continues to compute thrust commands equal to the sum of the thrust commands computed by active wind compensation in real time and the last valid steady-state thrust command computed from the error signal. At the same time, the controller signals the operator that it has entered the dead reckoning mode.

The controller continues to hold the last valid steady-state thrust command from the error signal plus adding the active wind compensation updated commands from the real-time wind sensor data until the sensor processing logic signals the resumption of normal operation. Obviously, the longer the controller is in dead reckoning, the larger the error is in position holding accuracy by the dynamic positioning system.

However, since the wind is a major disturbance element acting on the vessel and active wind compensation continues during dead reckoning, the performance of the dead reckoning is potentially very good. Naturally, the better the wind drag model and wind measurement, the better dead reckoning performs.

When a transfer is made from one sensor to another, there will very likely be a difference in their outputs, both in the short-term and long-term components. As a result, when the transfer occurs, a step function in the sensor data is applied to the filtering of the sensor data processing and the control law computations. If the step function is of sufficient magnitude, a sudden change in the thruster commands from the controller will be generated which drives the vessel to a new location as indicated by the new sensor. A strategy which is used to smooth out the transfer between sensors is called bumpless transfer.

Bumpless transfer is implemented by constantly computing an average difference between the operational or on-line sensor and every other valid standby sensor. Then, when a transfer from the operational sensor to a standby sensor is generated by some problem with the operational sensor, the average difference between the outputs of the two sensors is properly combined with the new sensor's output so that there is no average difference between the outputs of the old and new sensor. However, as time passes, the average difference is decayed to zero so that eventually the system is using the output of the new sensor without any changes.

Sensor data processing can be even more complex than what has been presented here. For example, estimation schemes based on Kalman filtering techniques can be used to "screen" the data for the controller to use.[5] However, it is important in whatever techniques used that the "screening" method does not introduce excessive phase shift in the control loop and the technique does not overscreen the data to the point that the controller receives little or no real-time data from which to compute the thruster commands.

Many of the details of the implementation of the previously-described sensor data processing have been omitted because they are beyond the scope of this book. However, these details should not be underestimated in achieving good performance from the sensor data processing.

Likewise, it should be noted that for a controller to include all the features of sensor data processing, not only are multiple sensors required, but also a controller capable of significant logical decisions in real time is required. Such a controller is the modern digital real-time control computer.

In summary the following features of sensor data processing have been discussed:

1. Filtering of unwanted data and noise
2. Monitoring data for sensor failures, data validity, and physical realizability
3. Automatic transfer to a properly operating standby sensor
4. Bumpless transfer between sensors
5. Dead reckoning

These features are illustrated in Fig. 5–8 in block diagram form.

5.4 Fixed-axis thrust allocation logic

Thrust allocation logic is a basic part of the controller of a dynamic positioning system. Thrust allocation requires a set of equations to distribute the two force commands and one moment command from the control equations among the thrusters. The equations developed in Chapter 4 are primarily for fixed-axis thruster installation where all the thrusters in the same axis have the same maximum thrust capability.

In this section, thrust allocation logic for the more general situations of fixed-axis thrusters of unequal maximum thrust capability and longitudinal thrusters with significant moment generating capacity are discussed. Also, the strategies that are required when the thrusters reach the limits of their capability are considered.

When the thruster installation consists of the minimum number of fixed-axis thrusters, i.e., two lateral units and one longitudinal unit, the lateral thrusters are of equal capacity, and the longitudinal thruster is on the centerline of the vessel, the thruster allocation equations are given by Equations 4.16 through 4.18. A more common situation is with multiple lateral thrusters, forward and aft of the center of rotation of the vessel, and two longitudinal thrusters (main screws) equidistant from their center of rotation (Table 4–4).

Frequently, the lateral thrusters are grouped close together in one forward and one aft location. Furthermore, they are usually identical for purposes of lower unit price and common spare parts and maintenance procedures. Similarly, the main screws are identical, and even though they are not mounted on the centerline of the vessel, their lever arms are so much smaller than the lateral thrusters that they are not effective moment generators.

As a result, the equations developed in Chapter 4 can be applied to many of the dynamically positioned vessels currently in existence. However, in case the thrusters are not of equal rating or there are more thrusters in one group than another (a more common practice), certain adjustments should be made in the allocation equations given in Chapter 4.

For thruster installations of multiple thrusters of the same rating, all the thrusters of the group receive the same command. Thus, a common command cable from the DP controller to the thruster control systems for each thruster is possible, reducing the number of cables. However, from the standpoint of redundancy, a separate command cable for each thruster is currently in use. Multiplexing data transmission systems and reliable microelectronics will definitely affect this in the future.

If the main screws are mounted on the vessel so that they can generate a significant moment, the allocation of thrust to them must be adjusted.

Before considering the situation for a fixed-axis thruster installation where the main screws can generate a significant moment compared to the lateral thrusters, the fixed-axis thruster installation with two equally-rated, closely-spaced main screws is considered. This configuration is currently the most common because all dynamically positioned ship-shaped vessels fall into this category because of their length-to-breadth dimensions.

The other configuration of widely spaced main screws applies to semisubmersibles with two parallel submerged hulls. However, the steerable thruster installation should become the more common thruster installation for semisubmersibles because of their reduced requirement for installed horsepower (see Appendix A).

The first adjustment that is made to the allocation equations is to apply weighting factors in computing the equivalent lever arms given in Equations 4.25 and 4.26. The modified lever arms equations are given as follows:

$$\ell_F = \sum_{i=1}^{N_F} \frac{T_{Fi}(\text{max})\, \ell_{Fi}}{\sum_{j=1}^{N_F} T_{Fj}(\text{max})} \qquad (5.13)$$

$$\ell_A = \sum_{i=1}^{N_A} \frac{T_{Ai}(max) \, \ell_{Ai}}{\sum\limits_{j=1}^{N_A} T_{Aj}(max)} \qquad (5.14)$$

where:

ℓ_F, ℓ_A = the equivalent forward and aft lever arms for the active thrusters, respectively.

N_F, N_A = the number of thrusters, forward and aft of the center of rotation, respectively

ℓ_{Fi}, ℓ_{Ai} = the lever arm of the ith thruster, forward and aft of the center of rotation, respectively

$T_{Fi}(max), T_{Ai}(max)$ = the maximum thrust for the ith thruster, forward and aft of the center of rotation, respectively

The second adjustment is to the allocation equations to each thruster which appears as follows:

$$F_{Fi} = \frac{T_{Fi}(max)}{\sum\limits_{i=1}^{N_F} T_{Fi}(max)} F_F \qquad (5.15)$$

$$F_{Ai} = \frac{T_{Ai}(max)}{\sum\limits_{i=1}^{N_A} T_{Ai}(max)} F_A \qquad (5.16)$$

where:

F_{Fi}, F_{Ai} = the thrust allocated to the ith forward and aft active thruster, respectively.

As pointed out in the definitions of the quantities in Equations 5.13 through 5.16, the equivalent lever arms and thrust allocations to the individual thrusters are a function of the "active" thrusters. In a thruster installation where there are multiple thrusters in the forward and aft groups, all the thrusters may not be in use. They may be out-of-service for repair or maintenance or, more commonly, they may be out-of-service because the weather is less than the design value.

As a result, the operator places certain of the thrusters in a standby or off mode. The logic, then, automatically recomputes the equivalent lever arms and distributes the thrust to the active thrusters.

Neither of these adjustments change the basic allocation equations which distribute the moment and lateral force commands between the forward and aft thruster groups. Thus, the basic equations are as given in Chapter 4 and are as follows:

$$F_F = M_Z \left(\frac{1}{\ell_A + \ell_F}\right) + F_y \left(\frac{\ell_A}{\ell_A + \ell_F}\right) \qquad (5.17)$$

$$F_A = -M_Z \left(\frac{1}{\ell_A + \ell_F}\right) + F_y \left(\frac{\ell_F}{\ell_A + \ell_F}\right) \qquad (5.18)$$

With these allocation equations, the thrusters within a group, either forward or aft, reach their maximum outputs at the same time. However, one group may reach its maximum output before the other. Once the thrust to be allocated forward or aft, i.e., F_F or F_A, exceeds the maximum value for the group, then the allocation equations are no longer valid, even though only one of the two equations has reached its limit.

To avoid the invalidation of the equations, logic is included with the thrust allocation equations to detect whenever one of the thrust allocation equations reaches a maximum value. When a maximum is detected, the logic limits the outputs of the equations to their maximum values. If the force and moment inputs from the controller continues to increase, the logic continues to hold the output forces of the allocation equations at the values for which the equations are still valid.

However, when limiting does occur, the thrusters may not supply sufficient thrust to satisfy both the moment and force commands from the controller. As a result, the vessel will lose heading and position control.

In the case of heading, the consequences can be very serious, especially for a ship-shaped vessel, because most ship-shaped vessels do not have sufficient lateral thruster horsepower to hold station with the specified operating environment on their beam. Thus, as the weather builds, they will be holding their heading into the weather and loss of heading will result in the lateral thrusters being totally overpowered by the weather.

To counter the possibility of loss of heading, the logic gives priority to the command which is responsible for holding the heading of the vessel, i.e., the moment command, M_Z. Then, once the moment command is satisfied, the remaining thrust, if there is any, is allocated to holding position up to the maximum capabilities of the thrusters.

Naturally, the vessel will lose some position because the position command is only being partially satisfied. However, heading control will be maintained.

Once the maximum output for one thruster group is reached and heading is given priority, thrust is allocated to the forward and aft thruster groups in steps. The first step is the allocation of thrust to satisfy the moment command, which gives the desired priority to heading.

This moment allocation can be performed in at least two ways. The first way is a two-step process where, when the moment command can be satisfied without reaching the limit of one of the thruster groups, the following equations are used to allocate moment-generating thrust:

$$F_{FM} = M_Z \frac{1}{\ell_A + \ell_F} \tag{5.19}$$

$$F_{AM} = -M_Z \frac{1}{\ell_A + \ell_F} \tag{5.20}$$

where:

F_{FM} = the moment-generating thrust allocation for the forward thruster group

F_{AM} = the moment-generating thrust allocation for the aft thruster group

Then, when the moment command is so large that the maximum thrust for one group is reached, satisfying the moment command only, the command for that group is held at the maximum and the command to the other group is increased toward its maximum, using the following equation:

$$F[\;\;]_M = (M_Z - F_{(\;)M}(max) \, \ell_{(\;)}) \frac{1}{\ell_{[\;]}} \tag{5.21}$$

where the ( ) brackets are the quantities for thruster group at its maximum output and the [ ] brackets are the quantities of the thruster group which has not yet reached its maximum. If the thruster groups have equal moment capabilities, then the second step is not used.

Prior to either of the thruster groups reaching their maximums in the allocation of the moment command, the magnitude of the thrust commands to the forward and aft groups are equal. After one group reaches its maximum output, the thrust commands to the two groups may not be equal which creates a residual lateral thrust.

This residual lateral thrust acts like a lateral distrubance thrust and, as such, adds to the lateral thrust equation. The residual is equal to the following expression:

$$\Delta F_M = (M_Z - F_{(\ )M}(max) \; [\ell_A + \ell_F]) \frac{1}{\ell_{[\ ]}} \tag{5.22}$$

This residual can increase or decrease the total lateral forces, driving the vessel off station, depending on its sign.

The second way of allocating thrust to satisfy the moment command is to weigh the commands to the groups according to their maximum capability. Then the two groups reach their maximum at the same time. However, unlike the two-step allocation just given, there is a residual lateral thrust created if the thruster groups have unequal capabilities. The allocation equations for this second method are as follows:

$$F_{FM} = \frac{F_F(max)}{F_A(max)} \; M_Z \; \frac{1}{\ell_A + \ell_F} \tag{5.23}$$

$$F_{AM} = -\frac{F_A(max)}{F_F(max)} \; M_Z \; \frac{1}{\ell_A + \ell_F} \tag{5.24}$$

where:

$F_F(max)$ = the maximum thrust of the forward thruster group
$F_A(max)$ = the maximum thrust of the aft thruster group

The residual thrust generated by the weighted allocation of the moment command is given by:

$$\Delta F_M = F_{FM} + F_{AM} = \left[\frac{M_Z}{\ell_A + \ell_F} \right] \left[\frac{F_F^2(max) - F_A^2(max)}{F_F(max) \; F_A(max)} \right] \tag{5.25}$$

The allocation of the lateral thrust command adds to the moment command of one thruster group and subtracts from the other thruster group because the two moment-generating thrust commands are in opposite directions to create the maximum couple and the two lateral thrust commands are in the same direction.

For the thruster group whose thrust command is increased in magnitude by the lateral thrust commands, its maximum output is reached before the other thruster group. As a result, if the moment allocation forces a thruster group to its maximum in the direction opposite the direction of the lateral command, then that thruster group will not be at its maximum value after the allocation of the lateral thrust command because of the summation to obtain the total thruster command.

The allocation of the lateral thrust command for the two-step method uses the following equations:

$$F_{( )L} = F_{( )}(\max) - F_{( )M} \qquad (5.26)$$

$$F_{[ ]L} = F_{( )L} \frac{\ell_{( )}}{\ell_{[ ]}} \qquad (5.27)$$

where:

$F_{( )L}$ = the lateral thrust allocation for the thruster group which has reached its maximum output

$F_{[ ]L}$ = the lateral thrust allocation for the other thruster group

The lateral thrust command from the controller does not appear in either allocation equation because the condition for allocating the moment command separate of the lateral command is that one thruster is already at its maximum output. Thus, if the moment command is satisfied first, then the lateral thrust command cannot be satisfied and, therefore, does not enter into the lateral thrust allocation.

However, the remaining capability of the lateral thrusters is allocated to achieve the maximum usable output from the lateral thrusters. To achieve the maximum usable output, the thruster group which will reach its maximum output first when the lateral thrust command is added, is commanded to its maximum.

The other thruster group is commanded in such a way to create zero moment so that the previously computed moment commands to the thruster groups remain satisfied.

With the lateral and moment commands from the separate allocations, the total command to each thruster group is computed by simple addition. Then the thruster group commands are subdivided among the active thrusters in the group and transformed into the units of the thruster command variable and scaled for proper data transmission to the thrust control system.

Allocation of the force and moment commands from the controller into individual commands for each active thruster involves several logic tests and decisions, as well as computations. This processing is best summarized by a flow chart which shows the sequence of the decision process. Fig. 5–9 is the flow chart for the previous discussion for the allocation of thrust to a thruster installation of fixed-axis lateral thrusters.

The allocation for the x-axis force command is not included in Fig. 5–9. If the thrusters which satisfy the x-axis command are two equally-rated main screws, mounted equidistant from the centerline of the vessel, where their lever arm distance is too small to be useful in generating any moment except in a total loss of the lateral thruster, the allocation of x-axis command from the controller is:

$$F_{xj} = \frac{F_{xc}}{N_{MS}} \qquad (5.28)$$

where:

N_{MS} = the number of active main screws, i.e., 1 or 2
F_{xj} = the x-axis thrust command to the jth main screw
F_{xc} = the x-axis force command from the controller

Generally, the main screws will have the same rating. Therefore, they will reach their maximum output at the same time, making priority allocation rules unnecessary.

If only one main screw is in operation, either because the other is being serviced or repaired, or because the longitudinal load on the vessel does not warrant two screws, a moment is generated by the single screw. The magnitude of the moment is given as follows:

$$|M_{zx}| = |\ell_{xj} F_{xj}| \qquad (5.29)$$

where:

$|M_{zx}|$ = the magnitude of the moment about the z-axis which is generated by the x-axis thruster or main screw
ℓ_{xj} = the lever arm through which the main screw acts

This moment acts as a disturbance moment, tending to change the heading of the vessel. However, for a ship-shaped vessel, the disturbance is not serious because the lever arms of the main screws are small compared to the lever arms of the lateral thrusters. Likewise, the longitudinal demand on the vessel is not at its maximum when the moment demand is at its maximum.

For a semisubmersible with a fixed-axis main screw installation, such as the Aker H-series units, the separation of the longitudinal

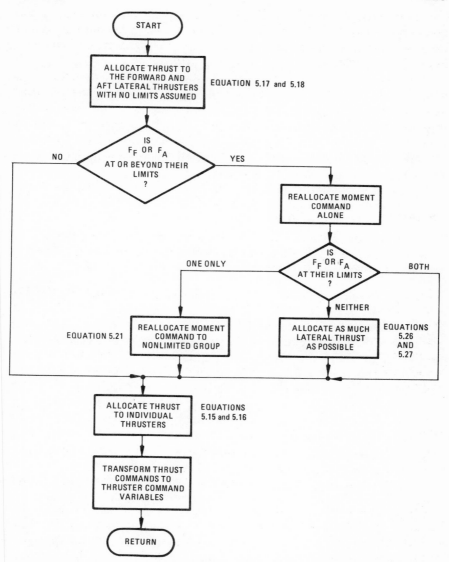

Figure 5-9. A Lateral and Moment Thrust Allocation Logic Flow Chart

or x-axis thrusters can be nearly equal to the separations of the lateral thrusters. Therefore, for such installation, the allocation equations must be different than developed so far in this section.

If viewed from the standpoint of the number of equations and unknowns, the situation with two lateral and two longitudinal thruster groups which are all capable of generating moment is one with three equations and four unknowns. Thus, either another equation or constraint is required to achieve a unique solution. As discussed in Chapter 4, a mandatory requirement for thrust allocation equations is that they always give a unique solution for every command input from the controller.

One strategy is to divide the moment command into two parts, one to be generated by the lateral thrusters and one to be generated by the longitudinal thrusters. Then, the allocation equations given in this section for the lateral thrusters can be used for the longitudinal thrusters as well as the lateral thrusters. The division of the moment command might be performed on the basis of moment-generating capability.

This strategy results in the thruster axis with the most active thrusters or largest equivalent lever arm generating the moment command, which is the case for the allocations equations described earlier in this section. However, this strategy loads the thruster groups producing thrust in the same direction as the thrust for moment commands so that they reach their maximum outputs first. Then, if the same equations given earlier are used to give priority to the moment commands, the corresponding force command will not be satisfied.

Of course, the priority equations can be modified so that once a thruster group in one axis reaches its maximum, the thruster groups in the other axis can be used to generate whatever moment cannot be satisfied in the axis with the saturated thruster group.

Another strategy divides the moment command so that the thruster axis with the larger force command produces the smaller amount of moment. This strategy also results in saturating at least one thruster group earlier than is necessary. As an alternative, the moment command can be divided in direct proportion to the unused moment-producing capability. This strategy does not assign moment command to a thruster axis if there is moment-generating capability in the other axis.

However, it does require more complex computations even when there are no thruster groups at their maximum output. Similarly, if any thruster group reaches its maximum output, the reallocation of either force or moment is quite complex because of the many potentially available options. A complete development of the allocation equations and logic for this strategy is beyond the intended goals of this book.

5.5   Steerable thrust allocation logic

In Chapter 4, the allocation of thrust to thrusters whose direction as well as their thrust output can be controlled is discussed briefly to show that at least two thrusters are required. Even then, there are more unknowns than equations. Thus, another equation or constraint must be introduced to give a unique allocation for every set of force and moment commands from the controller.

If there are more than two azimuthing or steerable thrusters, such as for the Sedco 709 with eight thrusters, even more allocation rules are required. In this section, the various aspects of allocation logic for two or more steerable thrusters are discussed. However, the discussion is limited to the introduction of various aspects without as much detail as given in the previous section.

Unlike with fixed-axis thrusters, the direction of the thrust is one of the degrees of freedoms for each steerable thruster which must be controlled to gain the full advantage of the steerable thruster. Otherwise, the allocation rules degenerate to those of the fixed-axis thrusters.

If the commands from the controller are only for a x-axis and y-axis force, then the logical choice for the direction of the thrust for each thruster is in the direction of the vector formed by the two force commands. Similarly, the logical thrust command for each thruster is of sufficient magnitude so that the total equals the vector magnitude of the two force commands. The distribution of the composite force command should be among the active thrusters in such a way as to create no moment. However, generally speaking there is a moment command from the controller.

There are many ways to command the thrusters to generate moment. One is to command each thruster in the direction to create the maximum moment, i.e., at right angles to the line connecting the center of rotation of the vessel and the thruster. The distribution of the moment can be on the basis of moment-producing capability. Thus, a thruster with a longer lever arm or a larger capacity receives a larger command than a less capable thruster. Then the two force commands from the controller can be allocated as described in the previous paragraph and as illustrated in Fig. 5–10.

Finally, the total command to each thruster can be calculated by the vectorial addition of the moment and composite force commands as Fig. 5–10 shows.

One feature of allocating the moment in a "pinwheel" pattern is that for at least one thruster, the moment command and the composite force command will be in the same quadrant. Thus, that thruster will have a larger total output than the other units and will reach its maximum output before the other units for most force command directions. Like the allocation logic for the fixed-axis thrusters, once a thruster reaches its maximum output, special allocation rules are required.

Another way to allocate the moment command is to first compute the direction of the thruster based on the composite force command from the controller. Then using the resulting directions of the thrust-

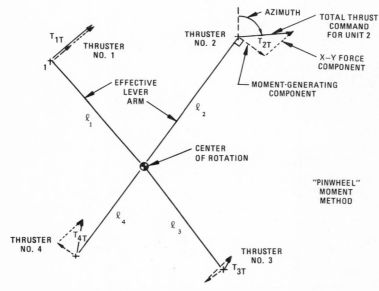

Figure 5-10. A Moment and Force Allocation for a Steerable Thruster Configuration

ers, the moment command is satisfied by increasing and decreasing the thrust magnitude to create a couple.

As Fig. 5–11 shows, the thrusters whose directions are aligned with the center of rotation cannot produce any significant moment. Consequently, the other thrusters must produce more moment-producing thrust than for the "pinwheel" moment allocation. Likewise, there is a tendency for at least one thruster of the group to reach its maximum output before the others, as with the "pinwheel" moment allocation. For the allocation example shown in Figs. 5–10 and 5–11, the same thruster, Number 1, will reach its maximum output first.

A third moment allocation scheme starts the same as the previous method by first allocating the composite force command. Then the moment command is allocated by commanding a thrust vector at right angles to the vector component of the composite force command, as illustrated in Fig. 5–12. In this scheme, the thrusters which generate the most moment are the ones that produce the least moment in the previous scheme.

However, as Fig. 5–12 illustrates, the problem of one thruster reaching its maximum as soon as in the previous two schemes is reduced, be-

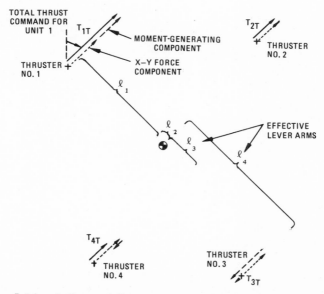

Figure 5-11. A Second Steerable Thruster Allocation Scheme

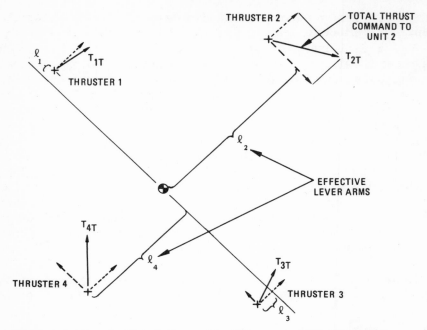

Figure 5-12. A Third Steerable Thruster Allocation Scheme

cause the moment-producing thrust is at right angles to the force-producing thrust.

With steerable thrusters, their directions change as the disturbance forces acting on the vessel change their angle of attack to the vessel. In the process of tracking the forces acting on the vessel, the direction of each thruster for some combination of disturbance forces lines up so that its output flow is directly into the input of another thruster.

As discussed in Chapter 4, the thrust output of the downstream thruster is reduced by an amount dependent on the inflow velocity from the upstream thruster. The inflow velocity into the downstream thruster is a function of the output thrust of the upstream thruster, the sea current, and the distance between the two thrusters.

Under many circumstances, logic can be included in the allocation logic to steer a thruster so that its output is not directly into the input of another thruster. The steering logic to avoid spoiling is only required to change the direction of the thrusters a small number of degrees and only at certain thruster-steering angles, as illustrated in

Fig. 5–13. It should be noted that another thruster must change its direction to compensate for the change in direction of the potentially spoiling thruster.

When the forces acting on the vessel are small and even varying from plus to minus, i.e., making reversals through zero, the direction

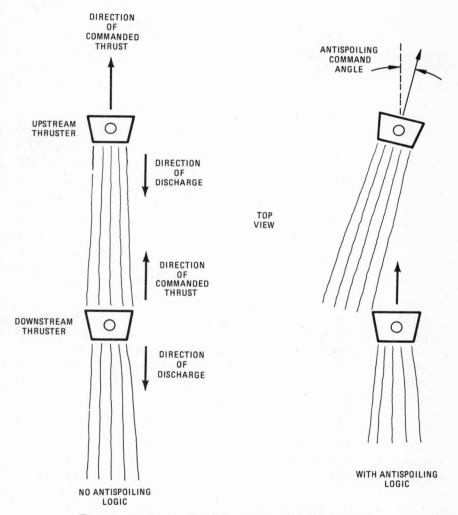

Figure 5-13. A Sketch of Antispoiling Logic

of the steerable thrusters will be making frequent large changes. Since the steering or azimuthing mechanism of the thrusters has a maximum rate of rotation, the position control may not be as good for small forces as when the forces acting on the vessel are large enough to give each thruster an average direction around which only small changes in direction are required.

To overcome this problem, a small bias thrust can be added to each thruster command so that, first, each thruster is operating in a definite direction and, second, the total forces and moment outputs from the bias thrusts for the overall thruster system are zero. Then as the forces acting on the vessel increase and the thrust commands to each thruster increase, the bias thrust can be reduced until at some level of thrust output the bias is zero.

After the force and moment commands from the controller are allocated to the thrusters, each thruster has a thrust magnitude and direction command. The thrust magnitude command is converted to a thruster-control variable command in the same manner as for the fixed-axis thruster discussed in the previous section.

The direction command does not require conversion to another thruster-direction control variable unless the thrust allocation logic is controlling the direction of the thruster directly. Then, the allocation logic calculates which direction and at what rate the thruster must rotate to execute the commanded direction from the allocation logic.

5.6   Control system modes

In an earlier discussion of the basic control problem, the thrust commands are computed from the difference between the measured and reference value of the variable being controlled. This difference is often referred to as the error signal because it represents the difference between the achieved and the desired or commanded value of the variable being controlled. When the reference or commanded value of the controlled variable is set by the operator at a desired value, the mode of control is often referred to as set-point control.

This is the most common automatic control mode used for dynamically positioned drillships where the desired position and heading are kept constant over long periods of time. However, there are other control system modes for drillships and other dynamically positioned vessels.

The next most common control mode for the drillship, and a very useful mode for support vessels, is referred to as joystick control. Joystick control is not an automatic control mode. Instead, a multiple degree-of-freedom control lever, usually two-axes, is manipulated by the operator to control the desired vessel parameter, such as the x-axis and y-axis position. This mode, although not automatic, is closed-loop control with the operator assuming the role of the controller.

To close the control loops, the operator observes a display of the measured value of the parameter to be controlled. He then mentally computes the difference between the measured and desired value of the parameter, i.e., the error. From the error he figures, based on a learned control law, the required manipulations of the joystick control lever to achieve the best control of the vessel parameter that he is capable of performing.

The learned control law depends on the display of the measured parameter, the dynamic behavior of the vessel, the thruster system capability, the relationship between control lever manipulations and thruster response, and the capabilities of the operator.

Generally, the best control is when the magnitude of the thrust output is proportional to the angular deflection of the control lever and in the direction in which the lever is deflected. Thus, by deflecting the control lever in a given direction, the operator can simultaneously command the thruster system to produce an x-axis and y-axis thrust.

To incorporate the joystick mode with the automatic set-point control mode, the x-axis and y-axis command from the joystick control lever are substituted for the x-axis and y-axis force commands from the PID control law given by Equation 5.4 whenever the joystick mode is selected. The joystick thrust commands are then summed with the active wind compensation thrust commands and transferred to the thrust allocation logic for distribution to the active thrusters in a coordinated manner.

A similar manual control of the heading of the vessel is possible. However, maintaining automatic control of the vessel's heading while the operator uses manual joystick control of the position reduces the operator's control task to a more manageable level. As a result, he is able to perform manual joystick control in heavier weather and for longer periods of time. Both of these factors are limiting factors for the joystick control mode and are the reasons automatic control generally outperforms the operator using joystick control.

In transferring between set-point control and joystick control, large differences in commands generally exist. These differences result in large changes in the commands to the thrusters and may result in driving the vessel away from a desired position.

To smooth the transition between control modes, a technique referred to as bumpless transfer is used. The way bumpless transfer works is that the last thrust command at the time of the transfer is maintained and decayed over time as the commands from the new control mode establish themselves to exercise proper control.

Another feature of the controller which prevents large sudden changes in the thrust commands is command rate limiting. This feature is used in the set-point command mode and limits the rate at which the set-point command changes from one operator command to the next. Normally, the rate is constant and is set so that the vessel moves from one command to the next at a rate which does not cause an appreciable increase in the thrust commands except for increases caused by environmental load changes.

There is another class of control modes where the set-point commands to the automatic closed-loop controller are changed automatically to achieve a given result. To perform an automatic change of the set-point commands, the controller contains a specific set of laws which operate on a given set of inputs to produce a unique set of set-point commands for the controller.

The inputs are often measured physical parameters, some of which may be used in the controller for other purposes. Thus, these control modes use the basic dynamic positioning control loop with the addition of an outer-loop which automatically computes a commanded value for the system. For proper control system performance, the rate of change of the set-point commands from these outer-loop systems must be slower than the inner-loops or else the inner-loops cannot follow the desired commands.

For vessels which have a preferred direction with respect to the total environment and the task-related load acting on them, the outer-loop mode referred to as automatic heading for minimum thrust (AHMT) is very useful. This mode is useful for drillships and other ship-shaped vessels which can change their heading freely.

In this control mode, the set point or reference value to the heading control loop is varied automatically, based on the average load to the thrusters which are in the axis of vessel which is perpendicular to the preferred direction of the vessel. For ship-shaped vessels, the average load to the lateral thrusters is used to change the heading command of the vessel.

If the change in the heading results in a reduction in the average load to the thrusters that are being monitored, then the heading command is changed by a preset increment in the same direction. If not, the heading is incremented by a preset amount in the other direction. The change in heading command continues in the direction which reduces the average load until the load goes through a reversal in sign.

Once the heading is found which results in an increase in the average load when the heading is incremented in either direction, the heading command will cycle back and forth around the preferred heading. To reduce this cycling, a smaller increment size can be used. Then, once the preferred heading is found again, a deadband can be introduced in the control mode logic. The deadband holds the heading command constant until the average load increases from the value at the preferred heading by an amount greater than that required to change the heading command by its larger preset increment.

For this control mode to perform properly, the rate at which the heading command is changed must not exceed the rate at which the thruster system can change the heading of the vessel and the rate at which the average thruster load can be computed accurately. Likewise, for this mode to achieve the minimum thrust, the vessel must have a single preferred heading without any local minimums.

In mathematical terms, the loading function which is being searched by the AHMT algorithm must be monotonic. However, even for a ship-shaped vessel there are two preferred directions: one with the bow into the weather, and a second with the stern into the weather. For either direction, the minimum lateral thrust is theoretically zero. However, the longitudinal load on the bow is generally different than the stern.

Thus, the absolute thrust minimum is with either the bow or the stern into the weather. But the aforementioned search algorithm, sometimes referred to as a "hill-climbing" technique, cannot determine which way to turn the vessel to get the absolute minimum without comparing the two minimums or a priori knowledge of the preferred direction. In either case the algorithm should change the heading of the vessel toward the closer local minimum to avoid passing through the maximum lateral loads which separate the two minimums and may surpass the lateral thrust capacity, causing the vessel to lose position.

Another control mode in which the set-point commands are changed automatically is used on a drillship when the marine riser is connected between the vessel and the sea bed equipment. In deeper water the riser may assume a shape as shown in Fig. 5–14 in an exaggerated sketch. By moving the drillship aft of its position shown in Fig. 5–14, the angle in the riser can be reduced. The ability to maneuver the drillship to reduce the riser angle at the lower ball joint is very important to the drilling process to reduce the rate of wear of the drill string or pipe at the lower ball joint.[6]

The set-point commands can be changed by the operator to move the ship. To decide how far to move the vessel to reduce the riser angle at the lower ball joint, the operator uses a display of the riser angle generated by a two-axis inclinometer-type sensor mounted to the riser just above the lower ball joint. More details regarding the riser angle are given in Chapter 3 on sensors.

Since the shape of the riser is a function of the current acting on it, the lower ball joint angle changes as the current profile changes. Generally, the rate of change of the current is slow and an operator can make the necessary adjustments to keep the riser angle nearly equal to zero. However, an automatic mode can be provided which slowly changes the set-point commands in such a way that the lower ball joint angle is reduced or changed to a desired angle which is normally zero.

To perform the automatic set-point adjustment, the output of the riser angle sensor, properly resolved into vessel coordinates, is compared to the desired riser angle. The difference is integrated slowly to create a set-point command which automatically moves the vessel until the error is zero or in a preset deadband around zero. Then, as the riser angle varies because of changes in the current, the position of the vessel automatically tracks to keep the riser angle to the desired value.

In certain applications, the dynamically positioned vessel tracks a subsea vehicle such as a submarine. If the position of the subsea vehicle is measured with respect to the dynamically positioned vessel, then the DP vessel can be made to automatically track the subsea vehicle by changing the X-Y set-point position commands at a rate equal to the speed of the subsea vehicle and in the direction of the subsea vehicle.

If the subsea vehicle is not moving at a speed or in a direction to overload the thrusters of the DP vessel, then the DP vessel will be able to track the subsea vehicle. To implement this control mode, the operator is furnished with X-Y control dials calibrated in speed. When he wants to station-keep over the subsea vehicle, the operator sets the

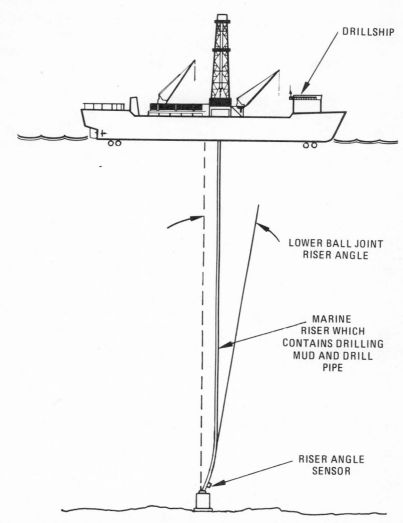

DRILLSHIP

LOWER BALL JOINT
RISER ANGLE

MARINE
RISER WHICH
CONTAINS DRILLING
MUD AND DRILL
PIPE

RISER ANGLE
SENSOR

Figure 5-14. A Sketch of the Static Riser Angle in Deep Water

commanded speed to zero. Then, as the subsea vehicle starts to move, the operator moves the speed controls to track the subsea vehicle.

The track mode can also be implemented if the position of the dynamically positioned vessel is measured by a position reference system which has a reasonable area of coverage. An example of a position reference system is a radio position reference. Then, the DP vessel can be made to track a given path at a given rate by changing the set-point commands corresponding to the desired path at the desired rate.

One method of changing the set-point commands along a desired path is to give the controller way points in the coordinates of position reference system. Then, the set point is varied as a straight line function between the way points at a rate equal to the command value.

The previously described modes are generally mutually exclusive and are, therefore, selected one at a time. The only mixing of control modes that is allowed is between position and heading control. For example, set-position control and automatic heading for minimum thrust can be used together. Whereas set-point control cannot be used with the riser-angle control mode.

5.7   Performance monitoring

Depending on the degree of automation required or desired and the capabilities of the controller, there are many auxiliary features that can be added to the dynamic positioning system controller which are not a part of the closed-loop control system. Their function is to monitor the performance of a particular element of the control system, such as the thrusters, and alert the operator of off-nominal performance.

As such, these features are not essential to the control system, but they add greatly to the usefulness of the system without being extremely costly in some cases.

The primary usefulness of the performance monitoring features is in the detection of failures or potential failures at the earliest possible moment. With early detection, problems caused by the failures can be minimized and maintenance and repair can be started immediately.

An automatic controller is particularly well suited for performance monitoring, because once the method of monitoring is developed and the criterion upon which off-nominal performance is identified, then the controller performs the operation continuously without tiring.

With "complete" performance monitoring, the operator is free to perform maintenance and repair, especially if the controller is equipped with an audible alert which is sounded when off-nominal performance is detected.

Thruster performance is very critical to the overall performance of the dynamic positioning system. First, knowledge of when the control thrust reaches an output level near its maximum capacity for each thruster is important. Then the operator can activate another thruster if one is available or he can reorient the vessel to reduce the loading on the most heavily loaded thruster is possible.

To provide this function, the controller monitors the thrust magnitude command to each active thruster. When the command reaches a preset threshold after a sufficient length of time, a thrust level alert is given. These alerts usually take the form of an audible and visual alert where the audible portion of the alert can be silenced by the operator acknowledging the alert through a control panel pushbutton.

A second thruster performance monitor compares the thruster command to the feedback of the achieved thruster control variable. If the comparison varies beyond a preset threshold for a sufficient number of times, then the "thruster compare" alert can be given. When the thruster compare alert occurs, the operator must first acknowledge the alert and then decide what action should be taken.

If the alert persists or if there is obviously a large difference between the command and the feedback, the operator should replace the suspect thruster with a standby unit if one exists. Then he can verify that the thruster has malfunctioned or that the feedback path is operating properly.

In either case, some form of troubleshooting is required which, if the operator can isolate the element requiring repair, reduces the time and effort required to return the thruster to the ready mode.

Thruster performance monitoring can be combined with logic which automatically activates a standby thruster when an active thruster reaches a preset thrust output. The logic can also include a time log on each thruster so that it selects the most eligible standby thruster, if a choice exists, based on the least amount of operating time.

The same logic can deactivate a thruster when the thrust demand increases to a sufficiently low value. This value is determined so that when the thruster is deactivated, the output of the remaining thrusters do not exceed the preset level which activates a thruster. The deactivation logic is more risky than the activation logic because erroneously activating another thruster is not a serious mistake, but erroneously deactivating a thruster in the midst of a storm is a serious mistake.

Equally important to thruster performance monitoring is monitoring the performance of the sensors. One form of sensor performance monitoring has already been discussed in Section 5.3. Another form is the comparison of inputs from sensors which measure the same quantity. As in the case of the comparison of thruster commands to thruster feedbacks, the "sensor compare" logic alerts the operator visually and audibly when the difference between two sensor inputs is greater than a preset threshold for a sufficient amount of time.

Upon the sensor compare alert, the operator, after acknowledging the alert, must decide which of the two sensors is in error and disable that sensor input into the controller. Once disabled, the sensor can be repaired.

In the case of systems with more than two sensors measuring the same quantity, more sophisticated schemes of averaging and voting can be used. However, if the sensors are of different types so that their dynamic response characteristics are different, all comparison or averaging techniques must be carefully used. A comparison is not meaningful if the quantities being compared are not alike under "normal" conditions.

Malfunction detection of the controller is also a form of performance monitoring. For example, by monitoring the average absolute magnitude of the error signal in the two-position and one heading-control loop, a failure in the control system can be detected because for a properly performing system with PID control, the average error is zero.

Similarly, in the case of a digital computer, if the failure for the computer to cycle through all of its calculations is detected for one or two cycles by a monitoring device external to the computer, then the computer failure alert should be given to the operator. If a second or redundant computer is in the system, the computer failure alert should cause an automatic transfer to the standby or back-up computer.

Commands from the operator or an automatic control algorithm can also be monitored to check that they are consistent with the presently selected control mode, or that they are equal to or less than any preestablished limits. The limits may come about because of operational considerations.

An example is a drillship in the AHMT mode. Limits on the variation of the heading may exist because of the subsea equipment in use. As a result, the AHMT logic should not be allowed to change the heading set-point command to any value beyond the given limits. By monitoring the heading command, the controller can alert the operator that the heading command is equal to one limit and at the same time stop the heading command from changing any more.

The operator can then decide to hold the present limit or increase it so that the thruster system can either reach a minimum thrust condition or maintain a minimum thrust condition. The heading command alert also serves as a warning that the loading on the vessel has shifted its direction to the vessel.

The same heading command monitor will alert the operator in the set-point control mode that he has commanded the vessel to turn to a heading beyond the limits he had previously set. In addition, it can be set up so that it prevents the heading from exceeding the preestablished limits or merely as a warning to the operator.

5.8   System architecture

Previous sections contain descriptions of the various functions that the controller must perform to be a dynamic positioning system controller, plus the functions that the controller can perform to provide a high level of usefulness. The structure of the various elements of the controller and their interconnections has been given to some degree, but the effects of safety and reliability on system structure or architecture have not been discussed to any extent.

Similarly, the effects of the physical layout of an offshore vessel and the environmental considerations have been given very little attention. In this section these factors are considered to arrive at several system configurations which meet differing system requirements.

A basic fact of life with a dynamic positioning system is that its parts are dispersed at many different locations around the vessel. The thrusters are located below the waterline near the bow and the stern. The sensors are located above and below the surface and at various locations on the vessel. For example, the wind sensor must be located in an elevated area which is as free of masking by other structures as possible.

Thus, the controller must accept sensor data from different locations and transmit thruster commands to other locations separated by hundreds of feet. The cables from the sensors and to the thrusters must pass through several bulkheads and are exposed to various electrical and mechanical environments. This dispersion of system elements is illustrated in Fig. 5–15.

Fig. 5–15 indicates that the central control room is forward on the vessel, and generally speaking, on vessels with a forward bridge it is. However, there are exceptions, such as on the *Discoverer 534,* where the DP control room is at the stern of the vessel.

Other variations from what is shown in Fig. 5–15 are that the op-

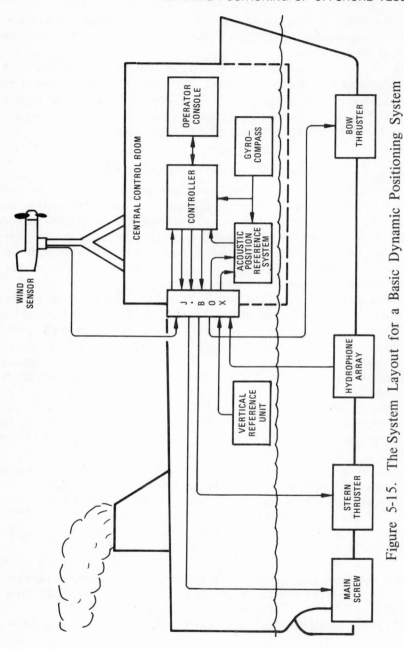

Figure 5-15. The System Layout for a Basic Dynamic Positioning System

erator console and the gyrocompass are not in the central control room. The console can be a part of the bridge installation or another room which provides good operator visibility to the surroundings of the vessel. The gyrocompass may be a part of the bridge installation or another room near the bridge because heading information is required for normal steerage from the helm when the vessel is underway and for radar.

On larger vessels, distances to sensors and thrusters are greater and the thruster system includes more units separated by greater distances. Likewise, on larger vessels the controllers generally provide the full range of features described in the previous sections along with other computing functions such as power management.

These functions involve more instrumentation which is fed into the controller's computing system which is performing multiple tasks in a time-shared manner.

As the system increases in size to perform additional tasks, the number of inputs and outputs increase. However, the system architecture remains reasonably unchanged as the system size varies, if it is represented as shown in Fig. 5–16. In the basic system in Fig. 5–15, the controller is the computer system, the operator console is the control and display system, and the interconnection cable is the data acquisition and transmission system.

In larger systems there is a tendency for data inputs and outputs to the computing system to reach a significant level of concentration.

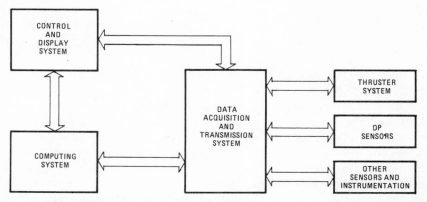

Figure 5-16. The System Architecture for a Dynamic Positioning System

For example, in the drillship application where there are as many as 14 thruster units, the commands from the DP controller go from the DP control room to the central engine control room. This room is generally near the main engine room where all the vessel's power is generated.

For a ship-shaped hull, the engine room is usually aft near the main screws. Similarly, if the DP controller is being combined in the same computing system with the power management system, there are several hundred inputs from the central engine room to DP control room.

Another area on the drillship which may produce a significant level of input/output concentration is the moonpool area amidships. In the moonpool area, inputs from various sensors and outputs to remote displays create a concentration of cabling to the DP control room. Also, the capability to give set-point commands or requests from the drill floor can increase the data transmission load to the DP control room.

With this concentration of data, the data acquisition and transmission system in Fig. 5-16 may logically become a serial multiplex system. At the locations on the vessel where there is a sufficiently large concentration of inputs and outputs, a remote data acquisition terminal can be located to collect and distribute the information for the central control room over a minimum number of interconnecting cables.

This approach can potentially save the cost of cable and cable installation.

One aspect of system design which has not been discussed which affects system architecture is reliability. As discussed several times earlier each application requires a different amount of reliability. In the cases requiring the highest reliability, such as drilling, the concept of dual redundancy has been applied with great success.

This concept applied to the basic DP control system results in the system architecture shown in Fig. 5-17. All the basic sensors are supplied in pairs, along with two controllers. At the operator console, dual displays are provided and the thruster system includes multiple units to supply thrust for each of the three control axes.

The one difficulty with a dual parallel architecture is that ultimately the command paths must be made singular. For that reason, the automatic switching logic shown in Fig. 5-17 is included to switch the command source to the thrusters from one controller to the other. However, the commands from the operator can feed both controllers in the

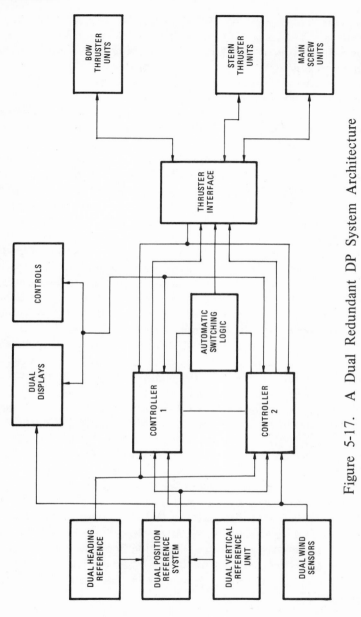

Figure 5-17. A Dual Redundant DP System Architecture

architectural scheme shown in Fig. 5–17, because they perform the same computations in parallel with interconnections between the two controllers to eliminate any long-term computational drift which may occur in the integrators in the PID control laws.

The automatic switching logic can be actuated by a detected failure in the controller which is furnishing commands to the thrusters at the time of the failure. Also, schemes which monitor other critical system functions could be used to generate a transfer of thruster control from one path to the other.

However, if all the sensor information is fed into both controllers and each controller is selecting the best input for signal and control law processing and executing performance monitoring, the failure of the controller commanding the thrusters is the only critical function which requires switching. But the switching function must be external to the controller system.

Additional redundancy features which can be and are provided with many DP control systems are individual manual control sources for each thruster and a joystick command system, independent of the controllers. The former is implemented by providing at the operator console a control dial for each thruster which can be switched in place of the command outputs from the controller or controllers.

The difficulty involved with this arrangement is that in the case of a thruster system with more than three thrusters, the operator has difficulty in giving the coordinate commands to hold the vessel in position. This is especially true if there is any weather acting on the vessel. However, the manual controls do function usefully to check out or test thrusters that have given indications of malfunctioning or that have been supposedly repaired prior to returning them to service. Also, the manual controls can be used to regulate the load on the power system.

For the parallel joystick control independent of the DP controller or controllers, the joystick command unit is supplied so that it can be switched from the controller path to the redundant path. The redundant path also contains some form of thrust allocation logic independent of the controllers. The allocation logic is required to give coordinated thrust commands to the thrusters in the axes where the thrusters are supplying translational thrust as well as moment-producing thrust.

For a dual redundant system as shown in Fig. 5–17, a dual failure is required in the controllers before independent joystick control is needed. So this redundancy scheme is more meaningful for the single controller DP system where a failure of the controller results in a total loss of dynamic positioning control.

In fact, since three-axis control is difficult in any degree of sea state, the addition of a simplified automatic heading controller, independent of the normal DP controller, should be considered.

5.9   Types of controllers

In this book, two types of controllers are considered — analog and digital. An all-analog system can be provided, but the digital system includes analog components along with digital components because of the analog nature of the physical world in which the DP system operates.

To narrow the discussion of controller types to a part of a chapter as opposed to a book, the topic is limited to the basic control problem for the lateral and heading control axes of a ship-shaped vessel with one bow and one stern thruster.

The three basic sensors, i.e., position, heading, and wind, supply analog outputs. The heading and wind direction angles are assumed to be each a single channel. Some sensors can be obtained which provide digital outputs and, as digital technology continues to expand, the number of sensors which provide digital outputs will increase. However, these devices will necessarily contend with the problems associated with digitizing information.

The outputs of the sensors are analog voltages which are proportional to the units the sensors are measuring. For example, the wind speed sensor may produce 10 volts for 100 knots of wind or 10 knots per volt output. The signal processing of the sensor outputs in the analog system is standard analog circuits which in the most complicated form may be a multistage high-gain operational amplifier.

The complexity of the circuit depends on the desired filtering characteristics and the characteristics required by the closed-loop control system. For example, the signal processing for the wind speed is simply a first-order low-pass filter described by the following transfer function:

$$H_w(s) = \frac{1}{\tau_w s + 1} \tag{5.30}$$

where s is the Laplace operator. In contrast, for the position sensor signals, the signal processing has the form of at least a second-order low-pass filter given by the following transfer function:

$$H_p(s) = \cfrac{1}{\dfrac{s^2}{\omega_n^2} + 2\zeta \dfrac{s}{\omega_n} + 1} \qquad (5.31)$$

where ω_n is the natural frequency of the filter and ζ is the damping factor.

For the analog controller, the error signals are generated in summing amplifiers where one input is the processed position or heading sensor signal and the other input is a voltage proportional to the command. For example, the position command might be scaled so that 1 volt equals 2% of water depth. As indicated in Equation 5.4, the error signal is used as the input to the control law which for this discussion is that given in Equation 5.4.

The implementation of the control law is straightforward for proportional and integral parts of the control equation. A combination of one amplifier in parallel with an integrator feeding a summing amplifier implements the proportional plus integral terms of the control law. The rate term must be implemented with a high-pass filter combination to attenuate any high-frequency noise in the system. Then the output of the high-pass circuit is summed with the proportional plus integral terms.

Fig. 5-18 shows an implementation in the form of a circuit diagram of analog PID control law and wind compensation prior to thrust allocation logic. At the various outputs of the analog devices shown in Fig. 5-18 there are voltages proportional to the appropriate units. Thus, a standard voltmeter can be used to monitor the various signals.

Outputs of the control law circuitry when summed with the counter wind outputs are then fed into a switching network where, in the automatic control modes, the combined outputs feed into the thrust allocation logic. When the joystick mode is selected, the inputs to the thrust allocation logic are the analog voltages from the joystick command assembly. In some systems the counter wind outputs are summed with the joystick commands, also.

The joystick command assembly generates analog voltages proportional to the deflection of the control lever by mechanically coupling the lever to potentiometers which are supplied by voltages scaled to a given number of units of thrust per percent deflection of the lever.

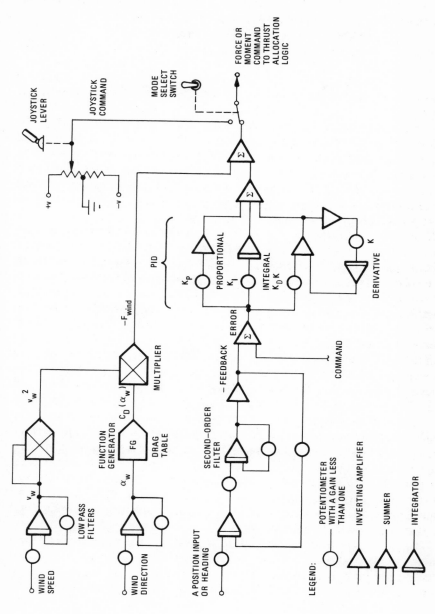

Figure 5-18. A Single Loop Analog Controller Without Thrust Allocation Logic

Thrust allocation logic for the analog controller in this simple case of one lateral bow thruster and one lateral stern thruster involves implementing the equations in Section 5.4. One such implementation is shown in Fig. 5-19. The final stages of the thrust allocation logic are function generators which convert thrust commands to analog voltages which are proportional to the thruster command variables.

The total analog controller for the lateral position and the heading is a combination of the analog circuitry in Fig. 5-18 taken twice plus the analog circuitry in Fig. 5-19.

For the digital controller varying amounts of the controller can be digital. The following discussion considers the system with digital implementation from the output of the sensors to the input of the thrusters. Also, for the controller described in this book, all the inputs are sampled at approximately the same time at a periodic rate and all the outputs are updated at approximately the same time.

The system does not need to be implemented in this manner, but it does make the initial understanding and the analysis of the system easier.

A major difference between the analog and the digital system is that the former is a continuous process and the latter is discontinuous, involving the sampling of input data, as illustrated in Fig. 5-20. If the rate of sampling is significantly faster than the highest frequency of the data used for control, the system behaves like the analog system.

For dynamic positioning systems, a sampling rate of once per second is still significantly faster than the critical frequencies of the control system which are less than 1/100 of a hertz.

Another difference between the two types of controllers is that analog systems compute with scaled voltages, but digital systems compute with scaled numbers. In the basic digital hardware the numbers are represented in binary form, i.e., a sequence of ones and zeros. Thus, the digital controller is well suited to perform arithmetic and logic operations.

However, to perform as a closed-loop controller, the digital computer must execute the required calculations sufficiently fast to minimize the delay time between when the sensor inputs are sampled and the thruster commands are given. With modern digital technology, a digital controller mechanized in a minicomputer can execute thousands of calculations and logical operations in a fraction of a second. Likewise, software systems have been developed especially for real-

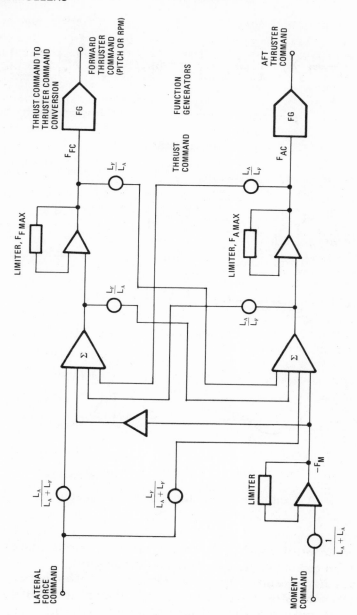

Figure 5-19. Analog Thrust Allocation Logic

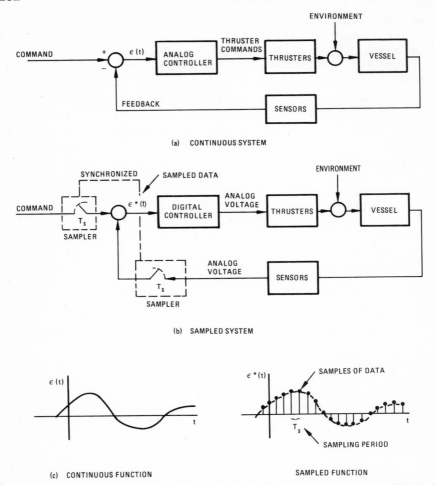

Figure 5-20. A Comparison of the Data for Analog and Digital
Controllers

time process control, which makes real-time closed-loop digital control possible with commercially available minicomputers.

Since the digital controller uses numbers instead of analog voltages, the controller requires a device to convert the analog outputs of the sensors into numbers. Such a device is called an analog-to-digital converter or A/D converter. With an A/D converter on the output of each sensor, the controller can periodically pass sequentially through the sensor outputs, converting their output voltages to binary numbers and storing the numbers in its memory.

With modern A/D converters with high-speed conversion and multiplexing, hundreds of inputs can be converted in a fraction of a second.

Once the sensor output is converted and stored, the digital controller can operate on the sensor outputs to compute thruster commands in much the same manner as previously outlined for the analog controller. A major difference, however, is that the continuous operations of filtering, integration, and differentiation require numerical implementation. In the case of filtering, difference equations of the following form are used for a second-order operation:[10]

$$Y(n) = AY(n-1) + BY(n-2) + CX(n) + DX(n-1) + EX(n-2)$$

$$(5.32)$$

where:

Y = the output of the filter
X = the input to the filter
(n) = the present value or sample of the variable
$(n-1)$ = the value of the variable on the previous computational cycle or sample
$(n-2)$ = the value of the variable on the second past computational cycle
A, B, C, D, E = the filtering coefficients

To perform the operations in Equation 5.32, five multiplications, five additions, and at least ten memory locations are required. If five of these filters were used in each control loop of which there are three for the basic control system, then the total processing is 75 multiplications and 75 additions with less than 150 memory locations. Less than 150 memory locations are required if the filters are in series because the output of one filter is the input to the next filter.

Even if double precision arithmetic is required, this computational and memory loads are small for a modern minicomputer. Typically, within an enclosure with less volume than 5 cu ft, a 16-bit digital computer with over 32,000 words of 16-bit memory is available. Thus, the basic digital filtering load for dynamic positioning is insignificant.

For the digital computer to perform the PID function, numerical approximations of integration and differentiation are used. The simplest approximation for integration is the rectangular rule which is given in difference equation form as follows:

$$I(n) = I(n-1) + T_s E(n) \qquad (5.33)$$

where:

$I(n)$ = the output of the integration on the present sample
$E(n)$ = the input to the integration or the error signal on the present sample
T_s = the sampling period.

For systems where the sampling period is significantly shorter than the period of the information being sampled, rectangular integration is sufficiently accurate.

The derivative portion of the PID control law can be approximated by computing the slope of the error signal as given by the following expression:

$$D(n) = \frac{1}{T_s}[E(n) - E(n-1)] \qquad (5.34)$$

where $D(n)$ is the output of the derivative term. However, as in the analog controller, the derivative term should be further approximated with a filter which does not amplify the high frequencies to control the noise in that thrust commands from the control laws.

The simplest implementation of the derivative filter is a first-order lead-lag, discussed in Chapter 7, which can be expressed in difference equation form as follows:

$$D(n) = \frac{K_{D1}}{T_s}[E(n) - E(n-1)] + K_{D2} D(n-1) \qquad (5.35)$$

where K_{D1} and K_{D2} are constants related to sampling period and the time constant of the filter. More complex lead-lag filters can be implemented with Equation 5.31 where the choice of the filter coefficients determine the lead-lag characteristics of the filter.

The total PID control law implementation for the digital controller is the sum of the terms which, in equation form, is:

$$C(n) = K_p E(n) + K_I I(n) + K_D D(n) \qquad (5.36)$$

or:

$$C(n-1) = K_p E(n-1) + K_I I(n-1) + K_D D(n-1) \qquad (5.37)$$

If Equation 5.37 is subtracted from both sides of Equation 5.36, the value of Equation 5.36 is not changed and can be expressed as follows:

$$C(n) = C(n-1) + \left[K_p + \frac{K_D\,K_{D1}}{T_s} \right][E(n) - E(n-1)]$$

$$+ K_I T_s E(n) + K_D(K_{D2} - 1)D(n-1) \tag{5.38}$$

With Equation 5.38, the computation of the command variable is somewhat simpler than Equation 5.36 because the integration variable is not calculated directly. If the command variable is to be initialized by bumpless transfer from the joystick mode, the first past value of the command variable is set to the initializing value.

Active wind compensation in the digital controller is quite straightforward to implement. The low-pass filters can be implemented by the following difference equation:

$$Y(n) = Y(n-1) + K \quad [X(n) - Y(n-1)] \tag{5.39}$$

where:

$Y(n) =$ the output of the filter
$X(n) =$ the input of the filter

$$K = 1 - \exp\left[\frac{-T_s}{\tau}\right]$$

$\tau =$ the equivalent time constant of an analog filter

The wind drag functions are stored in the memory of the digital computer either as arithmetic functions which curve fit the drag functions or as tables of drag coefficients for discrete wind angles. Then a curve-fitting routine can be used to determine the drag coefficient when the measured and filtered wind angle is between two discrete angles in the table.

Normally, linear interpolation is sufficiently accurate if the tables give drag coefficients for every 10 degrees of the wind angle. The square of the wind speed and multiplication required to complete the computation of the compensating wind command are trivial for the digital computer. Also, the summation of the wind command and the control law command to form the total automatic feedback command to the thrust allocation logic is trivial for the digital computer.

The digital controller also excels when doing the thrust allocation logic because of its arithmetic and logical capacity. As shown in Fig.

5–9, the thrust allocation logic is most easily represented in a logical flow diagram where decisions are made based on tests. This is especially true for more complex configurations of thrusters than are being considered in this section.

For example, a thruster configuration of as many as eight steerable thrusters with anti-spoiling logic and null environment biasing involves many decisions and computations of vector quantities. Implementing this logic on a digital computer is involved, but on an analog computer it is extremely difficult and impractical.

Once the thrust is allocated to the individual thruster units, an arithmetic function or a table can be used to convert the thrust commands into thruster control variable commands in the same manner as the wind drag coefficients. Then, the computer converts the thruster control variable commands which are in the form of numbers into analog voltages properly scaled to command the thruster controllers.

Previously, the point was made that the digital computer samples all the sensor inputs as shown in Fig. 5–20. Then the computer sequential performs the filtering, command computation, and thrust allocation. When the computer completes one cycle of computation it waits until the next sensor sampling time to read the control computations again. As a result, each calculation is performed and the computer then goes to a new calculation.

If the calculations at any one point in the sequence are not stored in memory they are lost. For this reason the converters, called digital-to-analog converters, which convert the thruster commands to analog voltages automatically hold each command until the next command is computed and converted. This feature is referred to as a zero-order hold and is an important element of a digital control system which affects its control characteristics. The transfer function for a perfect zero-order is given as follows:[10]

$$H_{OH}(s) = 1 - \frac{\epsilon^{-T_s s}}{s} \tag{5.40}$$

where s is the Laplace operator.

The operation of the digital controller described in this section can be represented in the normal block diagram form as shown in Fig. 5.21. Also, the operation can be represented in a flow chart form. If a designer is a control system engineer, he prefers the former for the purposes of his analysis. In addition, instead of difference equations for the filters and control laws, he prefers transfer functions which are

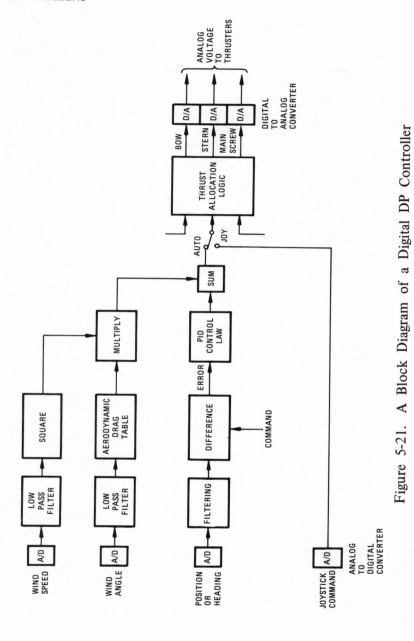

Figure 5-21. A Block Diagram of a Digital DP Controller

expressed with z-transforms or some approximate Laplace transform form.

If a designer is a software engineer or programmer, he prefers the flow chart because he can write the computer program from it which implements the controller. The program is then transferred into computer instructions and data to be stored in the memory of the computer and executed when the computer is properly initialized and running.

In the simplest case illustrated in the flow chart in Fig. 5–22, the computer starts at the top of the flow chart and serially executes the program. When it has completed the computations for one pass, the computer waits, as Fig. 5–22 shows, until the next sample time occurs. An external clock is used to give the computer the signal to start each sample cycle. In the time that the computer is waiting, it can be doing other tasks such as performance monitoring, data logging, display formatting, etc.

If the computer is being used to do more than dynamic positioning control, a different program execution method is more appropriate. An example of expanded computer usage is shown in Fig. 5–23. For this system, a task-oriented software system is more appropriate than the cyclic method shown in Fig. 5–22 where the executive routine services various subroutines on a priority interrupt basis. Those functions having the highest priority are real-time control and include dynamic positioning as number one priority.

Since the mid-60s, digital controllers have been used successfully in dynamic positioning systems (Tables 1–1 and 1–2). Today the only application of purely analog hardware to dynamic positioning or coordinated thruster control is for a joystick plus thrust allocation logic control combination. This is not dynamic positioning in the sense used in this book, but a separate joystick and allocation logic combination, independent of the DP controllers, is being advocated as a backup to the DP controllers.

This back-up approach certainly makes sense for a single controller DP system where transfer to standard helm control is not acceptable. However, for a dual redundant controller system, the independent joystick path may be of questionable value, because multiple failure must occur before it is needed. Likewise, unless the vessel is in very light weather, joystick control is of questionable worth and transferral of control back to the helm may be a more sensible strategy if both controllers fail.

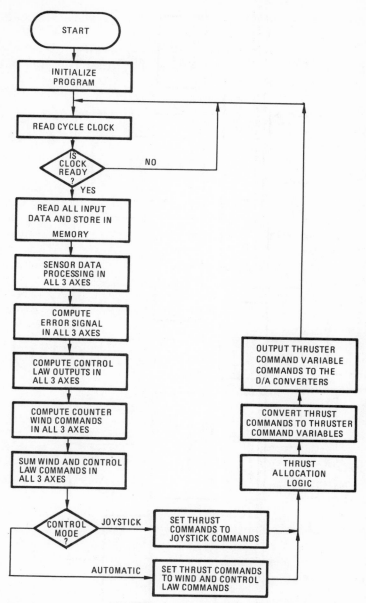

Figure 5-22. A Flow Chart of the Digital DP Controller

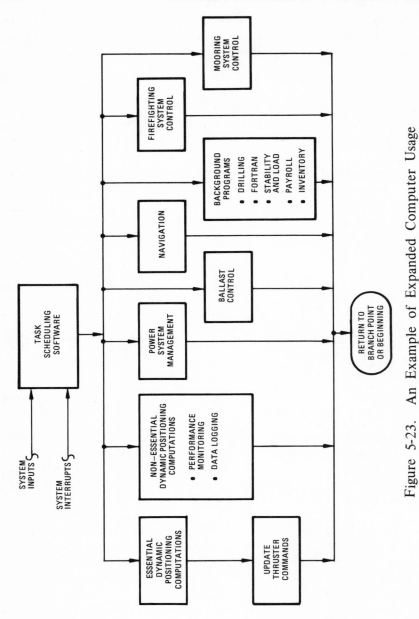

Figure 5-23. An Example of Expanded Computer Usage

The reasons for changing from the analog to digital controllers are the same as other areas of control changing from analog to digital controllers. Included in the list of reasons are the following:

1. Modern real-time control digital computers have the capability to perform complex numerical and logical operations with constant accuracy at very high speed. Increasing accuracy with a digital computer is less expensive than with analog devices and digital accuracy is not a time-varying function as it is with analog devices (no drift or aging).

2. With digital computer software programs, modifications in a controller to improve performance, accept increased capabilities, or correct a problem discovered during actual operations can be performed more easily and cheaply than for analog controllers. The feature of expansion of use in the most cost-effective manner is very important in today's changing marketplace so that vessels can change operational roles as market demands change.

3. Digital technology is a dynamic field with a high degree of competitiveness. As a result, innovations that improve system performance occur monthly and costs are remaining competitive if not decreasing.

4. Self-test and diagnostic schemes are implemented in digital computers because of their logical, sequential nature backed up with a large memory capacity.

There are certain drawbacks to the digital system that analog systems do not possess. The list of drawbacks includes:

1. Lack of independent signal paths so that the loss of a digital controller is likely to cause the total loss of control.

2. Probing the flow of signal or data processing requires special software and display features as opposed to the more conventional oscilloscope or voltmeter techniques available to the analog controller designer.

3. Digital controllers suffer with the standard set of problems of numerical computations with digital computer, i.e., truncation, overflow, etc. However, with good programming practice and thorough software verification using dynamic tests, these problems can be reduced to an insignificant level.

4. There is a delay between the time that sensor inputs are sampled and when thruster commands are updated. This delay can be mini-

mized to a fraction of the sampling period with proper programming structure where the essential dynamic positioning computations are executed before the less essential computations are performed.

The previous discussion is at least a sketch of the dynamic positioning system controller. In the areas of analog and digital controllers, more discussions of the control system design aspects related to them are given in Chapter 7. Also, several of the references give more information about actual controllers.

For example, Reference 7 gives some details of an analog controller for the original dynamic positioning system on the Terebel. Likewise, more information on analog computers and on digital control systems is given in References 9 and 10, respectively.

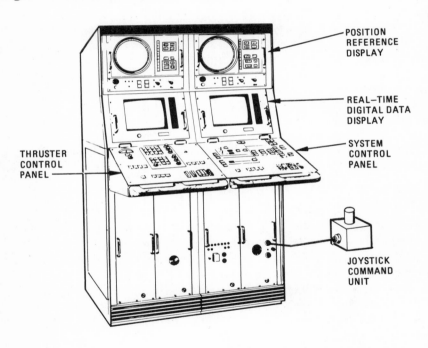

COURTESY OF HONEYWELL
MARINE SYSTEMS

Figure 5-24. A Currently Used Control Console for Dynamically Positioned Drillships

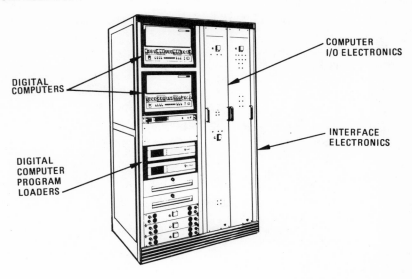

DIGITAL COMPUTERS

DIGITAL COMPUTER PROGRAM LOADERS

COMPUTER I/O ELECTRONICS

INTERFACE ELECTRONICS

COURTESY OF HONEYWELL
MARINE SYSTEMS

Figure 5-25. A Currently Used Computer System and Interface
Electronics for Certain Dynamically Positioned Drillships

Unfortunately, there are other topics related to controllers which
cannot be justified at this time. For example, the man-machine interface between the controller and the operator is important. Likewise, a
discussion of the equipment and its size might be of interest to some
readers. As a partial coverage of these areas, Figs. 5–24 and 5–25 show
a currently used control console with its man-machine interfaces and
a controller equipment suite, respectively.

There are other units on the market which are different in appear-
ance than that shown in Figs. 5–24 and 5–25. However, the basic
elements shown in the two figures are in every system. Included in the
list of basic elements are:

• Control console:
    Position reference display
    Real-time digital data display (alphanumeric display)

Keyboards
Control panels
Joystick command unit
• Equipment suite:
Digital computer
Program loader
Interface electronics
Computer input/output electronics

REFERENCES

1. Rules for Building and Classing Offshore Mobile Drilling Units 1973, American Bureau of Shipping, New York, 1973.

2. John Sjouke and George Lagers, "Development of Dynamic Positioning for IHC Drillship," OTC paper 1498, 1971.

3. H. J. Zunderdorp and J. A. vander Vlies, "How to Optimize a Dynamic-Stationing System," Soc. of Petroleum Engineers of AIME, SPE-European Spring Meeting, Paper # SPE 3757, 1972.

4. J. S. Sargent and J. J. Eldred, "Adaptive Control of Thruster Modulation for a Dynamically Positioned Drillship," OTC 2036, 1974.

5. J. G. Balchen, N. A. Jenssen, and S. Saelid, "Dynamic Positioning Using Kalman Filtering and Optimal Control Theory," Automation in Offshore Oilfield Operation, IFAC/IFIP Symposium, Bergen-Norway, June, 1976.

6. M. A. Childers, D. Hazlewood, and W. T. Ilfrey, "An Effective Tool for Monitoring Marine Risers," Journal of Petroleum Technology, March, 1972, pp. 337–346.

7. Jacques Harbonn, "The Terebel Dynamic Positioning System — Results of Five Years of Field Work and Experiments," OTC 1499, 1971.

8. P. M. Derusso, R. J. Roy, and C. M. Close, State Variable for Engineers, John Wiley and Sons, New York, 1965.

9. J. J. D'Azzo and C. H. Houpis, Feedback Control System Analysis and Synthesis, McGraw-Hill Book Co., 1966.

10. Julius, T. Tou, Digital and Sampled-Data Control Systems, McGraw-Hill Book Co., 1959.

11. W. P. Harthill and H. Van Calcar, Adaptive Filter, U.S. Patent, Number 3,867,-712, Feb. 18, 1975.

12. AMSOC Committee of the Division of Earth Sciences, Experimental Drilling in Deep Water at LaJolla and Guadalupe Sites, Pub. Nbr. 14, National Academy of Sciences — National Research Council, Washington, D.C., 1961.

6

The Power System

6.1 Introduction

The final element of the dynamic positioning system — and probably the one most likely to be overlooked — is the power system. The dynamic positioning power system not only supplies power for the thrusters but also for the sensors and the controllers. Naturally, the thruster power demands are significantly higher than the other two demands.

In fact, for most dynamically positioned vessels, when the thruster system is operating at its maximum output, it requires more power than any other task being performed by any other system on the vessel. Notable exceptions to this might be pipe burying or mining vessels where significant horsepower is required to bury the pipe or pump the mined ore from the bottom to the surface.

On smaller vessels, the power system usually consists of separate power-generating units for separate functions. Thus the main screws, each lateral thruster group or unit, and shipboard power for lights, DP sensors, DP controllers, etc., are driven from independent power-generating units. On larger vessels, centralized power-generating systems are used so that system can be jointly shared by the various users and redundancy can be provided for increased reliability.

In the small systems the design of the power system is straightforward once the sizing requirements are established. The design is more complex for the centralized power system because of the possible interactions among users, the size of the system, the importance of reliability to avoid the total loss of power which is tantamount to the loss of dynamic positioning control in heavy thruster load situations, and the variations in the power required by the users.

In a manner similar to Chapter 4 on thrusters, this chapter does not discuss the details of power system design. Instead, the discussion is limited to the aspects of the system that relate to dynamic positioning. The effects of other user systems on the ability of the power system to continue to meet the dynamic positioning power needs are considered under varying load and failure conditions.

From the discussion the merits of computer-based power management are illustrated. Then, two possible approaches to computer-based management systems are discussed.

6.2   Power system elements

Like all the other systems discussed in previous chapters, the power system is composed of several elements. The heart of the power system is the power generator. The output of the power generator feeds the user network through a matching network.

The power generator generally requires a control system to regulate its operation. Finally, since the power generator is simply an energy converter, a source of primary energy is required such as sunlight, petroleum fuel, or chemicals. These basic elements of a power system are illustrated in Fig. 6.1.

Included in the list of power generators are diesel engines, gas turbines, steam turbines, and batteries. The first is the most common for dynamically positioned vessels and the second is gaining popularity. Batteries do not have sufficient capacity and must be charged from a source of electrical power. Therefore, batteries are only used as backup power sources for small users such as the controller electronics.

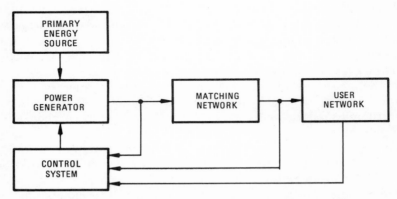

Figure 6-1.   The Basic Elements of a Power System

Steam turbines, which are frequently used on large ships moving from one port to another, are not normally used on dynamically positioned vessels because their power generating characteristics are better suited for a continuous load which does not change rapidly or frequently.

The nature of the matching network depends on the form and the amount of power required by the users and the extent of the user network. If a user requires mechanical power and the generator is mechanical, then the matching network is a mechanical coupler such as a gear train. If a user requires electrical power and the generator is mechanical, then the matching network is a conversion device commonly known as an electrical generator.

The kind of generator depends on whether the user requires alternating or direct current. However, alternating current (ac) generators are currently used for supplying either kind of power. If direct current is needed, the output of the ac generator is rectified to form dc. The advances in silicon-controlled rectifier (SCR) technology has made this economical and reliable.

The remaining combinations involving an electrical power generator are restricted to users requiring battery back-up. In a later section these power systems are discussed in relationship to supplying power to the dynamic positioning controller electronics.

Power systems using mechanical power generators which are mechanically coupled to the users are generally the ones which are most efficiently implemented with an independent generator for each user. This allocation strategy is usually the best policy where:

1. The users can be supplied power from standard marine generator units.

2. The total power required by any one user can be supplied by a single standard marine generator unit.

3. There is not a strict requirement for continuous dynamic positioning or work task operation such as in the case of the drillship.

If a vessel fits the foregoing criteria, it can be outfitted with equipment which is in common supply and is known to marine engineering personnel.

Smaller dynamically positioned vessels meet the above-mentioned criteria because their users are restricted to thrusters, ship service, and small standard marine work tasks (see Table 6–1 for the *Arctic Surveyor,* the *Seaway Falcon,* and *Kattenturm*). The thrusters, which are smaller than those used for drillships, can be driven individually

TABLE 6-1. SOME TYPICAL POWER SYSTEMS FOR SUPPORT VESSELS

VESSEL	TYPE POWER SYSTEM	NO. OF GENERATORS AND RATING	MAIN BUS VOLTAGE, VAC	TOTAL GENERATING CAPACITY	DP REQUIREMENTS			DRILLING REQUIREMENTS				VESSEL
					NO. OF MOTORS	TYPE	HP RATING (EACH)	NO. OF MOTORS	TYPE	HP RATING (EACH)	COMBINED USER REQUIREMENTS, ▲ KW	
SUPPORT VESSELS												
ARCTIC SURVEYOR	DIESEL AND DIESEL-ELECTRIC	4	NO COMMON BUS		2 (MAIN SCREW) / 2 (LATERALS)	DIESEL / DIESEL-ELECTRIC	1000 / 600	–	N/A	–		ARCTIC SURVEYOR
SEAWAY FALCON	DIESEL AND DIESEL-ELECTRIC	6	NO COMMON BUS		2 (MAIN SCREWS) / 4 (LATERALS)	DIESEL / DIESEL-ELECTRIC	1000 / 2 @ 574	–	N/A	–		SEAWAY FALCON
KATTENTURM	DIESEL	4	NO COMMON BUS		2 (MAIN SCREWS) / 2 (STEERABLES)	DIESEL / DIESEL	965 / 513	–	N/A	–		KATTENTURM
DRILLSHIPS												
GLOMAR CHALLENGER	DIESEL-ELECTRIC	10 x 800 HP	***	6000 KW	4 (LATERALS) / 6 (MAIN SCREWS)	DC / DC	750 / 750	7	DC	750	11000	GLOMAR CHALLENGER
SEDCO 445	DIESEL-ELECTRIC	5 x 2100 KW / 2 x 1050 KW	4160	12600 KW	11 (LATERALS) / 12 (MAIN SCREWS)	DC / DC	750 / 750	9	DC	750	19400	SEDCO 445
SAIPEM DUE*	DIESEL-ELECTRIC	6 x 2000 KVA	6300	12000 KVA	4 (BOW)* / 6 (STERN)	AC / AC	550 / 850	9	DC	800	12300	SAIPEM DUE*
HAVDRILL	DIESEL-ELECTRIC	5 x 3000 KVA		15000 KVA	5 (LATERALS) / 4 (MAIN SCREWS)	AC / AC	1750 / 1750	8 / 2	DC / DC	800 / 400	18600	HAVDRILL
DISCOVERER 534	DIESEL-ELECTRIC	6 x 2500 KW / 1 x 1050 KW	4160	16050 KW	6 (LATERALS) / 8 (MAIN SCREWS)	AC / DC	3000 / 2000	6 / 4	DC	750 / 1000	33180	DISCOVERER 534
SEDCO 471	DIESEL-ELECTRIC	5 x 2100 KW / 2 x 1400 KW	4160	13300 KW	12 (LATERALS) / 12 (MAIN SCREWS)	DC / DC	800 / 750	9	DC	800	20730	SEDCO 471
SEDCO 709	DIESEL-ELECTRIC	7 x 2500 KW / 1 x 1400 KW	4160 / 480**	23900 KW	8	AC	3000	9	DC	800	24760	SEDCO 709

* ONE OF THE BOW THRUSTER UNITS HAS BEEN REPLACED BY A 2800-HP AC MOTOR DRIVER AZIMUTHING PROPELLER.

** THIS GENERATOR CAN BE TIED TO THE 4160 BUS THROUGH A TIE TRANSFORMER.

*** EACH DIESEL ENGINE DRIVES A DC GENERATOR WHICH SUPPLIES POWER DIRECTLY TO A MATCHED DC MOTOR. THE MOTORS ARE CONTROLLED BY A MODIFIED WARD-LEONARD SYSTEM. [4]

▲ A COMMON SHIP'S SERVICE LOAD OF 1500 KW'S IS USED IN THESE REQUIREMENTS.

by diesel engines whose couplings to the propellers are standard mechanical installations which were developed long before the first dynamically positioned vessel. Likewise, the ship's service can be supplied by a standard diesel engine electrical generator set sized to meet the needs of the smaller vessels.

Similarly, diesel engines to drive winches, lifting machines, or hydraulic pumps are standard marine equipment. The controls for these power generators are also standard. Although they are potentially a common power system for dynamically positioned vessels because of the possible number of these vessels in the future, their design is better discussed by a marine engineer and is, therefore, not considered in this book.

For larger dynamically positioned vessels the amount of power required plus the number of users increases to the point that a centralized power system is the most cost- and space-effective approach. With the centralized power system matching the network to the user network includes not only several electrical generators being driven by diesel engines but also distribution, switching, and transforming systems. In the next section these large, centralized power systems are discussed in detail.

As Fig. 6–1 shows, the power system includes a control system. To achieve the proper regulation, the control systems are feedback systems. In the simpler systems the feedback is only from the output of the power generator. More complex control occurs when feedback is added from the matching and user network.

The final section of this chapter describes such a system which uses the same minicomputer used to implement the dynamic positioning system controller to supply feedback control to the power system.

6.3   Centralized power systems

Currently, the drillship uses the most power of any dynamically positioned vessel. Typically, as Table 6–1 shows, the installed power plant of today's drillship has a maximum capacity of 10 to 20 Mw.

The thrusters are the largest user in maximum output situations. However, there can be a large range of power demands from very little in light weather and no drilling, to maximum output in heavy weather. To meet these wide range of demands and provide power reliability in the face of system failures and maintenance requirements, the power systems are configured with multiple units and redundant features.

A typical centralized marine power system for a dynamically positioned vessel consists of several diesel engines, each driving a three-phase alternating current (ac) generator. The diesel engines are from 2,000 to 4,000-hp units and number from as many as seven to as few as five for the vessels shown in Table 6–1.

Often they are turbocharged to achieve the large horsepower output in the most compact physical dimensions because the space and weight of such a large power system are significant factors in the overall vessels layout. With each diesel engine there are controls to automatically start, stop, and control the throttle as the load varies on the electrical generator that it is rotating. Also, each engine is instrumented to monitor temperature, oil pressure, etc.

Ideally, for the most effective power system operation the power generator should be able to operate at zero or very low loads as efficiently as it can at medium or maximum loads. Such is not the case for the diesel engine which is best operated at mid-range to maximum output. This requirement affects the operation and management of the overall power system.

The electrical generators that the diesel engines drive are connected to a common busbar system through a circuit breaker, as shown in Fig. 6–2. Line diagrams are used to represent the structure and interconnections of the power systems. In reality, these systems are three-phase systems where each line in the figure is three or four conductors. Similarly, the circuit breakers are multiple-element devices.

Synchronizing controls are provided so that the generators can operate simultaneously with their outputs connected together. The synchronization is so important that if a generator loses synchronization within a certain tolerance, the controls remove the generator from the busbar by opening the circuit breaker. Synchronization can be lost due to a malfunction in the diesel-electric generator system, the control system, or an overload on the generator output for a certain length of time.

To generate the power required for the larger offshore dynamically positioned vessels, each generator is rated at several thousand kilovolt amperes (kva) with a terminal voltage of 4,000 to 6,000 volts (Table 6–1). The high kva rating is required to supply the necessary thousands of kilowatts for dynamically positioning and the work tasks. The high terminal voltage is used to reduce the required common busbar amperage which affects the size and construction of all the electrical equipment in the matching network.

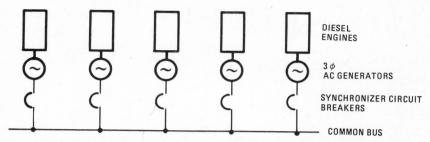

Figure 6-2. A Centralized Power System Common Bus Arrangement

The multiple diesel-electric generator sets give the power system the capability to adjust to load fluctuations and keep the loading on the individual diesel engines at a better operating level. Likewise, with multiple units, failures and overloads can be better tolerated and maintenance performed without the loss of the entire system.

Naturally, this inherent redundancy in multiple generating units is important for the power system of a dynamically positioned drillship and special measures must be taken to prevent the total loss or blackout of the power system. For that reason the common busbar shown in Fig. 6–2 is at least split in two parts, connected by a tie circuit breaker. In most redundant configurations two busbars are provided with connection circuit breakers so that either busbar can be connected to any diesel-electric generator set as Fig. 6–3 shows. A more complete discussion of busbar arrangements is given in Reference 2 at the end of this chapter.

From the common busbar the various loads are connected either through circuit breakers or through further coupling devices. The large three-phase ac induction motors which drive the controllable pitch thrusters are produced with rated terminal voltages equal to the high common ac busbar voltages so that they can be connected directly through circuit breakers to the busbar system.

For other ac loads which require lower voltages such as ships service, they are connected to the bus through transformers and circuit breakers. In fact, since there are usually several ac users, the transformer from the high-voltage busbar system feeds a low-voltage busbar system. For redundancy the low-voltage busbar system can be split with separate transformers from the redundant high-voltage busbar system feeding each of its parts and with a tie circuit breaker as shown in Fig. 6–4.

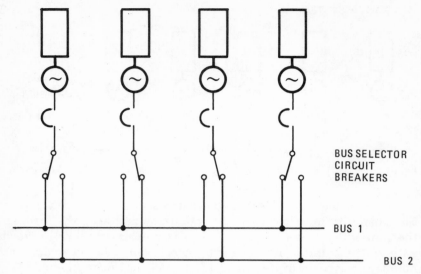

Figure 6-3. A Dual Busbar Centralized Power System

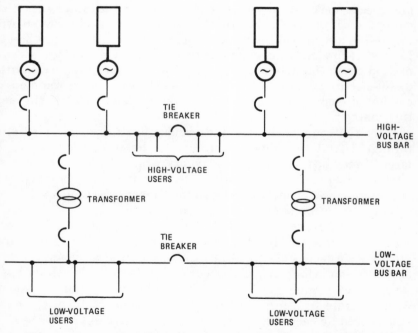

Figure 6-4. A Split High- and Low-Voltage Busbar System

For the dual-redundant high-voltage busbar system shown in Fig. 6–3 a dual-redundant high-voltage busbar system can be achieved as shown in Fig. 6–5.

The range of voltages used in various DP vessel power systems for the low-voltage busbar is 400 to 600 volts. There are users, such as lighting and small motors, which require a further reduction in voltage to 110 or 220 volts. These reductions are achieved with transformers connected to the low-voltage busbar system.

For vessels with users which require significant dc power, such as

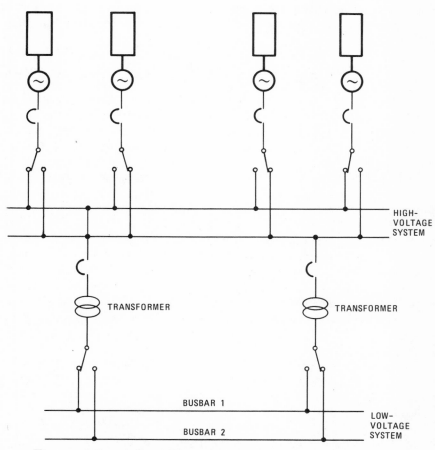

Figure 6-5. A Dual High- and Low-Voltage Busbar System

drilling and controllable speed thrusters, the most common method of supplying this power is with silicon-controlled rectifiers (SCR) connected to the low-voltage busbar system. Currently available three-phase SCRs can supply currents of several thousand amperes at 400 to 600 volts so that one SCR can easily drive a 1,000 to 2,000-hp dc motor. Thus, for a vessel like the *Discoverer 534* with eight 2,000-hp dc motors driving the main screw shafts, eight SCRs are required.

An SCR is a solid-state diode which rectifies ac voltage in a controlled manner. The basic principle of rectification of single phase ac power to obtain dc power is illustrated in Fig. 6–6. If the diode shown in Fig. 6–6 is an SCR, then by changing the phase or timing of when the diode conducts, the dc output of the diode can be controlled as Fig. 6–7 illustrates.

If more dc power than shown in Fig. 6–6 is required, then the conduction time of the diode can be increased by triggering the diode earlier in the positive cycle of the ac voltage.

The rectification shown in Fig. 6–6 and 6–7 is referred to as half-wave because the diode does not conduct on half of the ac voltage cycle. Full-wave rectification is achieved with a bridge arrangement and full-wave three-phase rectification is achieved with the arrangement shown in Fig. 6–8.

As Fig. 6–8 illustrates, a timing system is required which properly starts the diodes conducting in each phase and cycle of the ac voltage wave according to a control signal based on the required dc power. The control signal can be operator adjusted or automatically from a controller such as the dynamic positioning system controller which is commanding a thruster driven by a dc motor.

Since all the dc loads (usually motors) are not used at the same time, the SCR system includes switching circuits so that one SCR can be time-shared by more than one load. For example, an SCR which can supply 2,000-hp dc thruster motor with power can be switched to supply two 800-hp drilling motors when the thruster motor is not required.

Redundancy for an SCR system can be achieved from either low-voltage busbar system shown in Figs. 6–4 and 6–5. In the split busbar configuration in Fig. 6–4 the SCR system is split between the two low-voltage busbar systems. Furthermore, the loads are allocated to the SCRs so that the loss of a busbar system does not totally cause the loss of a critical load such as a thruster group or drawworks motor. Such a matrix arrangement is illustrated in Fig. 6–9.

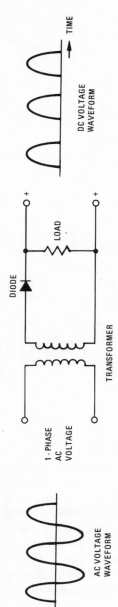

Figure 6-6. The Basic Principle of Diode Rectification of AC Voltage
to Produce DC Voltage

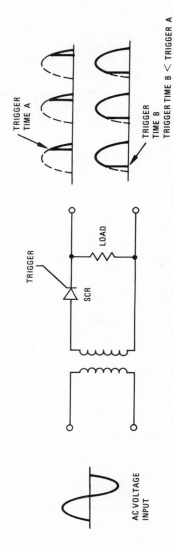

Figure 6-7. Variable Rectified DC Output by Varying
the Triggering Time of the SCR

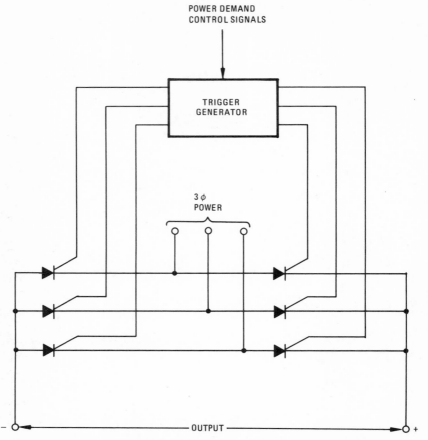

Figure 6-8.   A Three-Phase SCR Full-wave Bridge Circuit

A final element of an offshore centralized power system is an auxiliary diesel-electric generator set used as an emergency power supply. Normally, this emergency generator can be connected to the low-voltage busbar system with a tie circuit breaker which is closed manually. The emergency generator capacity is usually no greater than 500 kw and the start-up and synchronizing controls are all manually operated and independent of the main power system.

In this way if the main system fails, the emergency system can be started manually to supply the critical functions needed to restore or repair the main power system.

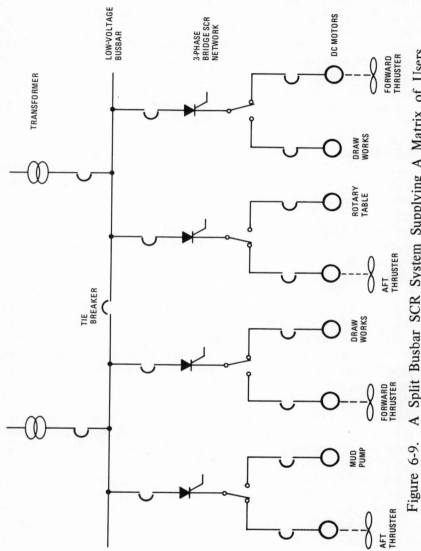

Figure 6-9. A Split Busbar SCR System Supplying A Matrix of Users

6.4 Operational considerations

The next topic of discussion after establishing the general layout of a centralized power system is the operational considerations of such a system. The discussion is only general and purposely develops those aspects which are pertinent to the next section of this chapter on the automation of a centralized power system. The drillship application used in the previous section is used in this section and in the next section of automation.

Basic to the operation of the power system is the start-up of the diesel engine and synchronization of the output of the corresponding electric generator with the other generators already connected to the high voltage busbar system. The start-up of a diesel engine is a standard operation and is performed before the circuit breaker which connects the electric generator to the busbar system is closed. Standard equipment for engine starting is available that requires only a start command from the engine control room or a local control panel at the engine.

For modern centralized power-generating system installations an enclosed central control room is used which is located adjacent to the diesel-electric generators and the busbar systems. In some installations viewing of the generators is provided for the power system operators.

The diesel starting equipment upon receiving the start command executes a preset sequence of operations to start the diesel engine. Since the electric generator is not connected to the busbar system, there is only an insignificant load on the diesel engine. Once the diesel engine is running at the proper speed, the starting equipment transmits a signal to the control room indicating the diesel engine is ready.

Modern start-up equipment also monitors engine parameters such as oil pressure so that if off-nominal performance is detected, an alarm is given, the "ready" indication is inhibited, and the engine is shut down if the fault is serious. The ready inhibit can be overridden by the operator if he feels that the detected off-nominal performance is not serious enough to harm the engine if it is used or if the diesel engine is needed for an emergency.

The operator can make such a judgment from the readouts of diesel engine parameters provided in the engine control room. A list of typically instrumented parameters is given in Table 6–2.

As a separate process or as a part of the diesel engine start-up, the

TABLE 6-2

TYPICAL MEASUREMENTS FOR A CENTRALIZED POWER SYSTEM

Diesel engine:
  RPM
  Oil pressure and temperature
  Cylinder temperature
  Cylinder pressure
  Cylinder wear
  Oil level
  Filter vacuum
  Fuel pressure
Electric generator:
  Terminal voltage
  Current
  Winding temperature
Busbar systems:
  Circuit breaker status
  Busbar voltage
  Power factor
User system:
  User status
  User current
  User voltage (if different from busbar voltage)
  SCR assignment (if dc power required and the SCR are shared among users)

diesel engine's speed and the generator's output voltage are changed so that the circuit breaker which parallels the generator's output with the other operating generators can be closed.

Before this closure can be performed, the frequency and amplitude of the output voltage of the generator must be within a small percentage of the frequency and voltage on the busbar system. The adjustment of the speed of the diesel engine regulates the frequency of the generator output.

Thus, synchronization requires either a manual or automatic control of the speed of the diesel engine. Both are generally provided with the manual control as a back-up. The automatic control of engine speed may be part of the start-up system. Likewise, the closure of the parallel circuit breaker may be automatic with a few seconds of delay after synchronization is detected. The speed control will make further adjustment after the breaker is closed because of the increase in load as the generator produces its share of the load demands on the centralized busbar system.

Ideally, the loading of the power systems should be of no concern to the users. When a user needs power, he prefers to only throw a switch. To provide such service, the power generators must operate efficiently at small as well as large load values. Operational efficiency includes not only fuel economy but also wear-and-tear on the generating equipment.

The more often that a generating unit must be maintained or repaired, the more costly the unit is to operate and the more likely that the power system will not be able to supply the user network with power under heavy load conditions. The operational costs are, of course, a serious consideration. But down time can be disastrous when either the weather demands maximum thruster system output or when the work task (drilling in this example) requires maximum power for a critical or emergency operation.

If the power-generating units are not ideal, which is the case with a diesel engine, the power system should be operated with the minimum number of on-line units to minimize wear-and-tear on the system. Maintenance of sufficient capacity with the minimum number of on-line units requires coordination among the users to achieve a reasonably efficient loading schedule. The coordination involves notifying the power system operating personnel that a change in the load is required.

For example, the driller may require more power for drilling or the DP operator sees that the weather is increasing and, therefore, he will need more thrusters and power capacity. Once notified, the power system operator decides whether he must change the make-up of the generators that are running. If the increase in the load is within the present on-line capacity, the user can immediately make the change in his load. If not, the user may have to wait until the number of on-line generators match the new demand.

If a user requires power for an emergency, he must still coordinate the change to prevent overloading the power system to the point that it fails.

Another complication in minimizing the number of on-line generating units is the potential failure of an on-line generating unit. To overcome this complication, the power system operating personnel must operate with sufficient reserve so that if a generating unit does fail, the total system does not fail. This precaution would not be necessary if a diesel generator set could be started and synchronized instantaneously, but in fact many seconds are required to start and synchronize a generator set. In that time the entire power system can black-out.

Some of the difficulties in operating the power system can best be understood from a graphic representation of the generating capacity as a function on the on-line units (Fig. 6–10). As Fig. 6–10 shows, generating capacity is a discrete multiple of the number of on-line units. If all the units are of the same capacity, then the stair-step function of the generating capacity has equal step size. Generating capacity can be expressed as follows:

$$\text{Generating capacity} = P_c = \sum_{i=1}^{N_G} k_G(i)\, P_G(i) \tag{6.1}$$

where:

N_G = the total number of generating units which can be connected to the common busbar system
$k_G(i)$ = 1 if the ith generator is on-line; 0 otherwise
$P_G(i)$ = the rated generating capacity of the ith generator unit

If the total power load on the system is represented by P_T, which is a time-varying function, then the following relationship must be true to avoid a power system black-out:

$$P_c \geq P_T\,(t) \tag{6.2}$$

where t is time indicating the functional dependence of the power system load on time. The maximum possible load on the power system for a particular user situation can be expressed as:

$$P_{UMAX} = \sum_{i=1}^{N_u} k_u(i)\, P_{UMAX}\,(i) \tag{6.3}$$

where:

N_u = the total number of users that can be connected to the common busbar
$k_u\,(i)$ = 1 if a user is connected to the busbar system; 0 otherwise
$P_{UMAX}\,(i)$ = the maximum power rating of the ith user

At any particular moment the total power load used in Equation 6.2 is less than or equal to the maximum possible load given by Equation 6.3.

The maximum power given by Equation 6.3 is only the total maximum user power when all the users are connected to the busbar system. As Table 6–1 shows, all the example centralized power systems

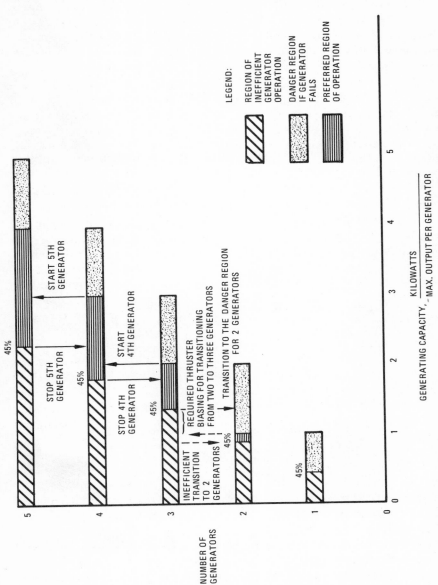

Figure 6-10.   Generating Capacity as a Function of On-Line Generating Units

do not have sufficient total capacity to supply the maximum requirements for all users at the same time.

To avoid the possibility of all users being connected to the common busbar system, most systems are organized so that certain users share the same connection to the busbar system through a mutually exclusive switch such as shown in Fig. 6–9.

For a diesel engine there is a minimum power level below which the engine does not operate efficiently. In fact, prolonged operation below a minimum loading level can result in damage or, at least, a decrease in the time before the engine requires an overhaul. If the engines share the load proportional to their maximum output and the minimum loading is the same percent of the maximum output for all the diesel engines, then the minimum loading is simply a percentage of the on-line generating capacity given by Equation 6.1.

The minimum loading of the power system can also be represented as a line on the generating capacity chart shown in Fig. 6–10.

For a particular number of on-line generators the operating personnel watch the total power system loading. When the loading reaches the minimum loading level, the operators must decide to either take a generating unit off-line or increase the load to a more efficient level of the total on-line generating capacity. To increase the load on the power system, the operator must have one of the users increase their power usage. The user which has the most control over its power usage is the dynamic positioning system through its thruster system.

The dynamic positioning system can increase the load on the power system without changing the position of the vessel by increasing the steady-state thrust output from two thrusters in such a way that their combined force and moment outputs are unchanged. The increase can be achieved by the operator of the dynamic positioning system using one of the thrusters which is not being used to automatically control the position of the vessel.

The operator commands the off-line thruster using a manual control to a given thrust output. Since the thruster is being manually controlled, its output remains constant at the commanded value. If the direction of the manually commanded thrust is such to add to the load already acting on the vessel, then the thrusters under the automatic control of the dynamic positioning system controller will necessarily automatically increase their outputs to counter the new force and moment tending to move the vessel off station.

As a result, the load on the power system is increased by the power used by the manually commanded thruster plus the power used by the automatically controlled thrusters to counter the thrust created by the manually commanded thruster, nearly doubling the manually commanded power.

If the manually commanded thrust is in the direction to subtract from the load already acting on the vessel, then the automatically commanded thrust from the dynamic positioning controller will only decrease to where the net increase in the power system load may actually decrease slightly. Not until the manually commanded thrust is greater than the load already acting on the vessel will the load on the power system increase in proportion to the power used by the manually commanded thruster. As a result, the use of a manually commanded thruster to increase the load on the power system must be performed in the proper direction.

Another difficulty in using a manually commanded thruster to load the power system is that if the operator suddenly changes the command to the thruster under manual control, the resulting sudden increase in force acting on the vessel may move the vessel from the commanded position by a significant amount. As a result, the operator should increase the command to the manually controlled thruster slowly over a period of several minutes.

Another method of creating an artificial thruster system load or bias to increase load on the power system is discussed in the next section where a computer is used to provide automatic power system supervision and management.

The method involves the dynamic positioning system controller automatically increasing the output of two thrusters to increase the power system load. The automatic biasing of thrusters has the added advantage that the thrust increase from the two thrusters can be performed in a coordinated manner so that there is no variation in the position of the vessel as the power system load is created.

There are several other advantages to using a computer to regulate the load on the power system and they are given in the next section.

Naturally, as the real load on the power system increases to an acceptable level, the pseudo-load created by the thruster system can be decreased appropriately toward zero. As the real load on the power system increases further, a loading factor threshold is reached where another generator should be started to avoid an overload situation on the existing on-line generators. The value of the loading factor threshold depends on the rate of increase of the load on the power system, on the time required to start another generator, and on the precautions used in operating the power system against failures of the generators and the users.

If the power system load is increasing because of thruster system loading caused by environmental effects, the rate of average increase will normally be sufficiently slow that the threshold for starting a new generator can be as high as 95% of the total on-line generating capacity.

At high thrust usage levels, such as those which exist in storms, the threshold may have to be less because the modulation on thrust commands may be of sufficient amplitude to make a 95% threshold overly risky. A sudden squall or an internal current wave, such as those which occur in the Andaman Sea, require a lower generator start-up threshold.

When the power system load increase is caused by a work-related task, the load changes as a function of the task. For example, if the task is drilling, which involves the use of the mud pumps and the rotary table, the load suddenly increases as the pumps and table start and continues at a nearly constant level of several thousand horsepower for several hours with intermediate short decreases to zero horsepower as another joint of drill pipe is added to the drill string.

In another drilling task of tripping, the load varies from zero to several thousand horsepower for a few minutes and then returns to zero as a section of drill pipe is added or subtracted from the drill string. This up-and-down variation in the power system load may continue for five to ten hours.

As a result, when the load is work task related, the power system operator must know the nature of the potential load variations or set-up the power-generating configuration to handle sudden significant load changes. This can result in a conservative number of on-line generators with inefficient loading factors simply to cope with an unknown maximum load which may or may not ever occur.

The thruster system influences the configuration of on-line generators in another way than direct demand for power. The number of on-line thrusters to cope with thruster failures also influences the number of on-line generators. Generally speaking, at least two thrusters or two drive motors to a main screw are kept on-line in each thruster group even in slack weather conditions. In a dynamic positioning system with the DP controller computer in control of the use of thrusters, this condition can be relaxed to one on-line thruster or drive motor and one thruster or drive motor in the ready condition in each group.

With two thrusters in each group the maximum thruster power load is the sum of rated horsepower for four thrusters plus two main screws drive motors for a fixed-axis thruster system, or four thrusters for a steerable thruster system. Normally, in minimum environmental conditions the thrusters will be using minimum power. If a thruster fails so that it produces zero thrust, the other thruster in the group automatically picks up the load of the failed thruster. Neither the normal or failure to-zero-output situation has any influence on the power system configuration other than in the way it satisfies normal demand.

However, if a thruster fails with maximum thrust output or if an erroneous command is generated by the controller for full thrust, the thrusters will suddenly be demanding full power. For larger thruster units full power demand can be several thousands of horsepower more than their normal demand. Unless the power system is configured to supply this sudden demand, or to limit the thruster outputs to within its capacity, a blackout will occur.

In the first case of configuring the power system to cover upsets in the thruster system, a minimum number of on-line generators results as a function of the minimum number of thrusters. This minimum can be computed for the worst case for a power system composed of equal sized generators in the following manner:

$$N_{GMIN} \geq \frac{\sum_{j=1}^{N_T} P_{TMAX}(j)}{P_{GMAX}} \tag{6.4}$$

where:

N_{GMIN} = the minimum number of on-line generators which is an integer

N_T = the number of on-line thrusters

$P_{TMAX}(j)$ = the maximum rated power of the jth on-line thruster

P_{GMAX} = the maximum rated power of each generator

As a result, if a power system has 2000-kw generators and 1000-kw thruster and main screw drive motors, the minimum number of generators to protect against the worst-case thruster upset, from Equation 6.4, is three generators. In slack weather the thruster load may be very small yet three generators are needed to cover the worst-case upset situation.

Also, thruster biasing is needed to bring the load on the power system up to an acceptable level to establish a minimum power usage for the power system and a minimum fuel consumption rate.

Obviously, the kind of failure required to make Equation 6.4 valid is very unlikely and as a result the power systems in slack weather are operated with a minimum configuration, less than given by Equation 6.4.

In the example just given the power system is most likely operated with only two generators in slack weather. Two generators give protection against a sudden thruster upset and a generator failure which must also be taken into consideration. In the next section of this chapter the power system can be operated at even smaller minimum configurations of thrusters and generators because a computer system keeps a constant vigil over both the thruster and generator system. A human operator cannot effectively keep track of both systems on a day-in and day-out basis.

The configuring of the power system to protect against the potential failure of one on-line generator is achieved by maintaining the load on the power system less than or equal to the on-line generating capacity minus one generator. This failure precaution can be expressed mathematically as follows for generators with the same rated capacity:

$$P_c \leq P_{GMAX} [N_{GT} - 1] \tag{6.5}$$

where N_{GT} = the total number of on-line generators

The combination of a minimum allowed load for generator operating efficiency and a maximum allowed load to protect against a generator failure gives rise to operating constraints and a minimum number of on-line generators, as illustrated in Fig. 6–10. As Fig. 6–10 illustrates, there is no region of preferred operation until two generators are on-line. Then, the preferred region is very narrow and only exists for two generators because a minimum allowed load is less than 50%. If the minimum allowed load was greater than 50%, then a preferred region of operation would only exist for three or more generators.

Fig. 6–10 also shows that until three or more generators are on-line a transition to more or less generators cannot be made so that the new on-line configuration is operating in its preferred region. The two generator situation in Fig. 6–10 can be changed to a three-generator configuration if thruster biasing is used to increase the load on the three-generator combination.

Likewise, the three-generator combination can be changed to a two on-line generator combination once the thrust biasing load on three generators increases to the point, at which, when one generator is taken off-line (stopped) with the thruster bias decreased to zero, the two generators are operating in their preferred load region. To make this transition from three generators to two generators unfortunately requires temporary operation of either two generators in their danger region or three generators in their inefficient region (Fig. 6–10).

Again, if an automatic system was available to limit thruster commands and other user demands to the available generator capacity, then the difficulties caused by having to protect against a generator failure with a reserve on-line generator could be avoided. This automatic system is the topic of the next section.

6.5   Power management system

For the successful manual operation of a power system on a dynamically positioned vessel the power system requires constant attention by operating personnel. They must not only configure the power system to achieve the lowest operating costs, but even more important, they must maintain sufficient capacity plus reserve to prevent blackouts.

Since operating costs include fuel consumption plus maintenance and repair demands, the operating personnel must maintain efficient loading of the generators plus performance monitoring and operational records on each generator so that maintenance can be performed at the proper time. As can well be imagined, this is asking too much of nearly all operating personnel on the round-the-clock schedule used for most dynamically positioned vessels.

However, a computer-based system can be added to supplement operating personnel to achieve a highly effective power management system.

There are at least two possible computer configurations that can be used to implement power management. The first places the power management functions co-resident with the dynamic positioning system controller in the same computers. The second separates the power management functions from the dynamic positioning system controller in another computer.

The separate power management system computer can be at the same location as the dynamic positioning system computers or at another location more convenient to the power system — e.g., the power engineering control room — if the DP system computers are at a location on the vessel remote from the power system. The more convenient location to the power system for the power management computer is favored.

Dynamic positioning can be performed by either an analog or digital controller. The digital controller implemented on the modern high-speed digital minicomputer is presently the best choice for the dynamic positioning control problem. The digital computer controller is superior because it has the capacity to process large volumes of data very quickly with complex logic and arithmetic.

This capability allows the digital computer to not only perform the real-time control law computations but also monitor various performance parameters so that the controller can alert the operator to current or impending problems and execute preset procedures to correct system difficulties such as sensor and thruster failures.

As will become obvious later, the power management functions involve large volumes of data which must be processed for performance monitoring, record keeping (data storage), and logical decisions in real-time. Consequently, power management favors the digital computer even more than dynamic positioning because of its logic performing and memory capabilities.

Power management, whether co-resident or independent of the dynamic positioning system controller, is the same. In either case there is an interchange of data between the two systems. The only difference is the method of interchange. For the co-resident situation the interchange is performed in the common computer memory. For the independent power management system the interchange is by inter-computer communication channels. As a result, in the following, power management is discussed independent of the computer configuration to begin with.

The power management scheme discussed in the following paragraphs is not the only way to perform this function. However, it does illustrate the various elements of a system which are required to perform power management.

To effectively manage the power system, the management system must be able to perform the following functions:

- Automatically start or stop any generator unit
- Detect a malfunction in the power system
- Limiting of selected user loads in critical situations
- Regulate the loading of the power system

The first function is important to proper power system management in at least three ways. First, it allows the management system to control the generating capacity of the power system. Second, it makes it possible to change the on-line generating unit configuration to maintain an even amount of running time on all units. Third, the starting and stopping capability is necessary to react to emergency situations, such as generator failures, as fast as possible.

The power management system (PMS) controller, if implemented in a digital computer, can perform whatever sequences of logic required to start or stop a generator unit. However, a simpler arrangement is for the PMS controller to issue a start or stop command to a sequencer supplied with the generator unit. This not only simplifies the cabling interface between the PMS controller and the generating unit, but also most likely assures a better match of equipment if the sequencer is provided by the supplier of the generator unit.

The second required function of the power management system is needed primarily so that the controller can react at the earliest possible moment to an actual or impending failure. In some cases a malfunction in the power system does not mean that some critical element of the power system has failed, thus causing a partial or total loss of power. However, in any case, operating personnel should be alerted to the malfunction of off-nominal performance so the appropriate action can be performed.

When a malfunction or sequence of malfunctions is detected that indicates a critical failure can occur in the near future, the management system can start a new unit using a normal starting sequence. Then, when the new unit is on-line, the unit suffering the malfunction can be stopped for repair. In some cases a malfunction may indicate an immediate critical failure in the power system. For these cases the PMS controller immediately signals a new generating unit for an emergency start. If a new unit cannot be placed on-line soon enough to pick up the load lost by the failure, then the management system must automatically decrease the load on the system.

Loading can be decreased by first eliminating any thruster bias being used to regulate the load on the generator system. Next, non-dynamic positioning system loads can be limited by the management system. Finally, if neither of these reductions is sufficient to prevent the power system from blacking out, then the commands to the thrusters can be limited for the length of time required to restore the necessary generating capacity.

With a digital computer all of the decisions can be made in a manner of seconds. However, the power management system requires the proper data from the power system from which to make its decisions and execute its response.

For the detection of generating system malfunctions the outputs of performance monitoring sensors installed in the various elements of the power system are interfaced to the digital computer of the PMS. Interfacing the sensors to the computer is performed with the same devices used to interface the sensors to the digital computer for the dynamic positioning system controller, i.e., analog-to-digital converters for analog signals and digital-to-digital converters for on-off signals.

As a result, if the power management system controller is being combined with the dynamic positioning controller in the same digital computer, the input/output interface used for the dynamic positioning system need only be expanded to include the power management inputs. With modern real-time control minicomputers and high-speed input converters, the number of inputs that can be interfaced to the computer and still allow sufficient time for processing and decision-making by the computer for the real-time control of the dynamic positioning and power system is several hundred analog signals and 500–600 on-off signals.

The capability to process many inputs is necessary for power management because there is a significant number of inputs from each element of the power system and there are many elements in the power system. For example, each diesel engine and electrical generator combination can have as many as 10 to 20 parameters measured by a sensor with an analog output. Since a typical power system for a dynamically positioned vessel consists of seven to nine diesel generator combinations, the analog signal inputs to the digital computer may number as high as 180 and as low as 70.

The input data from the power system include measures of various performance parameters and indicators that certain events have occurred. The indicators may in some cases be alarms generated by monitoring devices supplied with the power system equipment. For example, a diesel engine may include self-monitoring systems for oil pressure and temperature, the outputs of which are alarm signals for high or low oil pressure or high temperature. The thresholds for these alarms are generally preset by the manufacturer of the equipment.

When the indicators are not alarms, they still denote that some pre-established criterion has been satisfied. The criterion could be that the diesel engine has reached a certain rpm or that the diesel engine-electric generator combination is synchronized with the on-line generators or that a circuit breaker has closed.

In the case where the inputs are analog measures of performance parameters, such as oil pressure, temperature, or engine speed, the digital computer performs a monitoring function with a preset criterion for the detection of off-nominal performance. When off-nominal performance is detected, the computer alerts the operator with an audible and visual warning. Often, the computer is programmed to log all alerts on a hard copy printer to give operating personnel a permanent record of the power system's performance. Likewise, when the operator corrects the conditions that generated the alert, the computer logs that the alert is cleared.

The detection of certain indicators, off-nominal performance alerts, or combination of the two for a sufficient length of time indicates that the power system is failing or will fail soon unless prompt action is taken. In short, certain indicators and alerts are more important than others. In these cases, the power management controller can take pre-programmed action to avert the loss of the system.

For example, if a diesel engine has failed, the computer can immediately issue a start command to a new generator and limit the load until the generator is on-line. If the combination of off-nominal performance alerts indicates that a generating unit may fail, the computer can start a new generator. When the new generator is ready and on-line, the computer can stop the ailing generating unit.

With any automatic system which uses sensor data to make decisions there is a likelihood that the sensor data are erroneous. Naturally, if the data are erroneous, the computer may make the wrong decision and take the wrong action. The consequences of the computer erroneously warning the operator of off-nominal performance are not serious.

Erroneously starting a new generator and stopping a suspect generator is not a serious problem. However, erroneously stopping a generator without starting a new generator can be disastrous. Since the only time that such action is necessary from the power management system is when the load is decreasing, the decision to stop a generator does not need to be made immediately to save the power system.

Therefore, the decision to stop a generator without starting another generator can be made fail-safe by requiring that the computer requests the operator's permission to issue a stop command to a generator. This allows the operator to double check the computer with a final veto power if he detects that the computer is using erroneous data to decide to stop a generator.

In addition to the computer logging alerts on a hard-copy printer, the computer keeps records on the generating units so that operating personnel can determine when a particular unit should be maintained. Also, these records can be used to prioritize the selection of the next generating units to be started and stopped. As the power system is operating and events are occurring which result in a changing generating-unit configuration, the computer continuously updates which generator is best to start next if an emergency occurs or which generator should be stopped for a normal rotation of units under the uniform usage criterion.

The priority established by the computer is always displayed to the operators so that they can override or modify the priority, if they desire, to reflect information that the computer does not have or a more pertinent criterion which overrides the criterion being used by the computer.

Protection against total power system loss or blackout requires not only the first two functions of starting and stopping generating units and of detecting power system faults but also control of power usage. Without control of power usage, the generating system must be operated very conservatively with extra generators on-line to provide redundancy.

With extra generators on-line more wasteful thruster biasing is required to regulate the efficiency of the power system. Control of power usage to protect against blackouts need only be the capability of setting limits, based on existing generating capacity. With this strategy the management system is exerting no control of power use under normal operating conditions. Then, when a crisis exists in the generating system, the management system generates limits to the user network.

The limits are based on a priority criterion which first eliminates unnecessary loads such as thruster biasing and then cuts back power to other users based on their importance to the overall operation.

For a dynamically positioned drillship, the cut back of power to the drilling operation is not appreciated. However, power to the drilling users should be limited before the thruster system because drilling cannot be performed if the position of the vessel is not maintained over the wellhead. The limiting of power to work tasks may not always be first priority over thrusters if the relative importance of position control is less than that of work task.

However, generally the loss of position control results in a termination of the prescribed work task of the vessel and in some cases can result in significant damage, such as in drilling and pipe laying.

Cut back or power limiting can be performed by the power management system controller in several ways. For complete cut backs the controller can open circuit breakers. Partial cutbacks require proportional control signals from the management controller to a device which controls the power usage of the user to be limited.

In the case of drilling users the controller can issue a control signal to the SCR trigger generator shown in Fig. 6–8. Thruster commands can be limited by the power management system controller sending a power usage limit to the dynamic positioning controller. The DP controller, then, computes the thrust command limits used in thrust allocation logic based on the power usage limit established by the power management system. At the same time that the limit is issued to the DP controller, the DP operator is alerted so he can double check that the limit is necessary or should be overridden to prevent serious difficulty.

So far the discussion has concentrated on the management of the power system under conditions of malfunctions to prevent the loss of the power system. The management system also has important functions during normal operating conditions. First, the management system must maintain sufficient on-line generating units to supply the current power demand within the generating capacity of the overall power system.

This means that as the power demand changes, the management system must start or stop the proper number of generating units to satisfy the demand. Under conditions of normal operation, the starting and stopping of generator units is more straightforward because reaction time is not as critical as in the case of a failure.

To keep the power configured with sufficient on-line generating capacity, the power management system (PMS) controller performs certain computations, based on the instrumentation inputs from the power system. Since proper power management is essentially a balance between generating capacity and power demand to achieve efficient operation with sufficient reserve to prevent blackouts, the controller must keep track of the on-line generating capacity and the actual power demand.

The on-line generating capacity is computed using Equation 6.1 where the coefficients, k_G, are determined by interfacing the circuit breakers to the PMS controller which connect the generating units to the high-voltage busbar system. If the actual power demand is not measured directly, it can be computed from the total load current and the terminal busbar voltage using the following equation (The functional dependence on time represents the variations in the average power.):

$$P_u\,(t) = \sqrt{3}\,E_B\,(t)\,\cos\alpha_B\,(t)\sum_{i=1}^{N_G} I_G\,(i,\,t) \qquad (6.6)$$

where:

P_u = total user power, kw
E_B = RMS busbar voltage, kv
I_G = RMS current being supplied by the ith generating unit, amperes
  t = time

Generating capacity and actual user demand can be measured and compared in terms of kilovolt-amperes (kva), which is related to kilowatts by the cosine of the power factor angle, α_B. The power factor is a function of the reactive nature of the load. The larger the reactance of the load, the larger the power factor angle, and the fewer kilowatts that are consumed for the same kilovolt-amperes.

Since generators are rated for a given terminal voltage and current, they are kilovolt-ampere limited. As a result, using kilovolt-amperes is a better measure of power system capacity than the kilowatts. To compute kilovolt-amperes from Equation 6.1, the rated generating capacity for each generator unit is in units of kilovolt-amperes. Equation 6.6 gives the actual user power in kilovolt-amperes if the power factor is omitted.

The actual user demand is composed of thruster, work task related, and ship's service loads. The thruster load can include components for dynamic positioning control, thruster bias for power system load regulation, and manually commanded thrust. The power load caused by the manually commanded thrust is unpredictable because its purpose is not for control and it is controlled by a human operator. Thruster bias power loads are slowly varying, known, and under the control of the power management system controller.

Naturally, the DP thruster power loads vary with the disturbance forces acting on the vessel to which the DP system responds. This means that the load on the power system caused by the DP thruster system is composed of a slowly varying average value with a higher frequency component of less amplitude than the average superimposed on the average value. The composite of the three-thruster power load components can be represented as a function of time as shown in Fig. 6–11.

The power used for the work task being performed on the dynamically positioned vessel will vary depending on the task. For the drillship application the power usage for drilling depends on the particular drilling operation. When the drill bit is being replaced, the draw works is used to pull and run pipe. The resulting load variations during pulling and running pipe are from zero to several thousand horsepower for two or three minutes at each value over a period of five to ten hours, depending on the depth of the well. In a full-time drilling operation, pulling and running pipe occupies approximately 20% of the time.

A second drilling operation is pumping and rotating which consists of on-bottom drilling and making a connection. Pumping and rotating may continue one or two days before a drill bit must be replaced. On-bottom drilling is a steady load requiring 1,000 to 3,000-hp for approximately a half hour with a several minute period of zero hp. It occupies approximately 70% of the full-time drilling cycle. Making a connection is a transient load, like pulling and running pipe where the load goes from zero to several thousand hp for a couple of minutes as a new joint is added. Of the total drilling cycle, making a connection occupies about 10% of the time.

The last remaining power system load is the ship's service which supplies many different small loads on the vessel such as small pumps, hotel functions, and lights. The character of the ship's service load is a slowly varying average with small incremental variations around the mean (a small variance).

Thus, the total load on the power system is a composite, slowly varying average with short-term variations caused by disturbances in the thruster system plus significant sudden load changes resulting from manual thrust commands and work-related tasks. The load changes which involve a human operator are the most unpredictable and can involve several thousand hp which is equal to or greater than the capacity of one generating unit.

So that the power management system controller can properly configure the power system with sufficient capacity for the current demands plus provide reserve for unknowns, it continuously monitors

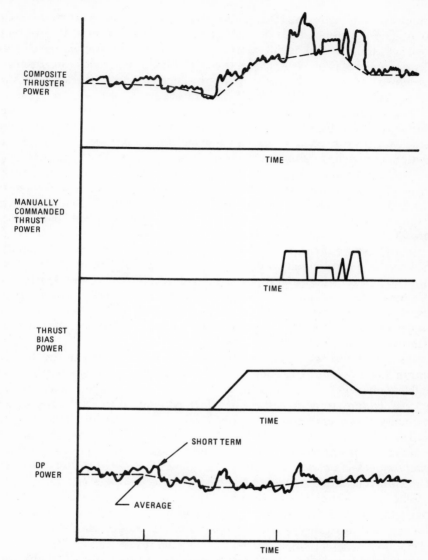

Figure 6-11. Power Consumption Resulting From the Thruster System

the present system load, the generating capacity, and the thruster bias load. The difference between the generating capacity and the system load is the instantaneous power system reserve. The sums of the instantaneous reserve and the thruster bias is the available reserve which can be reduced to zero without affecting any real users.

However, since the thruster bias requires a few seconds to be set to zero, the bias is not a truly instantaneous reserve. Therefore, start-up decisions must be made from the instantaneous reserve to give the quick reaction needed to prevent a system blackout, unless the power system can sustain short-term overloads.

Under normal conditions when the load on the power system is such that the generating units are operating in their preferred region (Fig. 6–10) the reserve varies between two thresholds, as shown in Fig. 6–12. When the reserve is less than the lower threshold for a preset length of time, a new generator is started (Fig. 6–12).

If the reserve increases to a preset percent of the generating capacity or threshold for a given length of time, the controller requests thruster bias to reduce the reserve to the level where the power system is operating more efficiently. The bias request is received and processed by the dynamic positioning controller to fulfill the request as closely as possible. The actual thruster bias achieved is transmitted back to the management controller.

As the thruster bias increases, the reserve stays near the lowest value for efficient operation of the power system, as shown in Fig. 6–12. However, at some value of thruster bias, the real load on the power system is at a sufficiently low percentage of the generating capacity to warrant the shutdown of one generating unit.

Therefore, at a preset thruster bias level (Fig. 6–12), the power management controller requests the stoppage of a generator. Once the request is acknowledged by the operator, the controller initiates the stop command to the next generator selected for stoppage.

The selection of the next generator to be shut down is based on a priority ladder established as the power system operates. The generator for shutdown is at the top of the shutdown ladder. At the bottom of the shutdown ladder is the next generator to be started. Thus, with one priority sequence, generators can be selected for start-up and shutdown.

The order of the generators in the priority ladder changes as on-line generators accumulate operating time and performance parameters

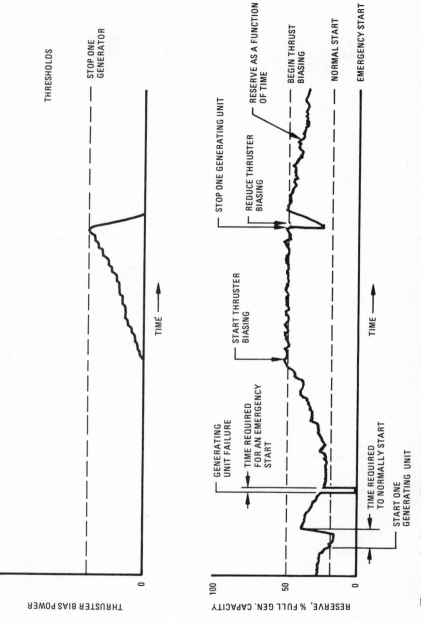

Figure 6-12. A Plot of Power System Reserve and Thruster Bias Power for Various Power System Events

change. The controller computes a new ladder on each cycle of new input information.

The power management system controller also monitors the status of users which are likely to show large variations in demand, both short-term and long-term. If the controller receives a signal from one of these users, it immediately reduces the reserve by the peak expected demand for that user. The reduction of the reserve below a certain preset threshold for a given number of consecutive seconds automatically initiates a normal generator start-up. If the reduction in the reserve exceeds a second lower threshold, an emergency generator start-up sequence is initiated, followed by load shedding if the reserve is reduced to zero or less. This start-up logic is the same as used under all situations.

For example, if a generator fails, the reserve is automatically reduced which may or may not require a new generator, depending on the system load factor. In the example shown in Fig. 6–12, the reduction in the reserve results in an emergency generator start command. Most likely the failure in Fig. 6–12 would result in power limiting to certain users.

One implementation of the previously discussed functions that the power management system controller must perform is illustrated in Fig. 6–13. The key element in the controller is the threshold detection logic which issues commands for starting generating units, limiting power, and requesting thruster biasing. The thresholds used for these decisions are parameters in the controller which are variable depending on the operating policy of the owner or operator of the vessel.

Likewise, the priority of shedding or limiting loads is variable. When the system first goes into operation, the thresholds and limiting priorities are set at some preconceived values. As the operator becomes more familiar with the operation of the overall system and the nature of the user loads, the system operating personnel may find that the thresholds and priorities should be changed. With the management controller implemented in a digital minicomputer, the adjustment of power management system parameters by operating personnel is easy and straightforward.

The record-keeping capacity of the minicomputer outfitted with mass memory and a hard-copy printer is also perfectly suited to track changes in the operating parameters so that operating personnel can evaluate the effect of their parameter selections. As a result of these

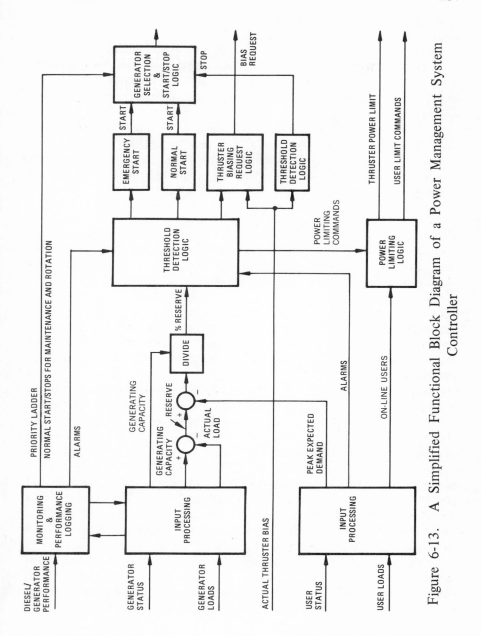

Figure 6-13. A Simplified Functional Block Diagram of a Power Management System Controller

evaluations they may find, for example, that the system parameters should change depending on the weather or the work tasks.

One difficulty in the threshold detection logic is setting the length of time that a threshold is exceeded before the corresponding command is issued. The difficulty arises because the time required to reach the decision to start a generating unit or to limit a load is critical to preserving the power system.

However, starting a generating unit or limiting a user unnecessarily is undesirable in many respects. As a result, the threshold detection logic must monitor the changes in the instantaneous reserve to screen transient surges and short-term changes from the decision process to avoid unnecessary start-ups or limiting of users.

The design of a fool-proof screening logic is, however, not trivial for all possible situations. In the final analysis, the designer will necessarily have to decide which is more serious, blackouts or false alarms.

Previously, it was mentioned that the power management system controller can be implemented in the computer with the dynamic positioning controller or in a separate computer.

Also, the fact that the power management system controller requires a significant quantity of information interchange with the power system was mentioned. In addition, it was mentioned that there is an interchange of data between the power management system and dynamic positioning system controllers. These interconnections are summarized in Fig. 6–14.

In deciding which computer system implementation is most appropriate for a particular dynamically positioned vessel, several factors need to be considered. First of all, the layout of the vessel and the location of the power system with respect to the dynamic positioning system controller.

Generally speaking, the location of the dynamic positioning system controller and the power system are not related. On most ship-shaped vessels the power system is aft. The majority of the DP control rooms which contain the DP controller are forward near the bridge. In a few cases, the DP control room is aft with the power system, e.g., the *Discoverer 534*. One reason that the location of the DP controller is not closely related to the location of the power system is that there is, at most, only limited interconnections between the two systems.

For example, in Fig. 6–9 if the thruster system uses controllable-speed propellers driven by dc motors, then the dynamic positioning controller interfaces to the trigger generator which controls the output

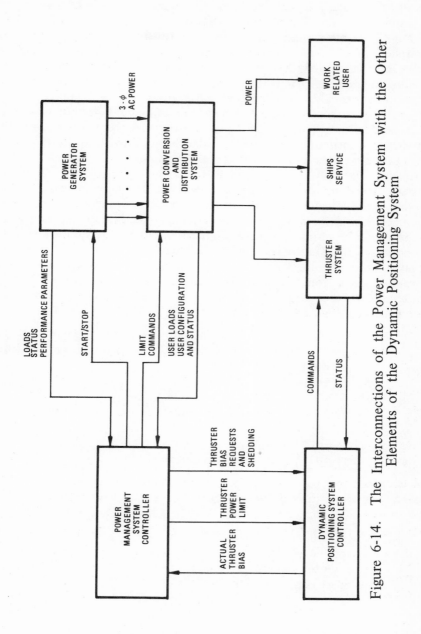

Figure 6-14. The Interconnections of the Power Management System with the Other Elements of the Dynamic Positioning System

of the three-phase SCR bridge network. With the lack of interconnections, the amount of cabling to transmit the signals between the two
systems is not significant nor is the cost.

For the power management system controller the situation of interconnections with the power system is totally different than for the
dynamic positioning system. Thus, if the management system controller is separated from the power system by a significant distance,
there are extensive cable runs which are costly.

For this reason locating the power management system controller
adjacent to the power system is advantageous. However, unless the
dynamic positioning controller is also located adjacent to the power
system, a separate computer for the management controller is required.

An alternate approach to the individual cables for each interconnection between the power system and the management controller
is a serial multiplex data transmission system. The multiplex system
requires only a single transmission path which may be an electrical
conductive pair or in the future, a fiber-optic light pipe. Thus, the
cost penalty for locating the power management system controller
some distance from the power system does not exist.

Then, the advantages of combining the two controllers in the same
computer can be exploited, including:

1. The information to be interchanged between the two controllers,
such as power limits on the thrusters, can be made directly within
one computer computational cycle.

2. Less overall input and output channels and computer memory
are required because there is common information used by both
controllers, such as thruster status and thruster power usage.

3. More complete information for the operating personnel of both
the power and dynamic positioning systems because of a centraliza-
tion of information for both systems at one point.

Locating the power management system controller in a separate
computer from the DP controller does have certain advantages, including:

1. The controller can be located in the power system control room
for more direct access by power system operating personnel.

2. The computer which implements the controller can be used for
other data processing functions.

The second advantage also applies for the DP controller computer and is discussed in more detail in the last chapter of this book.

REFERENCES

1. F. B. Williford and A. Anderson, "Dynamic Stationed Drilling SEDCO 445," OTC 1882, 1973.

2. O. Soyland, "Basic Design Principles for the Electrical Power Supply System," Det Norske Veritas Symposium on Safety and Reliability of Dynamic Positioning Systems, Oslo, November 17–18, 1975.

3. L. E. Owens, "Power Systems for Dynamic Positioning," Det Norske Veritas Symposium on Safety and Reliability of Dynamic Positioning Systems, Oslo, November 17–18, 1975.

4. J. R. Gramham, K. M. Jones, G. D. Knorr, and T. F. Dixon, "Design and Construction of the Dynamically Positioned Glomar Challenger," Marine Technology, April 1970, pp 159–179.

5. E. Swanson and S. Parlas, "Automatic Power Management for a Dynamically Positioned Drillship: An Integrated System," OTC 2416, 1975.

6. G. Holzenbein, "Automation of the Power Generating Plants On Board of the Saipem Offshore Vessels Scarabeo III, IV, and Castoro VI," Automation in Offshore Oilfield Operation, IFAC/IFIP Symposium, Bergen, Norway, June 1976.

7

Control System Analysis

7.1 Introduction

In previous chapters the four major elements of a dynamic positioning system were discussed from the standpoint of their characteristics and contribution to the overall system. The fact that the dynamic positioning system is a closed-loop feedback control system has been established. In its most common form the dynamic positioning system controls the heading and x-y position of the vessel on which it is installed. As a result, the dynamic positioning system is a multivariable feedback control system.

To better understand the design and operation of such a system, certain feedback control concepts, design techniques, and performance analysis methods are necessary. In this chapter, a basic collection of background information is presented. Further information is contained in References 1 through 5 at the end of this chapter.

The range of control systems discussed here includes linear, nonlinear, sampled-data, and multivariable system. Special consideration of types of control involved in dynamic positioning is important and the digital control system deserves emphasis because DP systems using a digital computer for a controller are common.

Performance measures are a key to the design and analysis of control because design is a compromise between performance measures. Also, the selection of analysis techniques affects the choice of the performance measures used in the design.

Two classical control system analysis techniques will be discussed here: the root locus and the Bode methods. These techniques were developed in the 1930's and 1940's and are still in common use for control system design. Since the 1940's, modern control techniques which are based on state-space formulations of the system, have been developed so that the most complex multivariable system can be analyzed. And powerful mathematical estimation and optimization techniques can be applied to the control system design.

7.2   Types of control systems

The broadest classification of control system types divides them into open-loop and closed-loop. The primary difference is that in open-loop control, the output has no influence on the control and in closed-loop control, the output does influence the control. An example of open-loop control is a dishwasher which cycles from one operation to the next based strictly on a preset timing plan. At no point in the operation does the output of a dishwasher i.e., the cleanliness of the dishes, have any influence on the control of the machine, until the cycle is complete.

Then, the operator of the dishwasher inspects the dishes and if they are not clean, he starts the machine again. At that point the dishwasher becomes part of a closed-loop control system. The other part, and the most important part, of the closed-loop system is the operator. In fact, the operator actually is responsible for closing the loop by inspecting the dishes, deciding they are clean enough or require further washing, and starting the dishwasher through another cycle.

Open-loop control is only effective when the necessary response of the control system is known or the required performance is not critical. In the dishwashing example, the length of the cycles has been determined experimentally to produce a clean dish if given the average dirty dish. The precision or degree of cleanliness is not important to the operator. He only becomes concerned if the dishes are not clean after he loaded what he considered to be a "dirty" dish.

In a closed-loop control system the output of the system does influence the control and, therefore, the feedback of the output to the controlling device is implied. This concept is illustrated in the block diagram in Fig. 7-1. Fig. 7-1 also defines the elements of the feedback control system to be used in the remainder of this book.

For the dynamic positioning control system, the plant is the thruster system and the vessel, the measuring devices are the sensors, and the controller is the controller discussed in Chapter 5. The disturbance and noise shown in Fig. 7-1 acting on the plant and the measuring

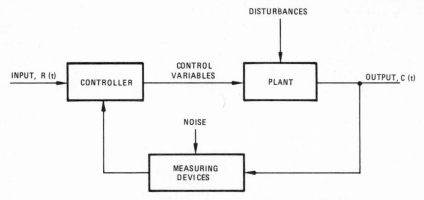

Figure 7-1.   The Basic Feedback Control System

devices, respectively, are unpredictable and unwanted system inputs which tend to reduce the accuracy of the control system.

The input shown at the left in Fig. 7–1 is the desired or commanded value of the output that the control system is attempting to achieve. By comparing the desired output with the measured output of the plant, the error in the control system can be computed. Since the error is a direct measure of the performance of the control system in achieving the commanded output, the controller uses the error signal to compute a command for some element of the plant which reduces the error signal.

The element of the plant which is commanded by the controller can be considered a part of the controller in some systems. However, for a dynamic positioning system the element is the thruster which is better considered as a part of the plant.

The computational processing of the error signal to obtain the controlling command is referred to as the control law. There are several standard control laws that are mentioned later. The element of the plant commanded by the controller has a certain amount of capacity to reduce the error which is referred to as its control authority. In the case of the dynamic positioning system, the element in the plant that reduces the error is the thruster system.

One concern that is inherent in a closed-loop system which does not exist in an open-loop system is stability. Stability relates to the response of the system to a disturbance or a change in the input command. If the system is stable, its response to a transient disturbance is bounded.

For a real control system, stability implies an even stronger condition of the transient response of the system decaying back to its original value prior to the occurrence of the next transient. Instability is, of course, a transient response which grows without bound. However, in most physical control systems some nonlinearity in the system limits the response, causing the control system to hold at some constant output or cycle between limits (limit cycle).

This nonlinear limiting action bounds the response of the control system which fulfills the first definition of stability. However, to the user of the system, the system response is totally unacceptable, especially if the amplitude of the response is large. Likewise, according to the first definition of stability, a system which oscillates at a constant amplitude is stable.

Again, the user will not agree unless he needs an oscillator. Thus, the degree of stability becomes an important practical aspect of the control system design. More discussion regarding stability and the degree of stability is given later.

There are many different types of closed-loop or feedback control systems, each requiring its own mathematical representation. In fact, the classification of the system type can be made on the basis of the mathematics required to represent and analyze the system.

The simplest and most straightforward feedback control system is the unity feedback, single-input, single-output, linear, time-invariant, continuous-data system. Because the system is linear and time-invariant, it can be described or modeled with linear differential equations with constant coefficients. A more convenient representation of this system is the block diagram where each block is the transfer function of a system element expressed in Laplace transform notation.

The block diagram is especially convenient from the standpoint of the control system designer because it not only gives him the mathematical models of each system element but also shows him the functional interrelation or "signal flow" among the elements of the system in a graphical form.

An example of a block diagram representing the nonunity feedback control system is shown in Fig. 7–2. Each block shown in Fig. 7–2 is a transfer function expressed in Laplace transform notation.

The conversion of a linear differential equation with constant coefficients to a transfer function in Laplace transform notation is

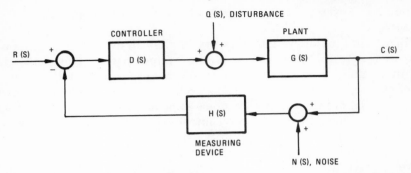

Figure 7-2.   The Single Input - Single Output Linear Continuous
Feedback System

straightforward. Given a nth-order linear differential equation with
a nth-order forcing function or input as follows:

$$\frac{d^n x}{dt^n} + a_1 \frac{d^{n-1} x}{dt^{n-1}} + \cdots + a_{n-1} \frac{dx}{dt} + a_n x = b_0 \frac{d^m y}{dt^m} + b_1 \frac{d^{m-1} y}{dt^{m-1}} \quad (7.1)$$

$$+ \cdots + b_{m-1} \frac{dy}{dt} + b_m y$$

where the coefficients a and b are time-invariant, x is the output and
y is the input. The conversion is performed by first expressing Equation 7.1 in Laplace transform notation with zero initial conditions.
Next, the ratio of the output to the input is formed to yield the following transfer function:

$$\frac{X(s)}{Y(s)} = \text{Transfer function} = \frac{b_0 s^m + \cdots + b_{m-1} s + b_m}{s^n + a_1 s^{n-1} + \cdots + a_{n-1} s + a_n} \quad (7.2)$$

where s is the Laplace operator.

In the analysis techniques described later, the singularities of
numerator and denominator of Equation 7.2 are important. The former
singularities are called zeros and the latter are called poles. They
are determined by finding the roots of the numerator and the denominator polynomials, respectively.

In its factored form Equation 7.2 is composed of the basic elements
given in Table 7-1. These elements have commonly referred-to names
which describe their effect on the input signal. These names are included as a part of Table 7-1 and are further discussed in the last

TABLE 7-1
BASIC LINEAR TIME-INVARIANT, CONTINUOUS-
DATA TRANSFORMS

Transfer function	Common name
1. K	Gain
2. $\dfrac{1}{\tau_i s}$	Integration
3. $\tau_d s$	Differentiation
4. $\dfrac{1}{\tau s + 1}$	First-order lag
5. $\tau s + 1$	First-order lead
6. $\dfrac{1}{\dfrac{s^2}{\omega^2} + \dfrac{2\zeta s}{\omega} + 1}$	Second-order lag
7. $\dfrac{s^2}{\omega^2} + \dfrac{2\zeta s}{\omega} + 1$	Second-order lead
8. $e^{-\tau s}$	Delay or transport lag

Note: All τ's have units of time and are constant.
All ζ's are constant, are referred to as the
damping factor, and are unitless.
All ω's are constant, are referred to as the
natural frequency, and have units of radians per second.

section of this chapter where their characteristics as a function of
frequency are examined.

Another reason for the use of block diagrams in the analysis of
linear feedback control systems is the ease with which a complicated
system with multiple feedback loops can be reduced to a system with a
single feedback loop. For example, in Fig. 7-3(a), the forward path of
the control system is composed of series elements in parallel with another system component, feeding forward a signal to an internal unity
feedback loop.

In Fig. 7-3(b) the series transfer functions and the unity feedback loop have been reduced to a single element. The next reduction
shown in Fig. 7-3(c), combines the parallel paths.

Finally, in Fig. 7-3(d), the two series elements in the forward loop
are combined by multiplication. If the transfer function of the total
system is desired, a final combination of the forward and feedback
blocks can be performed with the following equation:

$$\frac{C(s)}{R(s)} = \frac{G(s)}{1 + G(s)\ H(s)} \qquad (7.3)$$

However, in the analysis techniques to be described at the end of this chapter, only the product of the forward and feedback blocks, i.e., G(s) H(s), is used. This product is commonly referred to as the open-loop transfer function of the system[1]. The example given in Fig. 7-3 is not a complex system where inner loops and feed forward paths are interwoven. The rules for simplifying block diagrams and more complex examples are found in References 1 through 5 or nearly any basic feedback control theory publication.

Another topic of interest in control system analysis is the effect of disturbances, such as shown in Fig. 7-2, on control system performance. Equation 7.3 can be used to derive the system transfer function which operates on the disturbance. For example, in the system shown in Fig. 7-2, the closed-loop transfer functions which operate on the disturbance and the noise are given as follows by using equivalent terms in Equation 7.3:

$$\frac{C(s)}{Q(s)} = \frac{G(s)}{1 + G(s)\ D(s)\ H(s)} \qquad (7.4)$$

$$\frac{C(s)}{N(s)} = \frac{G(s)\ D(s)\ H(s)}{1 + G(s)\ D(s)\ H(s)} \qquad (7.5)$$

As can well be imagined by the difference in forms of the two transfer functions, the output resulting from the noise and the disturbance will be different even if the noise and the disturbance are equal. For example, when $G(s) \gg 1$, $D(s) \gg 1$, and $H(s) = 1$, the disturbance, $Q(s)$, will be divided by a large number or attenuated. Whereas the noise, $N(s)$, is not affected significantly because the system transfer function is approximately equal to unity.

Another type of linear feedback control system, and one which has become very prevalent, is the discrete-data system. In the discrete-data system, the signal or data in some part of the loop occurs at discrete instances of time. Two examples of devices which produce data at discrete times were discussed in previous chapters.

The first example is the acoustic position reference system discussed in Chapter 2. In this system, the measurement of position occurs each time an acoustic pulse is received at the hydrophones and transmitted to the position computer. Thus, the position measurements

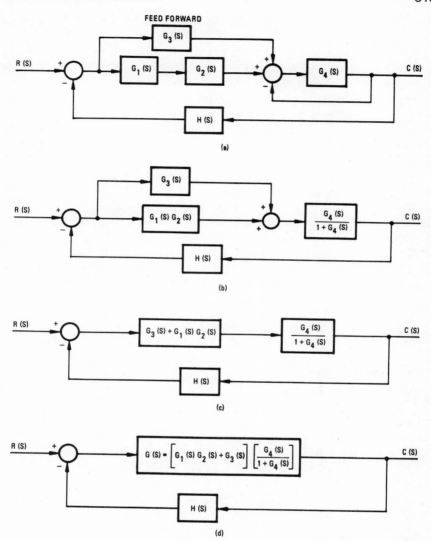

Figure 7-3. An Example of the Reduction of a Multilooped Block Diagram to a Single Loop

from this system appear at discrete instances of time or as a pulse train, as shown in Fig. 7–4.

Fig. 7–4 also illustrates that the discrete position measurements or position samples are delayed slightly from the actual position function. This delay represents the time required to process the acoustic pulse from the hydrophone and compute the position from the information in the acoustic pulse.

Not clearly shown in Fig. 7–4 is that the position samples from the acoustic position reference system occur in a nonperiodic pulse train. The lack of periodicity is because the hydrophone is moving with respect to the subsea acoustic source which results in a constantly changing distance between the two acoustic devices.

The second example of a discrete-time device is the digital computer used to implement the dynamic positioning controller in Chapter 5. The digital computer is a discrete-time device by its very nature. As it executes its software program, the computer reads all the inputs, converts them to digital numbers, and stores them in memory. Then the computer performs the preset computations prescribed by its software

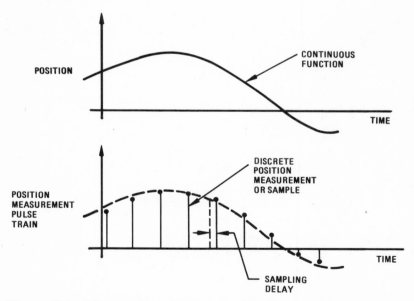

Figure 7-4. Discrete Data From an Acoustic Position Reference
System

program and outputs a short time later the preprogrammed output parameters.

During the time that the computations are being performed, the inputs to the computer can be changing, but until the computer returns to the point in its software program to read the inputs again, no new input data is used. As a result, the digital computer uses its input data in a sampled format. Also, the sampling by the computer is usually periodic which results in an equal length of time between samples.

A similar phenomena of an output pulse train results as the digital computer executes its software program. However, an output pulse train is generally unacceptable because the output of the computer is normally connected to a continuous data device like a thruster. As a result, the converter on the output of the digital computer not only converts the digital number from the computer to an analog voltage, but also holds the voltage constant until the computer outputs another number for conversion. The holding action of the output digital-to-analog converter is illustrated in Fig. 7–5.

As Fig. 7–5 shows, there is an error in holding a constant value from one output sample to the next from the computer. A more exact reconstruction of the output function from the computer could be ac-

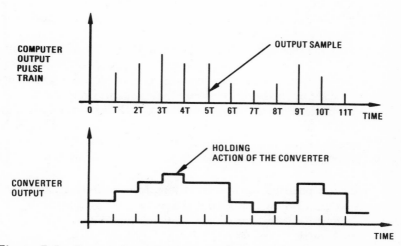

Figure 7-5. Conversion of Output Data From the Computer to an Analog Signal

hieved by a higher-order function such as a linear or polynomial fit between samples. However, the advantage in curve fitting may be lost in the gain and phase characteristics introduced by the sample-hold network.

Likewise, the zero-order hold (constant holding between samples) is easier to implement. As a result, most digital-to-analog converters use only a zero-order hold network.

Since a feedback control system with a digital computer is a sampled-data system, the discussion regarding the control system analysis is very important to the present day dynamic positioning system. Therefore, in the next section a more detailed account of the mathematical representation of a sampled-data control system is given.

Another type of linear feedback control system is one in which the coefficients of its representative differential equation are time-varying, as shown by the following equation:

$$\frac{d^n x}{dt^n} + a_1(t)\frac{d^{n-1}x}{dt^{n-1}} + a_2(t)\frac{d^{n-2}x}{dt^{n-2}} + \cdots + a_{n-1}(t)\frac{dx}{dt} + a_n(t)x = \quad (7.6)$$

$$b_0(t)\frac{d^m y}{dt^m} + b_1(t)\frac{d^{m-1}y}{dt^{m-1}} + \cdots + b_{m-1}(t)\frac{dy}{dt} + b_m(t)y$$

Although the law of superposition is as valid for the linear time-varying system as it is for the linear time-invariant system, there is no general method of solution to its time-varying differential equation. There are solutions for certain special cases, but generally, a computer simulation of the system is required to evaluate the total performance of the system.

Also, some systems with time-varying coefficients are studied at various instances in time where at the instances in time the equation is treated as time-invariant. One such system is an intercontinental ballistic missile control system where not only the forces acting on the missile change as the missile rises through the atmosphere, but also the mass properties and thrust efficiency change as a function of flight time.

In a dynamic positioning system there are changes in the coefficients of the equations which model the system as a function of time. However, generally, their rates of change are sufficiently small and the magnitude of the changes are within a tolerable range to consider the system time-invariant at least on the short-term. One parameter which does change sufficiently for certain vessels to require changing the coefficients of the modeling equations for those vessels is the draft.

If the changes are infrequent, the operator of the dynamic positioning system can make any necessary adjustments to the control laws in the controller to compensate for the changes in the vessel's model equations. If the changes in draft are more frequent, the draft can be measured and input into the controller to automatically compensate the control laws. The reason that compensation is necessary will be clearer after the section on control system analysis.

Another practical aspect of any control system is that it is a linear time-invariant system only approximately, or only over a limited range of the input variables to each element in the control system. In reality, all control systems contain nonlinear elements which makes the overall system nonlinear. Fortunately in many cases, the mathematical representations of the nonlinearities can be approximated by a linear function over some range of operating conditions. Then, the standard methods of analysis can be applied for each region of linearization.

When linearization is not possible or an evaluation of the performance of the system in a global sense is required, then the methods available to the designer are limited. A few nonlinear differential equations have known closed-form solutions. Describing function and phase plane analysis methods can be used for certain nonlinear systems. However, the use of computers to solve the nonlinear equations is by far the most powerful tool available to the designer. This is especially true with modern high-speed digital and hybrid computers. (A hybrid computer is a combination of a digital and analog computer.)

Even though dynamic positioning systems contain nonlinearities, the analysis of nonlinear control systems is not covered directly in this book.[4][6] Instead, one method of analyzing a dynamic positioning control system with its nonlinearities is presented in the chapter on models, simulation, and the basic design process. One step of the method is the approximation of the nonlinear characteristics of a control system component with a linear function.

If the characteristics of the control system component are defined by a mathematical function between its input and output (a transfer function) and the mathematical function can be expressed as a Taylor series in the neighborhood of the operating points of the system, then a linearization of the nonlinear function in the neighborhood of various operating points is possible as follows:

x = output of the component = f(input) = f(y)

$$= f(y_o) + \frac{df}{dy}\bigg|_{y=y_o} (y - y_o) + \frac{d^2f}{dy^2}\bigg|_{y=y_o} (y - y_o)^2 + \cdots \quad (7.7)$$

where y_o is the operating point around which the expansion is made.

Over some range of the input variable around the operating point where the expansion is made, the higher-order terms can be neglected which reduces Equation 7.7 to the following linear equation:

$$x = x_0 + K(y - y_0) \quad (7.8)$$

where:

$$x_0 = f(y_0)$$

$$K = \frac{df}{dy}\bigg|_{y=y_0}$$

The range of the input variable for which the function can be approximated by a linear equation depends to a great extent on the behavior of the higher-order derivatives of the function which define the curvature of the function around the operating point. In fact, for many control system components the higher-order derivatives are small or even negligible over nearly the entire range of values of the input to the nonlinear component.

Then, the component can be replaced by a single linear approximation or, at most, a small number of linear approximations. A common example of this degree of approximation is for the gain of an electronic amplifier whose output characteristics as a function of the input (the gain) and the resulting linear approximations are as shown in Fig. 7–6.

As Fig. 7–6 shows, the gain of the amplifier has three regions of operation. In the proportional region the output of the amplifier varies nearly linearly with the input. Only at the boundaries between the regions does the actual transfer function exhibit significant curvature.

For many amplifiers the transition region is sufficiently small to make the indicated three-segment piecewise linear approximation an excellent representation of the actual gain of the amplifier. This is because the approximation bounds the range of variation in the gain of the amplifier as it transitions from the proportional region (the normal region of operation) to the saturation region. Thus, as the

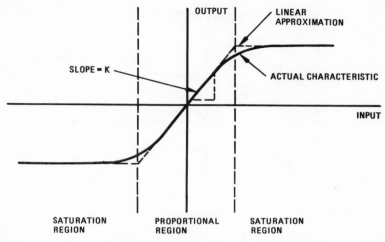

Figure 7-6.   Linearization of the Transfer Function of an
Electronic Amplifier

analysis is performed in the proportional region and the saturation region, the gain of the amplifier is never greater or less than the two values of gain given by the linear approximation.

If the amplifier gain or some other control device exhibits more regions of operation where the slope of the transfer function curve is sufficiently different from that represented by a three-segment linear approximation, then additional linear segments are necessary, such as shown in Fig. 7–7.

No matter how many linear segments are required to represent the transfer function of a nonlinear device, the nonlinearity is still classified as simple so long as the function has only one output value for every input value, i.e., a single-valued function. Even if the nonlinear functions contain discontinuities, they are still classified as simple. Thus, all the nonlinear characteristics shown in Fig. 7–8 are classified as simple.

If a nonlinearity is multivalued, then it is classified as complex. Several common examples of multivalued nonlinear characteristics are illustrated in Fig. 7–9. As can be imagined, the analysis of a system with a multivalued nonlinearity is even more difficult than for one with a single-valued nonlinearity. However, describing function techniques and phase plane analysis techniques can be used. Refer-

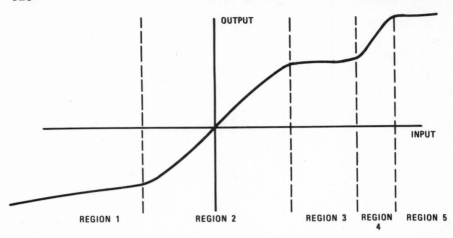

Figure 7-7. A Transfer Function of a Nonlinear Device with Multiple Regions of Operation

ence 6 contains an extensive table of describing functions for 23 common nonlinearities.

Nonlinearities can also be multivariable as well as multivalued. For example, for a dc motor the torque output is a nonlinear function of the armature current and the field strength. Thus, if both variables are controlled, a more complex nonlinear analysis results. For the particular example of the dc motor and for most control devices, one of the variables is held constant at some value. As a result, the device performs as a single variable nonlinearity. Also, the value selected for the variable which is held constant is one which gives the device the most useful performance from the standpoint of that variable.

In control systems certain mathematical operations are nonlinear and multivariable. These operations include multiplication of two variables, vector resolution, and coordinate transformation. Most often these nonlinearities are a part of the controller.

To this point in the discussion, the linear control system with one input and one output has been the dominant type of control system. The discussion of time-varying and nonlinear systems indicated that a few special classical analysis techniques exist, but the preferred approach was to approximate the complex system with a suitable linear representation. Fortunately, the classical approach is an effective method of analysis and design for many control systems.

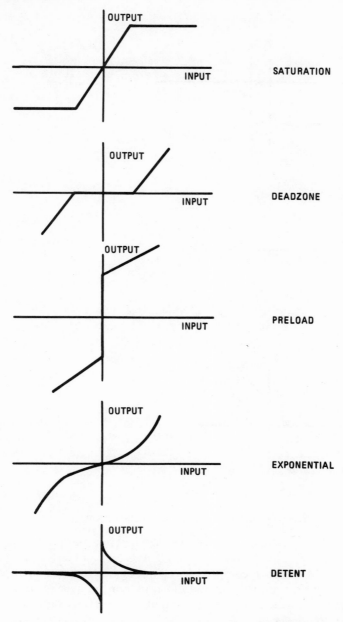

Figure 7-8. Five Common Simple Nonlinearities

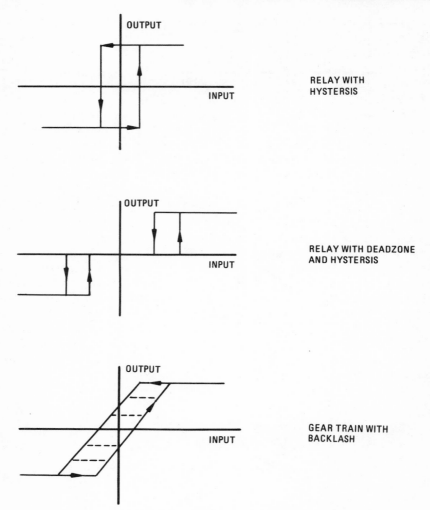

Figure 7-9.   Common Multivalued Nonlinear Characteristics

In fact, the classical approach can be and has been used for systems with multiple inputs and outputs. However, as control systems became more and more complex in terms of the number of inputs and outputs, time-varying components, and nonlinearities, the classical approaches became less and less capable of furnishing proper system analysis and designs.

As a result, complex computer simulations became more and more the only analysis and design tools. These simulations are not only costly to create and operate but also very time consuming to achieve a proven performance to wide variations in inputs and disturbances. As a result, a more powerful theory naturally started to develop to fulfill the demand.

The new theory, based on a very general mathematical formulation of a system, developed from existing mathematical techniques used in generalized mechanical systems, matrices, and optimization. A complete coverage of the topic of modern control is beyond the scope of this book. But a brief introduction into the representation of a system with the modern approach is possible.

The example used to illustrate the modern approach is also shown in the classical manner. For more details regarding the modern approach, see References 1, 3, and 7.

The mathematical formulation of a control system using the modern approach uses the concept of system state. The state of a system is no more than the minimum number of variables that must be known at some instant in time, $t = t_0$, which, when combined with the knowledge of the input for $t \geq t_0$, completely determine the behavior of the system for all $t \geq t_0$.[3] The variables are referred to as state variables.

In the most general case, a physical system can be represented by a set of nth first-order nonlinear ordinary differential equations expressed in the following form:

$$\dot{\bar{x}} = \begin{bmatrix} \dfrac{dx_1}{dt} \\ \dfrac{dx_i}{dt} \\ \dfrac{dx_n}{dt} \end{bmatrix} = \bar{f}(\bar{x}, \bar{u}, t) = \begin{bmatrix} f_1(\bar{x}, \bar{u}, t) \\ f_i(\bar{x}, \bar{u}, t) \\ f_n(\bar{x}, \bar{u}, t) \end{bmatrix} \qquad (7.9)$$

$$\bar{y} = \bar{g}(\bar{x}, \bar{u}, \bar{t})$$

where:

x_i = ith state variable
y_i = ith output variable
u_i = ith input variable
t = time
f_i = ith nonlinear function
g_i = ith nonlinear function

Modern control theory provides no general solution to Equation 7.9. However, it does provide powerful analytic techniques for linear systems, both time-invariant and time-varying. Linear systems result either from linear approximations of the nonlinear equations using series expansions (as described earlier for the classical approach) or from other formulations which result in linear equations.

For a system which is described by linear ordinary differential equations, the state equation formulation of the system is as follows:

$$\dot{\bar{x}}(t) = A(t)\bar{x}(t) + B(t)\bar{u}(t) \tag{7.10}$$
$$\bar{y}(t) = C(t)\bar{x}(t) + D(t)\bar{u}(t) \tag{7.11}$$

where:

$$\bar{x} = \text{nth order state vector}$$
$$\bar{y} = \text{pth order output vector}$$
$$\bar{u} = \text{mth order input vector}$$
$$A(t) = \text{n} \times \text{n matrix}$$
$$B(t) = \text{n} \times \text{m matrix}$$
$$C(t) = \text{p} \times \text{n matrix}$$
$$D(t) = \text{p} \times \text{m matrix}$$

The matrices, A, B, C, and D, given in Equations 7.10 and 7.11 are time-varying. In the case where these matrices are time-invariant, the state equations are as follows:

$$\dot{\bar{x}}(t) = A\bar{x}(t) + B\bar{u}(t) \tag{7.12}$$
$$\bar{y}(t) = C\bar{x}(t) + D\bar{u}(t) \tag{7.13}$$

The state equations can be represented in block diagram form as shown in Fig. 7–10 where the broad arrows connecting the blocks represent the multidimensional quantities of the indicated variables.

Since Equations 7.12 and 7.13 are time-invariant, linear ordinary differential equations, they can be expressed in Laplace transform notation in the following manner:

$$\bar{X}(s) = (sI - A)^{-1}B\bar{U}(s) + (sI - A)^{-1}\bar{x}(0) \tag{7.14}$$
$$\bar{Y}(s) = C\bar{x}(s) + D\bar{U}(s) \tag{7.15}$$

From Equations 7.14 and 7.15 the transfer function of the overall system can be derived by substituting Equation 7.14 into Equation 7.15 and setting the initial conditions to zero. The result of the substitution is:

$$\bar{Y}(s) = [C(sI - A)^{-1}B + D]\bar{U}(s) \tag{7.16}$$

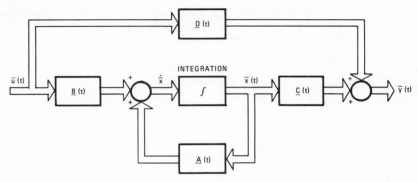

Figure 7-10. The Block Diagram Representation of the Generalized Linear State Variable System Formulation

where I is the unity matrix and $(\quad)^{-1}$ indicates the inverse of the matrix within the brackets.

The resulting multiple input-multiple output transfer function of the system is given as follows with $Y(s)$ as the output and $U(s)$ the input:

$$G(s) = C(sI - A)^{-1}B + D \qquad (7.17)$$

In order to illustrate the use of the modern approach and a comparison of the modern approach to the classical approach, the single input-single output feedback control system shown in Fig. 7-11 shall be reconfigured into state variable notation. The overall transfer function for the system shown in Fig. 7-11 is:

$$\frac{Y(s)}{U(s)} = \frac{K_1K_2 (s + b)}{s^3 + (a + c) s^2 + (ac + K_1K_2) s + bK_1K_2} \qquad (7.18)$$

which in differential equation form is as follows:

$$\frac{d^3y}{dt^3} + (a + c) \frac{d^2y}{dt^2} + (ac + K_1K_2) \frac{dy}{dt} + bK_1K_2\, y = K_1K_2 \frac{du}{dt} + K_1K_2bu$$

$$(7.19)$$

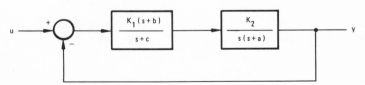

Figure 7-11. An Example Linear Feedback Control System

There is more than one choice of the set of state variables. One such choice which does ensure a proper solution of the state equation is as follows:

$$x_1 = y - \beta_0 u$$
$$x_2 = \dot{y} - \beta_0 \dot{u} - \beta_1 u = \dot{x}_1 - \beta_1 u \qquad (7.20)$$
$$x_3 = \ddot{y} - \beta_0 \ddot{u} - \beta_1 \dot{u} - \beta_2 u = \dot{x}_2 - \beta_2 u$$

where the β's are expressed as follows:

$$\beta_0 = b_0$$
$$\beta_1 = b_1 - a_1\beta_0$$
$$\beta_2 = b_2 - a_1\beta_1 - a_2\beta_0 \qquad (7.21)$$
$$\beta_3 = b_3 - a_1\beta_2 - a_2\beta_1 - a_3\beta_0$$

The a_i's and b_i's in Equation 7.21 are defined by the generalized form of the original overall system transfer function or differential equation given as follows:

$$\frac{d^n y}{dt^n} + a_1 \frac{d^{n-1}y}{dt^{n-1}} + \cdots + a_{n-1}\frac{dy}{dt} + a_n y =$$

$$b_0 \frac{d^m u}{dt^m} + b_1 \frac{d^{m-1}u}{dt^{m-1}} + \cdots + b_{m-1}\frac{du}{dt} + b_m u \qquad (7.22)$$

With this definition of the a's and b's, the β's given in Equation 7.21 are as follows for the example problem:

$$\beta_0 = 0$$
$$\beta_1 = 0$$
$$\beta_2 = K_1 K_2 \qquad (7.23)$$
$$\beta_3 = K_1 K_2 b - (a + c)K_1 K_2$$

Then, the state equations for the example problem become:

$$\begin{bmatrix} \dot{x}_1 \\ \dot{x}_2 \\ \dot{x}_3 \end{bmatrix} = \begin{bmatrix} 0 & 1 & 0 \\ 0 & 0 & 1 \\ -bK_1K_2 & -(ac + K_1K_2) & -(a+c) \end{bmatrix}\begin{bmatrix} x_1 \\ x_2 \\ x_3 \end{bmatrix} + \begin{bmatrix} 0 \\ K_1K_2 \\ K_1K_2(b-a-c) \end{bmatrix} u$$

$$(7.24)$$

$$y = \begin{bmatrix} 1 & 0 & 0 \end{bmatrix}\begin{bmatrix} x_1 \\ x_2 \\ x_3 \end{bmatrix} \qquad (7.25)$$

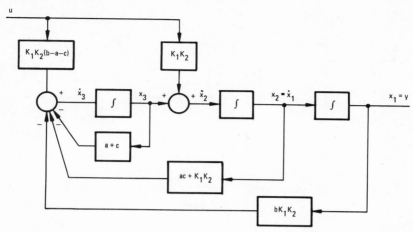

Figure 7-12. The State Equation Representation of the Example System in Figure 7-11

These state equations are shown in Fig. 7–12 in block diagram form. A comparison with Fig. 7–11 shows a significant difference in the two representations; however, the two systems are equivalent from input to output.

7.3 Sampled-data control systems

In this section, linear time-invariant sample-data feedback control systems are considered. The element of the system which makes it a sampled-data system is a sampler. The sampler converts continuous data into a train of very narrow pulses whose amplitudes are equal to the sampled function at the instants of sampling. The sampler may be located at different points in the feedback system.

Another element of most sampled-data systems is a holding device which converts the pulsed data into a form to drive control devices which require a continuous-data input to operate. A very common example of such a sampled-data system and the primary example to be used in this section is the control system which contains a digital computer.

A typical digital computer-based control system is shown in Fig. 7–13. The sampler in Fig. 7–13 is the analog-to-digital (A/D) converter and the holding device is the digital-to-analog (D/A) converter. Within

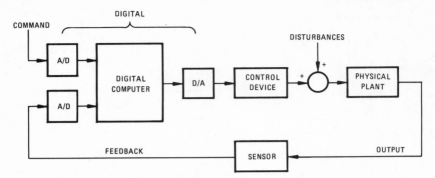

Figure 7-13.   A Digital Computer-Based Feedback Control
System

the computer the data sampled by the A/D converter is processed in
pulsed-data forms, one step at a time as the software program is executed in the computer.

A sampled signal can be formed by modulating a continuous signal
with a train of impulses as represented in Fig. 7–14. The impulse train
can be equally (periodic) or randomly spaced. The simpler of the two
situations to be expressed mathematically is the periodically-spaced
impulse train which is considered in the remainder of this section. In
equation form the periodically sampled signal, $y^*(t)$ is:

$$y^*(t) = \sum_{n=-\infty}^{\infty} y(t)\delta(t - nT_s) \tag{7.26}$$

where:

$\delta(t - nT_s)$ = an impulse at $t = nT_s$
T_s = the sampling interval or period, seconds

The pulse train series given in Equation 7.26 can be expressed as a
complex Fourier series which changes Equation 7.26 to the following:

$$y^*(t) = \frac{1}{T_s} \sum_{n=-\infty}^{\infty} y(t)e^{-jn\omega_s t} \tag{7.27}$$

where $\omega_s = \dfrac{2\pi}{T_s} = 2\pi f_s$. The Laplace transform of Equation 7.27 is
given as follows:

$$Y^*(s) = \frac{1}{T_s} \sum_{n=-\infty}^{\infty} Y(s + jn\omega_s) \tag{7.28}$$

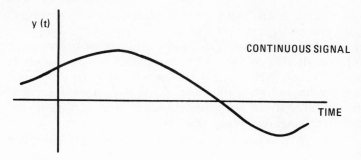

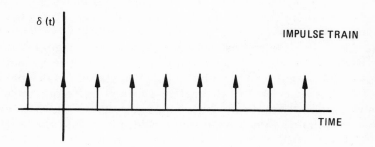

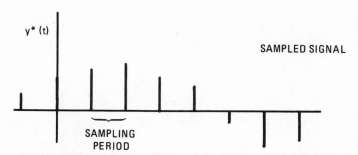

Figure 7-14.   A Sketch of the Sampling Process

Equation 7.28 is an important mathematical relationship from the standpoint of understanding the effect of sampling. In terms of frequency content, Equation 7.28 shows that the sampling process has taken the original spectrum of the sampled signal and spreads it over the entire frequency band (at least theoretically).

This spreading effect is shown in Fig. 7–15 where the original spectrum is compared to the resulting sampled spectra. Equation 7.28 also reveals something regarding the choice of the sampling frequency, ω_s, if the original signal is to be recovered at some point beyond the sampler.

If adjacent members of the series of which the sampled signal is composed have components at the same frequency, then these common components combine to form the sampled signal and separation following the sampler is not generally possible. This effect is illustrated in Fig. 7–16 where the original signal has significant frequency components above $\omega_s/2$.

Thus, for complete reconstruction of the original signal, the sampling frequency must be twice the highest significant frequency component of the signal being sampled. This is a statement of Shannon's sampling theorem. In actual practice a sampling frequency of five times the highest significant frequency component is a better choice.

When a sampled signal is fed into a continuous-data linear system

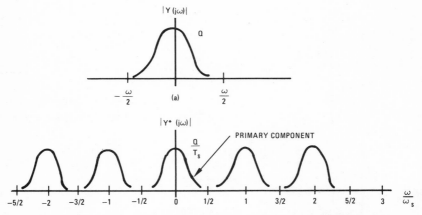

Figure 7-15.   The Frequency Spectra of a Continuous Signal and its Sampled Value

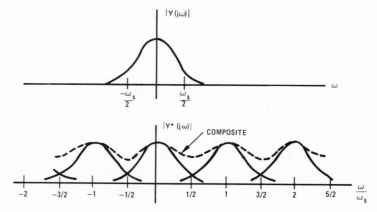

Figure 7-16. The Effect of Sampling at Less Than Twice the Highest Frequency Component

represented by the transfer function, G(s), the output of the system becomes the following product:[2]

$$X(s) = G(s) \; Y^*(s) \tag{7.29}$$

where X(s) is the Laplace transform of the output of the system. If the output of the system is sampled, then Equation 7.29 is:

$$X^*(s) = G^*(s) \; Y^*(s) \tag{7.30}$$

and the sampled data transfer function of this system is:

$$G^*(s) = \frac{1}{T_s} \sum_{n=-\infty}^{\infty} G(s + jn\omega_s) \tag{7.31}$$

If two transfer functions are not separated by a sampler, then the sampled-data transfer function for the overall combination of the two transfer functions is as follows:

$$G_1 G_2 {}^*(s) = \frac{1}{T_s} \sum_{n=-\infty}^{\infty} G_1(s + jn\omega_s) \; G_2(s + jn\omega_s) \tag{7.32}$$

or

$$G_1 G_2 {}^*(s) = \frac{1}{T_s} \sum_{n=-\infty}^{\infty} G_1 G_2(s + jn\omega_s) \tag{7.33}$$

This is not the same as if two transfer functions are separated by a sampler which is as follows:

$$G_1^*(s)\, G_2^*(s) = \frac{1}{T_s^2} \sum_{n=-\infty}^{\infty} G_1(s + jn\omega_s) \sum_{m=-\infty}^{\infty} G_2(s + jm\omega_s) \quad (7.34)$$

Classical methods of control system analysis rely on a plot of the open-loop transfer function as a function of frequency. The open-loop transfer function of a control system is the product of the various transfer functions around the entire loop.

Thus, for a sampled-data control system (Fig. 7–17) the open-loop transfer function is GH*(s). Unlike the nonsampled case, the sampled open-loop transfer function has an infinite number of components expressed as follows:

$$GH^*(j\omega) = \frac{1}{T_s} \sum_{n=-\infty}^{\infty} GH(j\omega + jn\omega_s)$$

$$= \frac{1}{T_s} [GH(j\omega) + GH(j\omega + j\omega_s) + GH(j\omega + j2\omega_s) \quad (7.35)$$
$$+ \cdots + GH(j\omega - j\omega_s) + GH(j\omega - j2\omega_s) + \cdots]$$

Each term of the series for a particular value of frequency is a vector of a given magnitude and direction (phase). For most physical systems the magnitude of vectors decrease as frequency increases. As a result, if the sampling frequency, ω_s is high in comparison to the most significant frequency components of the system or the bandwidth of the system, the dominant component of the series is the first term, GH($j\omega$). In fact very often, the frequency response of the sampled-data system can be approximated by no more than the following terms of the series:

$$GH^*(j\omega) \cong \frac{1}{T_s} GH(j\omega) + \frac{1}{T_s} GH(j\omega + j\omega_s) + \frac{1}{T_s} GH(j\omega - j\omega_s) \quad (7.36)$$

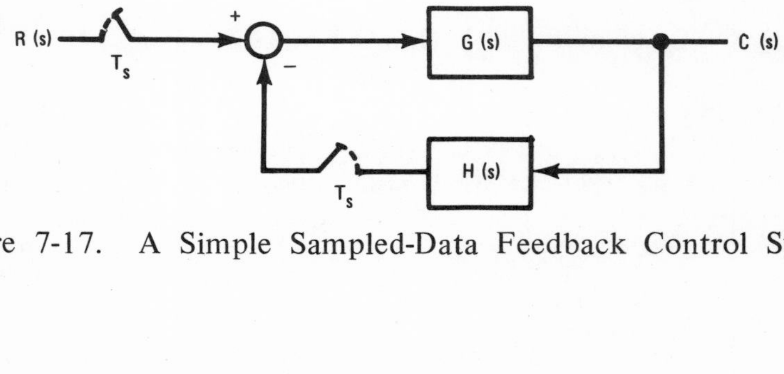

Figure 7-17.   A Simple Sampled-Data Feedback Control System

With only three terms of the series to compute and plot, the calculation of the open-loop frequency response of a sampled-data control system is not totally unreasonable as is the calculation of an infinite series. There are frequently cases where only the first term of the series is significant, at least in the critical regions of the frequency response. Then, the calculation and plotting of the open-loop frequency response of the system is no more difficult than that required for the continuous-data system.

The other element of the digital computer-based sampled-data control system is the holding device on the output of the computer. The output of the digital computer is a train of command pulses without the holding device which, if applied to a control device like a thruster, results in unsatisfactory performance. So the holding device is used to smooth between pulses so that the commands from the computer are approximate continuous functions.

The most common holding device used for digital computers is one which holds its output constant at a value equal to input pulse until the next input pulse arrives. As a result, the output of the holding device, called a zero-order hold, is a staircase function whose amplitude is equal to the input pulses, as shown in Fig. 7–5. The transfer function of a zero-order hold is given as follows[2]:

$$G_{ho}(s) = \frac{1 - e^{-T_s s}}{s} \tag{7.37}$$

which, when expressed as a function of frequency, has the following magnitude (gain) and direction (phase):

$$\text{Phase} = \phi_{ho}(\omega) \doteq \frac{-\omega T_s}{2} \tag{7.38}$$

$$\text{Gain} = |G_{ho}(\omega)| = \frac{T_s \left| \sin\left(\frac{\pi\omega}{\omega_s}\right) \right|}{\frac{\pi\omega}{\omega_s}} \tag{7.39}$$

Plots of the gain and phase functions for the zero-order sample hold, i.e., Equations 7.38 and 7.39, are given in Fig. 7–18. If the sample hold is placed in series with the sampler illustrated in Fig. 7–15, then the action of the sample hold can be seen to be a low-pass filter which transmits the primary component and attentuates or blocks the higher frequency spectra of the sampled function.

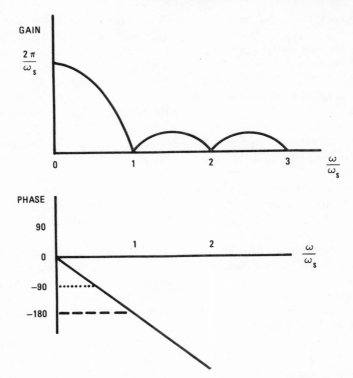

Figure 7-18.　The Gain and Phase Characteristics of the
Zero-Order Sample Hold Circuit

The more widely spaced the sampled spectra are, i.e., the higher the sampling frequency, the better the zero-order sample-hold device does in restoring the original input into the sampler.

Higher-order holding devices are possible. However, they generally introduce more negative phase shift than the zero-order devices. This is undesirable from the standpoint of the design of the dynamic performance properties of the system.

With the inclusion of the sampler and the sample hold into a feedback control system a more complete block diagram of a digital computer-based system can be given. Fig. 7–19 is such a block diagram, showing the digital computer enclosed by the dashed lines. In fact, the system illustrated in Fig. 7–19 is a good approximation to one of the control loops of a dynamic positioning system without active wind

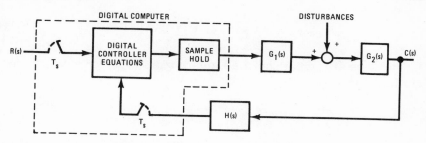

Figure 7-19. A Digital Computer-Based Feedback Control System

compensation. The equivalent parts of the dynamic positioning system are:

- $G_2(s)$ represents the vessel
- $G_1(s)$ represents the thruster system
- $H(s)$ represents the sensor in the feedback path
- $C(s)$ is the output which is either x-position, y-position, or heading
- $R(s)$ is the set-point command for x-position, y-position, or heading

As can be imagined, the control system designer, who is accustomed to linear continuous-data analysis techniques, is not ready to cope with transfer functions which are infinite series. As a result, he favors the approximation of the series given in Equation 7.36. However, this approximation is not valid for all systems.

Furthermore, the equations implemented in the software program of the digital computer are in the form of difference equations, not differential equations. Thus, some other mathematical representation of a discrete (sampled-data) system is needed if the powerful classical analysis techniques developed for linear continuous-data systems are to be applied. For this purpose the z-transform is used.

The z-transform of a sampled-data function is determined from the sampled Laplace transform of the function previously given in Equation 7.28 and given here in an equivalent form:[2]

$$Y^*(s) = \sum_{n=0}^{\infty} y(nT_s)e^{-nT_s s} \qquad (7.40)$$

Then, by defining $z = e^{T_s s}$ the z-transform of the sampled Laplace transform becomes:

$$Y(z) = Y^*\!\left(\frac{1}{T_s}\ln z\right) = \sum_{n=0}^{\infty} y(nT_s)z^{-n} \tag{7.41}$$

For purposes of comparison the Laplace transforms for a continuous-data function are:

$$Y(s) = \int_0^{\infty} y(t)e^{-st}dt \tag{7.42}$$

Further comparisons between Laplace transforms and z-transforms for several time functions are given in Table 7–2.

TABLE 7–2
TABLE OF TRANSFORMS

Time function	Transforms	
	Laplace	Z
$Y(t)$	$Y(s)$	$Y(z)$
$\delta(t)$	1	1
(Impulse)		
$\delta(t - KTs)$	$e^{-KT_s s}$	z^{-K}
$U(t)$	$\dfrac{1}{s}$	$\dfrac{z}{z-1}$
(Step)		
e^{-at}	$\dfrac{1}{s+a}$	$\dfrac{z}{z-e^{-aT_s}}$
$1 - e^{-at}$	$\dfrac{a}{s(s+a)}$	$\dfrac{z(1-e^{-aT_s})}{(z-1)(z-e^{-aT_s})}$
ta^{-at}	$\dfrac{1}{(s+a)^2}$	$\dfrac{T_s z e^{-aT_s}}{(z-e^{-aT_s})^2}$
$e^{+at}\sin \omega_n t$	$\dfrac{\omega_n}{(s+a)^2 + \omega_n^2}$	$\dfrac{z e^{-aT_s}\sin \omega_n T_s}{z^2 - 2z e^{-aT_s}\cos \omega_n T_s + e^{-2aT_s}}$

The z-transform method can also be applied to differences equations so that the control equations in the digital computer can be represented with the same operators as the other parts of the control system. A nth order time-invariant difference equation at the instant of time given by $t = kT_s$ is defined as:

$$y(kT_s + nT_s) + a_1 y(kT_s + (n-1)T_s) + \cdots$$
$$+ a_{n-1} y(kT_s + T_s) + a_n y(kT_s)$$
$$= b_0 u(kT_s + mT_s) + b_1 u(kT_s + (m-1)T_s) + \cdots$$
$$+ b_{m-1} u(kT_s + T_s) + b_m u(kT_s) \qquad (7.43)$$

where:

$y(kT_s)$ = the output of the system at $t = kT_s$
$u(kT_s)$ = the input of the system at $t = kT_s$

and can be represented by z-transforms as follows with the initial conditions set to zero as used previously in transfer functions:

$$(z^n + a_1 z^{n-1} + \cdots + a_{n-1} z + a_n)\, Y(z)$$
$$= (b_0 z^m + b_1 z^{m-1} + \cdots + b_{m-1} z + b_m)\, U(z) \qquad (7.44)$$

or in transfer function form as:

$$G(z) = \frac{Y(z)}{U(z)} = \frac{(b_0 z^m + b_1 z^{m-1} + \cdots + b_{m-1} z + b_m)}{z^n + a_1 z^{n-1} + \cdots + a_{n-1} z + a_n} \qquad (7.45)$$

Once the digital computer is represented by z-transforms, then the entire digital computer-based feedback control system can be modeled using transforms. For example, the control system illustrated in Fig. 7–19 becomes that shown in Fig. 7–20. The overall sampled-data transfer function of the system becomes the combination of pulsed transfer functions expressed in Laplace transforms and z-transforms.

By transforming the pulsed transfer functions expressed in Laplace transforms to z-transforms, the overall transfer function of the total

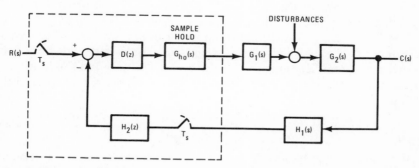

Figure 7-20. A Complete Block Diagram of the Digital Computer-Based Feedback Control System Using Transforms

system can be expressed in z-transforms. In the example system shown in Fig. 7–20, the overall system transfer function in z-transforms is:

$$\frac{C(z)}{R(z)} = \frac{D(z)\ G_{ho}G_1G_2(z)}{1 + D(z)\ H_2(z)\ G_{ho}G_1G_2H_1(z)} \tag{7.46}$$

where:

$$G_{ho}G_1G_2(z) = G_{ho}G_1G_2{}^*\left(\frac{1}{T_s}\ \ln z\right)$$

$$G_{ho}G_1G_2H_1(z) = G_{ho}G_1G_2H_1{}^*\left(\frac{1}{T_s}\ \ln z\right)$$

One of the methods of control system analysis emphasized in the next section, the Bode method, is valid for the Laplace transform representation of a continuous-data control system, but the Bode method cannot be used with z-transforms. Root-locus analysis techniques are valid for the z-transform representation of a control system. However, they are not discussed in this book.[2]

So that the Bode method of control system analysis can be used for sampled-data control systems, the following bilinear transformation is used to convert the z-transform representation of the system:

$$z = \frac{1 + w}{1 - w} \tag{7.47}$$

The resulting transform is referred to as the w-transform.

The Bode method involves plotting the magnitude (gain) and the phase of the open-loop transfer function of the control system as a function of frequency. For the continuous-data system represented by Laplace transforms, the Bode plots of magnitude and phase are made by substituting $j\omega$ for s in the open-loop transfer function.

Similarly, in the w-transform representation of the open-loop transfer function of a sampled-data system a substitution is made for w. The substitution is a fictitious frequency which is related to the frequency used for substitution into the Laplace transform representation in the following manner:

$$w = \frac{z - 1}{z + 1} \tag{7.48}$$

which is equal to the following when $z = e^{T_s s}$ is substituted into Equation 7.48:

$$w = \frac{1 - e^{-T_s s}}{1 + e^{-T_s s}} \tag{7.49}$$

or is equal to the following when $s = j\omega$:

$$w = \frac{1 - e^{-j\omega T_s}}{1 + e^{-j\omega T_s}} = j \tan \frac{\omega T_s}{2} \tag{7.50}$$

This implies that the fictitious frequency used to substitute for w in the Bode method is

$$v = \tan \frac{\omega T_s}{2} \tag{7.51}$$

Discrete system can also be represented in state space form which is very similar to the representation of the continuous-data system. The similarity is quite obvious by comparing the following equations to Equation 7.10 and 7.11 given earlier:

$$\bar{x}\,[(k + 1)\,T_s] = A(kT_s)\,\bar{x}(kT_s) + B(kT_s)\,\bar{u}(kT_s) \tag{7.52}$$

$$\bar{y}\,(kT_s) = C(kT_s)\,\bar{x}(kT_s) + D(kT_s)\,\bar{u}(kT_s) \tag{7.53}$$

where:

$$\bar{x}(kT_s) = \text{the state vector at } t = kT_s$$
$$\bar{u}(kT_s) = \text{the input vector at } t = kT_s$$
$$\bar{y}(kT_s) = \text{the output vector at } t = kT_s$$
$$A(kT_s), B(kT_s), C(kT_s), \text{ and } D(kT_s) = \text{the time-varying matrices at } t = kT_s$$

A block diagram of Equations 7.52 and 7.53 is like Fig. 7–10 except the integration block is replaced by a unit time delay equal to T_s seconds.

For the case of a linear time-invariant system the state equations given by Equations 7.52 and 7.53 simplify to the following:

$$\bar{x}\,[(k + 1)T_s] = A\,\bar{x}(kT_s) + B\,\bar{u}(kT_s) \tag{7.54}$$

$$\bar{y}\,(kT_s) = C\,\bar{x}(kT_s) + D\,\bar{u}(kT_s) \tag{7.55}$$

where A, B, C, and D are constant matrices.

There are many other aspects to sampled-data control systems which will not be covered in this book. Multirate sampling and modified z-transforms have not been mentioned. Complete details on these subjects plus much more on the entire topic of sampled-data control systems are given in Reference 2 for the classical techniques and Reference 7 for the modern techniques.

7.4 Control system performance

The design of a control system begins with a statement of the basic task of the system to be controlled. For example, the basic task of the dynamic positioning system for a drillship is to hold the drillship over the wellhead as the well is drilled. Ideally, the perfect dynamic positioning system would respond instantly to any input command or disturbance, resulting in zero position error.

However, every physically realizable control system contains elements which store energy and, therefore, cannot respond instantaneously. Likewise, control elements are power-limited. Thus, at the beginning the designer must face imperfection and aim for the best performance he can achieve within the physical limitations of the system.

Faced with imperfection, the designer must establish the limit of tolerable imperfection, or he must establish the required system accuracy in performing its basic task. Usually, some aspect of the basic task to be performed by the system will generate an accuracy requirement. In the drillship example the position-holding accuracy requirement is established by the amount of offset from the wellhead that can be tolerated before the drilling equipment can no longer perform efficiently.

Currently, the accepted position-holding accuracy requirement for drillships is a watch circle centered on the wellhead of a radius less than 5 and 6% of water depth for water depths less than 2,000 ft. Below 2,000 ft of water depth the requirement has yet to be definitely established because it depends on the riser system as well as the lower ball-joint angle.

However, generally speaking the accuracy requirement for water depths greater than 2,000 ft will be less than 5–6% of water depth.

A control system exhibits dynamic and static behavior which is a function of the input. Likewise, the control devices which compose the system are power-limited. As a result, the designer must specify the inputs to the system for which the accuracy must be satisfied.

The specification must include not only the time histories or spectral definition of the inputs but also the magnitudes of the inputs, because of the limitation of system power and nonlinearities in the systems. In this section only the linear aspects of the problem are considered within the limits of system power.

In the absence of a specification of inputs by the user or of the definition of physical inputs to the system, such as wind and waves, the designer selects certain inputs upon which to base his design. Even if the inputs are specified, the designer generally uses a well-behaved test input signal in his design.

The most commonly used design inputs are step, ramp, and parabolic functions of time. These three inputs are useful because they represent a sudden change in the input, followed by a steady input, and an acceleration-like input, respectively.

The step function is popular as a design input because it establishes the transient response characteristics of the system to a sudden change in the input. Key transient characteristics are delay time, rise time, peak time, maximum overshoot, and settling time. These are usually defined as shown in Fig. 7–21.

The first two transient characteristics shown in Fig. 7–21 are measures of how fast the system responds to a sudden change in the input. The peak overshoot characteristic defines the maximum dynamic error in the control system in reaching its final output value. In many systems the peak overshoot characteristic also defines the

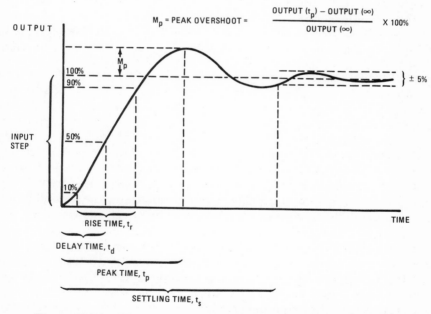

Figure 7-21. Unit-Step Transient Response Characteristics

dynamic error of the control system in reproducing the input step function. The final characteristic, t_s, is a measure of how quickly the system recovers from the sudden input and is ready for another input change.

Not all control systems have a step function response like Fig. 7–21. For example, a system whose transfer function is a simple lag, i.e.,

$$\frac{C(s)}{R(s)} = \frac{1}{\tau s + 1} \tag{7.56}$$

exhibits no overshoot response to a step function input. All systems with higher-order transfer functions can exhibit a step function response with overshoot. However, the response of a higher-order system depends on the parameters of the system.

The step function transient response of a second-order control system described by the following transfer function:

$$\frac{C(s)}{R(s)} = \frac{\omega_n^2}{s^2 + 2\zeta\omega_n s + \omega_n^2} \tag{7.57}$$

does exhibit the transient characteristics shown in Fig. 7–21, depending on the values of the natural frequency, ω_n, and the damping factor, ζ. This can be seen from the mathematical expressions for the peak time and peak overshoot of the system in response to a step function input:

$$\text{peak time, } T_p = \frac{\pi}{\omega_n \sqrt{1 - \zeta^2}} \tag{7.58}$$

$$\text{peak overshoot, } M_p = 100\% \exp\left[\frac{-\zeta\pi}{\sqrt{1 - \zeta^2}}\right] \tag{7.59}$$

For values of the damping factor between 0 and 1, the peak time varies from π/ω_n to infinity and the peak overshoot varies from 100% to 0%. Thus for damping factors greater than 1.0, the second-order system exhibits no overshoot and responds much slower than for damping factors less than unit.

For the damping factor equal to zero, the overshoot is equal to 100% and the peak time is π/ω_n. Further investigation of the time response for the damping factor equal to zero reveals that the response is oscillatory and never settles out, i.e., $t_s = \infty$.

In the case of the nth order linear feedback control system with the following closed-loop transfer function:

$$\frac{C(s)}{R(s)} = \frac{G(s)}{1 + G(s)\,H(s)} \tag{7.60}$$

the G(s) and H(s) are ratios of polynomials expressed as follows:

$$G(s) = \frac{N_1(s)}{D_1(s)} \tag{7.61}$$

$$H(s) = \frac{N_2(s)}{D_2(s)} \tag{7.62}$$

where the order of numerator polynomials is less than or equal to the order of denominator polynomials. When the closed-loop transfer function of the control system is expressed in terms of polynomial ratios given by Equations 7.61 and 7.62, the result is:

$$\frac{C(s)}{R(s)} = \frac{N_1(s)D_2(s)}{D_1(s)D_2(s) + N_1(s)N_2(s)} \tag{7.63}$$

which can be factored into the following forms:

$$\frac{C(s)}{R(s)} = \frac{K(s + z_1)(s + z_2)\,\cdots\,(s + z_m)}{(s + p_1)(s + p_2)\,\cdots\,(s + p_n)} \tag{7.64}$$

where $m \le n$, the z's are referred to as zeros, the p's are referred to as the poles and the poles and zeros are functions of the parameters of the system.

For example, in the second-order system represented by Equation 7.57, there are two poles given as follows and there are no zeros:

$$p_1 = \zeta\omega_n + \omega_n\sqrt{\zeta^2 - 1} \tag{7.65}$$

$$p_2 = \zeta\omega_n - \omega_n\sqrt{\zeta^2 - 1} \tag{7.66}$$

When the damping factor is less than 1, the poles form a complex conjugate pair with a real part equal to $\zeta\omega_n$ and an imaginary part equal to $\omega_n\sqrt{1 - \zeta^2}$. This conjugate pairing of poles and zeros happens for all systems. Thus, if the conjugate pairs of poles and zeros are collected, Equation 7.64 appears as follows:

$$\frac{C(s)}{R(s)} = \frac{K\displaystyle\prod_{i=1}^{L}(s + z_i)\;\displaystyle\prod_{j=1}^{M}(s + \sigma_{zj} + j\omega_{zj})(s + \sigma_{zj} - j\omega_{zj})}{\displaystyle\prod_{k=1}^{N}(s + p_k)\;\displaystyle\prod_{l=1}^{R}(s + \sigma_{pl} + j\omega_{pl})(s + \sigma_{pl} - j\omega_{pl})} \tag{7.67}$$

where:

Π = the product of the arguments
σ_{-j} = the real part of the jth conjugate pair
ω_{-j} = the imaginary part of the jth conjugate pair
$m = L + 2M$
$n = N + 2R$

The poles and zeros of the closed-loop transfer function and their relative values determine the transient characteristics of the system as the second-order system example demonstrates. However, without taking the inverse transform of Equation 7.64 for a particular input, i.e., a step function, the transient time response cannot be determined. Even with the mathematical expression for the time history of the closed-loop response, the determination of the transient characteristics may require a laborious task of plotting the time history.

For this reason, the analysis techniques given in the next section are more convenient because they use the open-loop transfer function, avoiding factoring of high-order polynomials, inverting the Laplace equation, and plotting time histories for each input of interest.

One interesting characteristic of the closed-loop transfer function of the control system as expressed in Equation 7.63 is the composition of the poles and zeros of the system. If the original polynomials, i.e., $N_1(s)$, $N_2(s)$, $D_1(s)$, and $D_2(s)$, are in factored form, which is frequently the case, then the zeros of the system are simply "the zeros of $N_1(s)$ and the poles of $D_2(s)$." However, the poles of the closed-loop transfer function are not the factors of the various polynomials, but the roots of a composite polynomial which must be factored independently of the factors of the polynomials of $G(s)$ and $H(s)$. This factor is re-emphasized in the root-locus analysis technique which is described in the next section.

For the nth order linear feedback control system with the closed-loop transfer function given in Equation 7.60, the transfer function between the error signal, e(t), and the input, r(t), is given as follows:

$$\frac{E(s)}{R(s)} = \frac{1}{1 + G(s)\ H(s)} \tag{7.68}$$

where e(t) is equal to the difference between the input and the feedback signals or is equal to the signal driving the forward transfer function of the system. The steady-state error, i.e., the value of e(t) as time approaches infinity, is given as follows by the final-value theorem for a stable system:

$$e(\infty) = \lim_{t \to \infty} e(t) = \lim_{s \to 0} \frac{sR(s)}{1 + G(s)\ H(s)} \tag{7.69}$$

If the input is unity step function, the steady state error, $e(\infty)$, can be derived from Equation 7.69 as:

$$e(\infty) = \lim_{s \to 0} \frac{s\left(\dfrac{1}{s}\right)}{1 + G(s)\ H(s)} = \frac{1}{1 + G(0)\ H(0)} = \frac{1}{1 + K_p} \tag{7.70}$$

where $K_p = G(0)\ H(0) =$ the static position error coefficient.

The general form of $G(s)\ H(s)$ is given as follows:

$$G(s)\ H(s) = \frac{K_{GH}\ \Pi\ (\tau_{zi}s + 1)\ \Pi\left(\dfrac{s^2}{\omega_{zj}^{\ 2}} + \dfrac{2\zeta}{\omega_{zj}}\, s + 1\right)}{s^N\ \Pi\ (\tau_{pi}s + 1)\ \Pi\left(\dfrac{s^2}{\omega_{pj}^{\ 2}} + \dfrac{2\zeta}{\omega_{pj}}\, s + 1\right)} \tag{7.71}$$

where τ is the time constant of the corresponding pole or zero. For $N \geq 1$, $G(0)\ H(0)$ or K_p is equal to infinity. As a result, the steady-state error for $N = 0$ (referred to as a type 0 system) is equal to:

$$e(\infty) = \frac{1}{1 + K_{GH}} \tag{7.72}$$

and for $N \geq 1$ (a type 1 or higher-order system)

$$e\ (\infty) = 0 \tag{7.73}$$

The results given in Equations 7.72 and 7.73 shows that for only a type 0 system is there any steady-state error driving the forward transfer function, $G(s)$, for a step input. Thus, to avoid steady-state error for a step function or constant input, the forward or feedback loop must contain an integration. (See Table 7–1 for the definition of an integration.) An integration in the forward loop is most commonly used, usually located in the controller part of the system.

If the input is a ramp function of unity slope, the steady-state error is:

$$e(\infty) = \lim_{s \to 0} \frac{s\left(\dfrac{1}{s^2}\right)}{1 + G(s)H(s)} = \lim_{s \to 0} \frac{1}{sG(s)H(s)} \tag{7.74}$$

which for the system defined by Equation 7.71 yields:

$$e\,(\infty) = \infty \qquad \text{for } N = 0$$

$$e\,(\infty) = \frac{1}{K_{GH}} \qquad \text{for } N = 1 \tag{7.75}$$

$$e\,(\infty) = 0 \qquad \text{for } N \geq 2$$

These results show that for a control system to track a ramp input with finite or zero steady-state error, the system requires an integration or more in the control loop. Likewise, it can be shown for a parabolic input, a double integration or more is required in the control loop to limit the steady-state error to a finite or zero value.

However, there is a significant price to be paid for each integration, which is a phase lag of 90 degrees per integration.

If the input specified by the user of the control system is a random variable or if the input is a physical quantity which is a random variable, then determining the time history of the output of the system by the technique of inverse Laplace transforms is not possible. Determination of the time history by computer solution is possible, but it requires a simulation whose results may not be very useful.

Another approach which is in keeping with the mathematical representations of the control systems developed so far uses the spectral density function of the input random variable to derive the spectral density function of the output. The variance of the output can be determined from the spectral density. The variance is useful in defining the behavior of the output of a control system which has been subjected to a random input.

If the spectral density of the input to the control system is defined by $S_u\,(f)$ and the input is wide-sense stationary, then the spectral density of the output of a stable linear time-invariant control system, $S_y\,(f)$, is given as follows:[10]

$$S_y\,(f) = |\text{Transfer Function }(j\omega)|^2\ S_u\,(f) \tag{7.76}$$

where f equals the frequency and $\omega = 2\pi f$. In the case of the control system described by Equation 7.60, the spectral density of the output becomes:

$$S_y(f) = \left| \frac{G(j\omega)}{1 + G(j\omega)H(j\omega)} \right|^2 S_u(f) \tag{7.77}$$

Once given the output spectral density, the variance or the square of the standard deviation of the output is determined by the following equation:

$$\sigma_y{}^2 = \int_{-\infty}^{\infty} S_y(f)df = \int_{-\infty}^{\infty} \left| \frac{G(j\omega)}{1 + G(j\omega)H(j\omega)} \right|^2 S_u(f)df \qquad (7.78)$$

where σ_y is the standard deviation of the output.

The most basic performance characteristic of a closed-loop feedback system which has been stated as a condition for certain previously given relationships is the stability of the system. Stability is such a basic issue that unless a control system is stable, it is not actually capable of controlling anything.

A common definition of stability explains why a control system must be stable to be a system capable of control. The definition states that a system is stable if the output of the system is finite in any response to a bounded input. Thus, if a system is unstable, the output of the system diverges to infinity in response to any input.

There are degrees of stability. Under the definition of stability given above, a system whose output is purely oscillatory with a bounded amplitude is stable. However, few, if any, real control systems with an oscillatory output are considered stable by the user unless the amplitude is very small. Likewise, even if a system exhibits an output with a decaying oscillatory response, the user of the system may demand that the oscillations effectively decay to an insignificant amplitude in less than two or three cycles (like the response shown in Fig. 7–21).

As a result, if the designer or the user specifies the transient response characteristics of the system, he has specified the relative stability of the system. The difficulty that the designer has in his design, in fact, is generally not stabilizing the system, but achieving the desired degree of relative stability within the constraints imposed by the physical limits of the system parameters. The analysis techniques given in the next section address this design problem of relative stability.

For linear time-invariant systems represented by Laplace transforms, stability is determined by the poles of the closed-loop transfer function. If any of the poles have negative real parts, then that pole gives rise to a transient response which increases indefinitely (or until the system becomes nonlinear and limits the response). This can be best understood by the simple example of a system with a first-order pole given as follows:

$$\frac{C(s)}{R(s)} = \frac{p_1}{s - p_1} \qquad (7.79)$$

The pole for this system has a value of $-p_1$. When the system is subjected to a unity step function input, the time response is:

$$c(t) = 1 - e^{-p_1 t} \qquad (7.80)$$

if p_1 is positive, then the time response of the output is bounded as shown in Fig. 7-22(a). However, if p_1 is negative, then the time response grows exponentially without bound as shown in Fig. 7.22 (b) and as is given in the following equation:

$$c(t) = e^{p_1 t} - 1 \qquad (7.81)$$

If a conjugate pair of poles has a real part equal to zero, then the pair gives rise to an oscillation in the time response which is not acceptable for a real control system. Likewise, if closed-loop transfer function has one or more poles with zero real parts and zero imaginary parts, the system is unstable. As far as zeroes of the closed-loop transfer function are concerned, a stable system can have zeros with negative real parts. These systems are referred to as nonminimum phase systems and do not generally occur. Thus, the boundary which determines the stability of a control system is when the real parts of the poles of the closed-loop transfer function are positive or zero.

A graphical way to represent the poles and zeros of the closed-loop transfer function is in a complex plane where the Laplace transfer operator is defined as a complex vairable which is given as follows:

$$s = \sigma + j\omega \qquad (7.82)$$

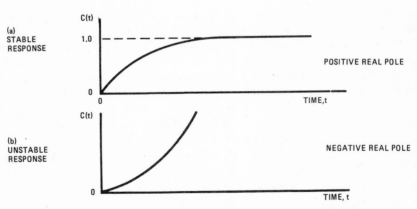

Figure 7-22. Plots of a Stable and Unstable System Response

Then, a transfer function of poles and zeros can be represented in the complex plane by plotting the poles and zeros at the locations where the following equations are satisfied;

$$s + p_i = 0$$
$$s + z_j = 0$$

(7.83)

or:

$$-\sigma_{pi} = Re\ [p_i]$$
$$-\omega_{pi} = Im\ [p_i]$$
$$-\sigma_{zj} = Re\ [z_j]$$
$$-\omega_{zj} = Im\ [z_j]$$

(7.84)

where:
RE [ ] = the real part of the argument
IM [ ] = the imaginary part of the argument

For example, the poles and zeros of the following transfer function can be represented graphically as shown in Fig. 7–23:

$$G(s) = \frac{(s + 5)\ (s - 2 + j3)\ (s - 2 - j3)}{s(s - 4)\ (s + 3 + j4)\ (s + 3 - j4)}$$

(7.85)

As Fig. 7–23 shows, the example transfer function has two poles and one zero in the left-hand part of the complex plane, one pole at the origin, and two zeros and one pole in the right-hand part of the complex plane. If G(s) is a closed-loop transfer function, the system which G(s) represents is unstable because of the one pole in the righthand part of the complex plane.

The s-plane representation of the closed-loop poles shows graphically whether a system is stable or not. A more useful purpose of the s-plane representation is establishing the relative stability of a control system. Simply because all the poles of a closed-loop transfer function are in the left-hand part of the s-plane does not result in a satisfactory transient time response.

As might be suspected on a intuitive basis, a system where the dominant poles display a more oscillatory or sluggish response the closer the poles are to the $j\omega$-axis.

If a transfer function possesses a dominant pole or pair of conjugate poles, the pole or poles are the nearest to the $j\omega$ axis by a factor of at least 5 in terms of the ratio of the real parts of the nearest and next nearest poles and there are no zeros near the dominant poles.

For example, the poles of the second-order system in the s-plane

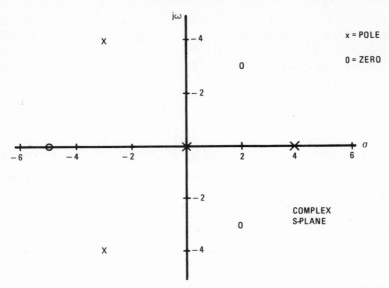

Figure 7-23. The Pole-Zero Locations for Equation 7-85

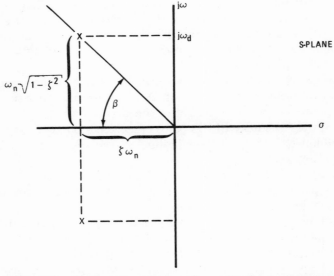

Figure 7-24. The S-Plane Representation of a Second-Order System

appear (Fig. 7–24) as a function of the damping factor, ζ, and the natural frequency of the system ω_n. To limit the system so that the damping factor is never less than some given percent which limits the maximum overshoot to less than some given percent of the step function input amplitude, the poles of the second-order system must be to the left of the line shown in Fig. 7–23.

Thus it appears that if the designer can position the poles of the closed-loop transfer function in the s-plane where he wants, he can obtain the desired transient response of the control system. This is true, but the designer cannot position the poles of the closed-loop transfer function precisely where he wants. As a result, the designer is faced with compromises.

7.5 Control system analysis

As is established in the previous section, one of the primary elements of the control system design is the transient response of the system or its relative stability. The key to the transient response is the poles of the closed-loop transfer function, which is repeated here for easy reference:

$$\frac{C\ (s)}{R\ (s)} = \frac{G(s)}{1 + G(s)\ H(s)} \tag{7.86}$$

The poles of the closed-loop transfer function are the zeros of the denominator of Equation 7.86, which is referred to as the characteristic equation. The zeros of the denominator can be determined by the following equation:

$$1 + G(s)\ H(s) = 0 \tag{7.87}$$

where $G(s)\ H(s)$ is the open-loop transfer function of the control system.

Solving Equation 7.87 to find the roots of the characteristic equation is not easy if its order is greater than three. Likewise, the roots of the characteristic equation only establish whether the system is stable or unstable and not its degrees of stability. To establish the transient response characteristics of the system once the zeros of Equation 7.87 are determined, the designer must invert the total system transfer function, Equation 7.86, and plot the result as a function of time.

As well can be imagined, this can be a tedious task. For this reason several analysis techniques were devised by control system designers to eliminate the necessity of this course of action and they are the topic of this section.

There are four primary classical analysis techniques; all of which are related, all of which use a graphical plot, and none of which are universally applied by control system designers. Almost always each designer has his own preference of analysis techniques with which he feels most comfortable and with which he has the most success in achieving his preferred design.

The four techniques listed here are often referred to by the plot they use, which is named after the man who originated the technique:

Analysis technique	Common name
Logarithmic gain and phase versus logarithmic frequency	Bode
Polar gain and phase	Nyquist
Logarithmic gain versus phase	Nichols
Root locus	Root locus (Evans)

In this section the discussion is limited to the Bode and root locus methods. The Nyquist and Nichols methods are essentially another method of presenting the information that is included in the Bode method. Each has certain advantages which certain designers prefer. There are many control system analysis texts which contain complete discussions of the Nyquist and Nichols methods[159]. The root locus method is considered first and only briefly. Most of the attention of the section is devoted to the Bode method.

In the root locus technique the zeros of the characteristic equation are determined by solving Equation 7.87 expressed in the following form and realizing that G(s) H(s) is a complex quantity:

$$G(s)\ H(s) = -1 \qquad\qquad (7.88)$$

Because G(s) H(s) is a complex quantity, the following relationships must be true to satisfy Equation 7.88:

$$|\ G(s)\ H(s)\ | = 1 \qquad\qquad (7.89)$$
$$\underline{/G(s)\ H(s)} = \pm\ 180°\ (2n + 1) \qquad\qquad (7.90)$$

where $|\ \ |$ is the magnitude of the open-loop transfer function, $\angle$ is the phase angle of the open-loop transfer function and $n = 0,1,2,...$ A plot of the phase-angle equation, Equation 7.90, in the complex plane results in the locus of roots of the characteristic equation or, in other words, the root locus. Then, the solution of the magnitude equation,

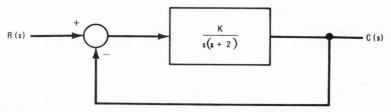

Figure 7-25. A Simple Second-Order System to Illustrate the Root-
Locus Method

Equation 7.89, gives the location of the zeros of the characteristic
equation or the poles of the closed-loop transfer function.

Very often the magnitude equation is expressed as a function of an
adjustable controller parameter such as gain. Then, the design can be
varied by changing the parameter of the controller.

The root locus method is most easily understood by plotting the root
locus for a simple example. The example system is shown in Fig. 7–25.
The open-loop transfer function of the example system is:

$$G(s) = \frac{K}{s(s + 2)} \tag{7.91}$$

where K is the gain of the system.

The first step in using the root locus method is to locate the poles
and zeros of the open-loop transfer function in the complex plane.
Some factoring of the polynomials which make up the open-loop transfer function may be necessary to determine the separate poles and
zeros of G(s) H(s).

For the example, there is one pole at the origin of the complex plane
and another pole at s = −2 in the left-hand part of the s-plane. Even
though the open-loop poles for the example are suitably located to
make the example system stable, the system is not necessarily stable
because these poles are not closed-loop poles.

Next the root locus or solution to the phase-angle equation is determined. There are many rules which assist the designer in determining the root locus. These rules are contained in most basic control
system analysis textbooks and are not included here [1] [5] [9]. There are
also devices, ranging from a manually operated protractor-like device
called a Spirule* to digital computer programs, which the designer

---

* Spirule is a trade name given by W. E. Evans to a transparent plastic combination
of a protractor and an arm. Its use is explained in Reference 5.

can use to plot the root locus. For the simple example under consideration, the root locus is as shown in Fig. 7–26.

The reader can convince himself that the root locus shown in Fig.
7–26 satisfies the phase-angle equation by considering any point on the
locus. The total of the angular contributions from the poles, θ_1 and
θ_2 (and the zeros if the open-loop transfer function includes them) for
each point on the locus sums up to 180 degrees. If some other locus is
tried, then the angular total is not 180 degrees.

For example, if the locus is hypothesized to be vertical from the pole
at $s = -2$, then the angular total from the two poles is greater than 180
degrees because $\theta_2 = 90$ degrees and $\theta_1 > 90$ degrees. Likewise, if the
locus is hypothesized to be vertical from the pole at the origin, the
$\theta_1 = 90$ degrees and $\theta_2 < 90$. As a result, the total angular contribution
is less than 180 degrees.

Once the designer plots the root locus as determined by the phase-
angle equation, he solves the magnitude equation to locate the zeros
of the characteristic equation. The solution of the magnitude equation

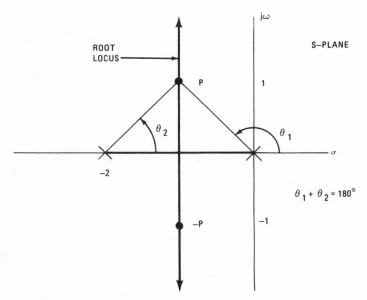

Figure 7-26. The Root Locus for the Example System Shown in
Figure 7-25.

can either be by mathematical equation, graphically with a Spirule, or with a computer program.

For instance, in the example shown in Fig. 7–26 where the gain of the system is unspecified, if the designer wants to locate one of the closed-loop poles at point P (the other pole is necessarily located at −P to form a conjugate pair), then he can solve the following equation for the gain, K:

$$\left|G(s)\ H(s)\right| = \left|\frac{K}{s(s+2)}\right| = 1 \tag{7.92}$$

or

$$|K| = |s(s+2)| \tag{7.93}$$

where $s = -1 + j1$.

For this very simple example the root locus is on the left-hand part of the s-plane for all values of positive system gain. Thus, the zeros of the characteristic equation or the poles of the closed-loop transfer function can never be in the right-hand part of the s-plane and the system is always stable if K is positive.

For higher-order systems the root locus is more complex and in many instances the locus moves into the right-hand part of the s-plane. As a result, for some value of system gain there are solutions or zeros to the characteristic equation in the right-hand part of the s-plane and the system is unstable for the corresponding value of gain.

In Fig. 7–27, the root loci for three separate systems are shown, all of which are capable of being unstable for certain values of system gain, K. For the first two systems, the systems are not unstable until the gain increases to the point where the loci cross the $j\omega$-axis. In the third case the possibility exists that the system is unstable for two separate ranges of gain.

If the task of the control system designer is merely to analyze a given system and he chooses to use the root locus technique, then he plots the root locus and uses the magnitude equation to find the closed-loop poles of the system. From this information he can determine the stability of the system. Also, the designer can formulate an estimate of the time response characteristics of the system, i.e., response time, overshoot, recovery time, etc., by the relative location of the closed-loop poles in the s-plane and the relative location of the closed-loop poles to the closed-loop zeros.

To a great extent, the accuracy with which the designer can estimate the resulting time response characteristics of the system depends on

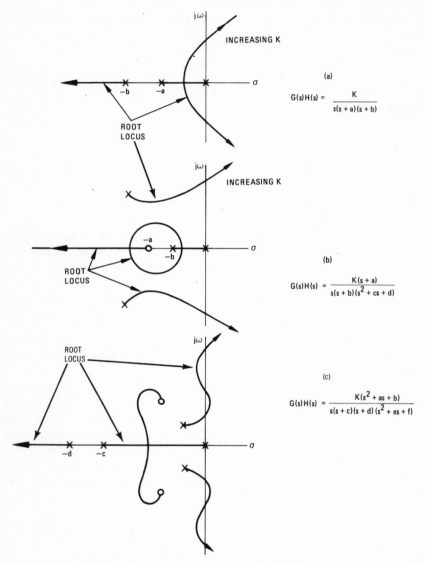

Figure 7-27. Three Possible Root Loci

the depth of the designer's experience and his intuition. However, he knows that the closer that the closed-loop poles are to the jω-axis of the s-plane, the more likely the system is to exhibit an oscillatory or sluggish response.

The presence of closed-loop zeros can affect this because of a tendency of the zeros to "cancel" the effect of the poles. In fact, pole cancellation using a zero is a technique used to compensate for undesirable effects of certain system poles.

Previously, the poles of the closed-loop transfer function of a system have been shown to be the zeros of the characteristic equation of closed-loop transfer function of the system, i.e., the zeros of $1 + G(s) H(s)$. The zeros of the closed-loop transfer function are shown in Equation 7.63 to be the zeros of the forward transfer function of the system, $G(s)$, and the poles of the feedback transfer function, $H(s)$.

As a result, for the case of a unity feedback system, i.e., $H(s) = 1$, the zeros plotted in the s-plane which are the zeros of the open-loop transfer function, $G(s) H(s)$, are also the zeros of the closed-loop transfer function.

To accurately determine the time response of the closed-loop system, the designer must invert the closed-loop transfer function using inverse Laplace transform techniques. Then, if the response does not meet the previously established specification, the designer must change some part of the control system to adjust the time response to the design specification.

The designer has several devices at his power to achieve the proper design. First and most convenient, the designer can adjust the gain of the system which is generally the parameter of the system which varies along the root locus. However, there are definite limitations to the degree of improvement that can be obtained by varying the system gain. Generally, either parameter in the control system or additional networks must be added to the system, usually in the controller to modify the shape of the original root locus. This process is called compensation.

The selection of the proper compensation network is based on certain established guidelines and the experience of the designer. Once the compensation network is added, a new root locus is plotted. The root locus may be plotted as a function of the control system gain or some parameter of the compensation network. Methods exist to shape the locus and position the poles of the closed-loop transfer function.[11] These methods are referred to as synthesis techniques.

However, at the completion of whatever method is used to reshape and reposition the closed-loop poles, the designer must invert the system transfer function to determine the time response of the system so that it can be compared to the design specification. If further reshaping and repositioning is required, the process iteratively continues in a cut-and-try fashion.

The number of required iterations generally depends on the skill and expertise of the designer. High-speed computers for determining the time response of a system, given the transfer function, have assisted greatly in this design process. More is said about the use of computers in determining the time response in the chapter on simulation techniques.

With the root locus analysis techniques, the poles of the closed-loop response are determined from which the time response of the system can be determined using inverse Laplace transform techniques. To plot the root locus, the poles and zeros of the open-loop transfer function are plotted in the complex plane. This means that the two transfer functions that make up the open-loop transfer function i.e., G(s) and H(s), had to be factored into their respective poles and zeros.

For the next analysis technique to be discussed factoring of G(s) and H(s) is not necessary. However, determination of the transient response of the system becomes more qualitative and indirect because the entire analysis is conducted in the frequency domain and the individual poles and zeros of the closed-loop system are not necessary. In fact, experimental measurements of the transfer functions of the elements of the control system can be used to perform the analysis.

As mentioned previously, the forthcoming discussion concentrates on the Bode method. However, all three methods, Nyquist, Bode, and Nichols, are based on the steady-state response of the system to a sinusoidal input. Then, because the systems are linear time-invariant systems, linear superposition applies and the response of the system to other inputs which are combinations of sinusoidal inputs can be estimated. For certain nonlinear systems, frequency response analysis techniques can be used. The interested reader is referred to References 1, 4, and 6.

If the input to a linear time-invariant system, represented by a transfer function, G(s), is a sinusoidal time function of a single frequency, ω, then the output is a sinusoidal of the same frequency. In general, the output signal has a different amplitude and is displaced in time by a fixed amount. Thus, the system acts as a linear filter with a gain and a phase characteristic which vary as a function of frequency.

To obtain the sinusoidal transfer function of a linear system expressed in terms of the Laplace operator, the Laplace operator, s, is replaced by $j\omega$. The gain and phase characteristics of the system are, then, found as follows:

$$\text{Gain} = |G(j\omega)| = \frac{|N(j\omega)|}{|D(j\omega)|} \tag{7.94}$$

$$\text{Phase} = \underline{/G}(j\omega) = \underline{/N}(j\omega) - \underline{/D}(j\omega) \tag{7.95}$$

Each of the three frequency response methods plot the gain and phase characteristics of the open-loop transfer function of the systems, $G(j\omega) H(j\omega)$, over some range of frequency. For the Nyquist method the gain and phase of $G(j\omega) H(j\omega)$ are plotted vectorially as a polar plot as the frequency is varied from zero to infinity.

The Nichols method plots logarithmic gain in terms of decibels versus phase in degrees with the frequency as the running variable along the gain-phase locus. The Bode method also uses logarithmic gain in terms of decibels, but the gain and phase are each plotted as a function of logarithmic frequency.

The reason that logarithms are used in the Bode method can be seen when considering the generalized frequency response expression for the open-loop transfer function, i.e.:

$$|G(j\omega)H(j\omega)| = \frac{K' \prod\limits_{i} |(1 + j\omega\tau_{ni})| \prod\limits_{k} \left| \left(1 - \frac{\omega^2}{\omega_{nk}^2}\right) + j2\zeta_k \frac{\omega}{\omega_{nk}} \right|}{|j\omega|^N \prod\limits_{i} |(1 + j\omega\tau_{di}) \prod\limits_{\ell} \left| \left(1 - \frac{\omega^2}{\omega_{n\ell}^2}\right) + j2\zeta_\ell \frac{\omega}{\omega_{n\ell}} \right|} \tag{7.96}$$

If the logarithms of Equation 7.96 are taken, the products given in Equation 7.96 become additions and the division of the numerator by the denominator is a subtraction as given below:

$$20 \log_{10}|G(j\omega)H(j\omega)|, \text{ decibels} = 20 \log_{10} K' + \sum_{i} 20 \log_{10}|1 + j\omega\tau_{ni}|$$

$$+ \sum_{k} 20 \log_{10} \left| \left(1 - \frac{\omega^2}{\omega_{nk}^2}\right) + j2\zeta_k \frac{\omega}{\omega_{nk}} \right| - 20 N \log_{10} \omega$$

$$- \sum_{i} 20 \log_{10} \left| 1 + j\omega\tau_{di} \right| \tag{7.97}$$

$$- \sum_{\ell} 20 \log_{10} \left| \left(1 - \frac{\omega^2}{\omega_{n\ell}^2}\right) + j2\zeta_\ell \frac{\omega}{\omega_{n\ell}} \right|$$

The phase of Equation 7.96 is given as follows:

$$\underline{/G(j\omega)H(j\omega)} = \underline{/K'} + \sum_i \underline{/1+j\omega\tau_{ni}} + \sum_k \underline{\left/\left(1 - \frac{\omega^2}{\omega_{nk}^2}\right) + j2\zeta_k \frac{\omega}{\omega_{nk}}\right.}$$

$$-N \underline{/j\omega} - \sum_i \underline{/1+j\omega\tau_{di}} - \sum_\ell \underline{\left/\left(1 + \frac{\omega^2}{\omega_{n\ell}^2}\right) + j2\zeta_\ell\frac{\omega}{\omega_n}\right.} \tag{7.98}$$

or:

$$\underline{/G(j\omega)H(j\omega)}, \text{degrees} = \underline{/K'} + 57.3 \sum_i \arctan(\omega\tau_{ni})$$

$$+ 57.3 \sum_k \arctan\left(\frac{2\zeta_k \dfrac{\omega}{\omega_{nk}}}{1 - \dfrac{\omega^2}{\omega_{nk}^2}}\right) \tag{7.99}$$

$$- 90N - 57.3 \sum_i \arctan(\omega\tau_{di}) - 57.3 \sum_\ell \arctan\left[\frac{2\zeta_\ell \dfrac{\omega}{\omega_{n\ell}}}{1 - \dfrac{\omega^2}{\omega_{n\ell}^2}}\right]$$

Thus, by taking logarithms of the gain expression of G (s) H (s), multiplication is changed to addition which simplifies the plotting of the gain function. If the logarithmic gain is plotted against logarithmic frequency, further plotting simplifications can be achieved.

First, the gain terms which involve frequency only, i.e., $20 \log_{10} \omega$, are straight lines with slopes of 20 decibels per decade of frequency which cross the zero decibel axis at $\omega = 1$. Second, the other terms of the gain expression can be plotted approximately as two straight line segments. These line segments are asymptotes to the exact gain curves for frequencies larger and smaller than some critical frequency.

For example, in the case of a simple pole or zero of the gain expression, the gain, when $\omega\tau \ll 1$, is given as follows:

$$20 \log_{10}|1 + j\omega\tau| \cong 20 \log_{10} 1 = 0 \text{ dB} \tag{7.100}$$

As ω increases, the relative magnitude of $\omega\tau$ to unity reverses until at $\omega\tau \gg 1$, the gain is given as:

$$20 \log_{10}|1 + j\omega\tau| \cong 20 \log_{10} \omega\tau, \text{ dB} \tag{7.101}$$

When $\omega = 1/\tau$, the gain given by Equation 7.101 is 0 dB. For one decade higher in frequency, i.e., $\omega = 10/\tau$, the gain is 20 dB per decade

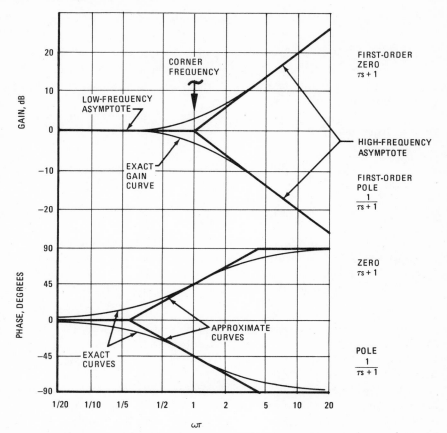

Figure 7-28. The Gain and Phase Curves for a Simple Pole and Zero

of frequency. The sign of the slope of the asymptote is positive for a zero and is negative for a pole. The low-frequency asymptote is the zero dB axis. The two asymptotes intersect at $\omega = 1/\tau$ which is called the corner frequency. The two asymptotes are shown in Fig. 7–28 along with the exact gain curve for purposes of comparison.

The exact phase characteristics for a simple pole and a simple zero are also shown in Fig. 7–28. The phase characteristic for either the pole or the zero can be approximated by three-line segments as shown in Fig. 7–28 where the breakpoint frequencies are defined as follows:

$$\frac{\omega_{CF}}{\omega_{lower}} = \frac{\omega_{upper}}{\omega_{CF}} = 4.8 \qquad (7.102)$$

where $\omega_{CF} = 1/\tau$ and ω_{upper} and ω_{lower} are defined in Fig. 7–28.

Better approximations can be achieved by using more line segments. However, the more line segments that are used, the less convenient the approximation becomes.

The gain and phase characteristics of the second-order pole and zero pairs in Equation 7.96, i.e.,

$$\text{Pole pair } \frac{1}{\left[\left(1 - \dfrac{\omega^2}{\omega_n^2}\right) + j\, 2\zeta\, \dfrac{\omega}{\omega_n}\right]} \qquad (7.103)$$

$$\text{Zero pair } \left(1 - \frac{\omega^2}{\omega_n^2}\right) + j\, 2\zeta\, \frac{\omega}{\omega_n}$$

can also be approximated with straight-line segments. However, the errors in the approximation are a function of the damping factor, ζ, as Fig. 7–29 illustrates. The peak of the gain function for a second-order system occurs at the following frequency with the following magnitude:

$$\omega_p = \omega_n \sqrt{1 - 2\zeta^2} \qquad (7.104)$$

$$M_p, \text{dB} = 20\, \log_{10}\left[\frac{1}{2\zeta\, \sqrt{1 - \zeta^2}}\right] \qquad (7.105)$$

These expressions are valid for the damping factor greater than or equal to zero and less than or equal to 0.707. For a damping factor greater than 0.5 the magnitude of the peak is not significant, as Fig. 7–29 shows.

The phase characteristic of the second-order pole and zero pair can be approximated by three-line segments like the simple pole and zero are in Fig. 7–28 using the following equation:

$$\frac{\omega_n}{\omega_{lower}} = \frac{\omega_{upper}}{\omega_n} = 4.8^\zeta \qquad (7.106)$$

This approximation is not very accurate. However, the phase approximation given in Equation 7.106 does express some dependence on the damping factor and is easy to plot. Better approximations (more line segments) reduce the error, but require more plotting time.

The advantage of the Bode method over other analysis techniques which comes about from the asymptotic approximation is most easily

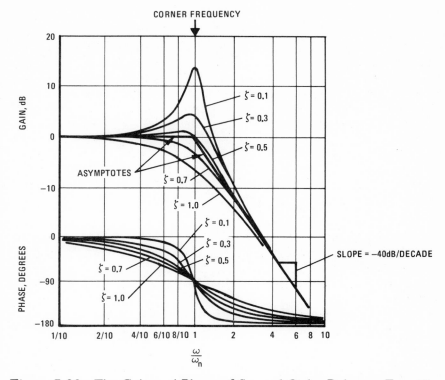

Figure 7-29. The Gain and Phase of Second-Order Pole as a Function of Frequency and Damping Factor

seen by the following example. Given the following open-loop transfer function,

$$G\,(s)\,H\,(s) = \frac{5(s+4)}{s(s+0.5)(s^2 + 1.6\,s + 4)} \qquad (7.107)$$

the first step is to reform the transfer function in a normalized form and a function of $j\omega$:

$$G(j\omega)H(j\omega) = \frac{10\left(\dfrac{j\omega}{4} + 1\right)}{j\omega(j2\omega + 1)\left(1 - \dfrac{\omega^2}{4} + j\,0.4\omega\right)} \qquad (7.108)$$

Next, the corner frequencies are identified and the asymptotic approximations for each element are sketched as a function of frequency on semilog paper. Finally, the approximation of the frequency response of the total transfer function is formed by totalling the separate gain and phase functions at the various corner frequencies and then connecting these points by straight lines. These steps are illustrated in Fig. 7–30.

Fig. 7–30 also illustrates the exact gain and phase curves for the example system. As is previously noted, the gain approximation is better than the phase approximation.

Once the Bode plots of gain and phase are completed, the next ques-

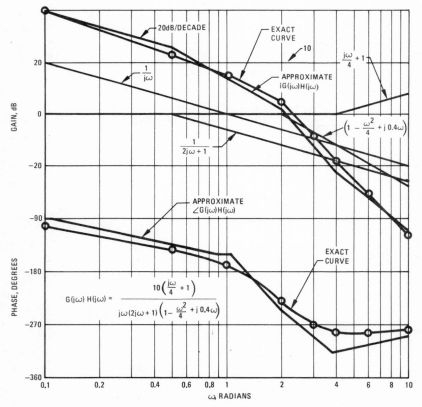

Figure 7-30. A Bode Plot of $G(j\omega)H(j\omega)$ for the Example Problem

tion is whether the system is stable. If it is stable, the designer asks what is its degree of stability and does that degree of stability satisfy the design requirements. If the plots predict that the system is unstable or that the system does not satisfy the design requirements, the designer then asks what modification to the gain and phase characteristics are necessary to achieve a stable system which meets the design requirements.

The determination of system stability from the Bode plots requires simultaneous consideration of both the gain and phase plots as a function of frequency. If the gain of the system is less than one or less than 0 dB on the gain curve when the phase of the system crosses the −180 degree line, the system is stable. Otherwise, the system is unstable. Fig. 7–31 shows sketches of each situation.

The degree of stability of the system is a function of the excess or margin with which the stability criterion is satisfied. For example, in Fig. 7–31 the gain curve for the stable case is a given number of decibels less than zero at the phase crossover point with the −180 degree line. Until the gain of the system is increased by that number of decibels, shifting the gain curve upward so that it crosses the zero decibel line at the same frequency as the phase crossover, the system is stable. That amount of gain is referred to as the gain margin of the system.

Relative stability is also a function of the amount of phase margin that exists when the gain curve crosses the 0 dB line just prior to the phase crossover frequency. Phase margin is illustrated in Fig. 7–31 and is determined by the following equation where ϕ is the open-loop phase angle at the gain crossover frequency:

$$\text{Phase margin} = \text{PM} = 180 + \phi \qquad (7.109)$$

A complete explanation of the reasons that relative stability is a function of both the gain and phase margin is most easily given using the Nyquist method. The interested reader is referred to Reference 1 or 5.

Another way of looking at gain and phase margin is that they represent allowable variations which can be caused by changes in system components before instability occurs. However, they also are correlated to the transient response characteristics of the system. For example, in the case of the second-order system which has the following open-loop transfer function:

$$G(s) = \frac{\omega_n^{2}}{s(s + 2\zeta\omega_n)} \qquad (7.110)$$

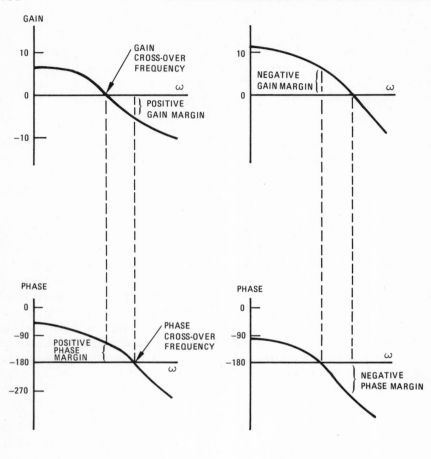

(a) STABLE SYSTEM   (b) UNSTABLE SYSTEM

Figure 7-31. Bode Plots for Stable and Unstable Systems Showing
Gain and Phase Margins

the phase margin of the system using the previously given definition
is found at the gain crossover frequency or when $|G\,(j\omega)| = 1$, i.e.,

$$\omega_n^2 = |j\omega\,(j\omega + 2\zeta\omega_n)| \tag{7.111}$$

Solving Equation 7.111 for gain crossover frequency, ω_c, yields

$$\omega_c = \omega_n[\sqrt{4\zeta^2 + 1} - 2\zeta^2]^{1/2} \tag{7.112}$$

The phase angle of the second-order system at ω_c is given as follows:

$$G(j\omega_c) = + \bigg/\!\frac{1}{j\omega_c} + \bigg/\!\frac{1}{j\omega_c + 2\zeta\omega_n}$$

$$= -90° - \arctan\left[\sqrt{\frac{\sqrt{4\zeta^2 + 1} - 2\zeta^2}{4\zeta^2}}\right] \qquad (7.113)$$

From Equations 7.109 and 7.113 the phase margin of the second-order system becomes:

$$PM = 90° - \arctan\left[\sqrt{\frac{\sqrt{4\zeta^2 + 1} - 2\zeta^2}{4\zeta^2}}\right] \qquad (7.114)$$

which is equivalent to the following:

$$PM = \arctan\left[\sqrt{\frac{4\zeta^2}{\sqrt{4\zeta^2 + 1} - 2\zeta^2}}\right] \qquad (7.115)$$

A plot of the phase margin and the peak overshoot for a second-order system, shown in Fig. 7–32, clearly illustrates that the greater the phase margin, the less overshoot the transient response displays.

A second-order system does not exhibit gain margin in the sense of the systems shown in Fig. 7–31 because the phase characteristic never crosses the −180 degree line. However, the gain crossover frequency of the system is correlated to the speed of response of the system in the following manner:

$$t_p = \frac{\pi}{\omega_c}\sqrt{\frac{\sqrt{4\zeta^2 + 1} - 2\zeta^2}{1 - \zeta^2}} \qquad (7.116)$$

where $\omega_p = \frac{2\pi}{t_p}$ and t_p = the time to the peak overshoot of the system.

As Equation 7.116 states, the higher the gain crossover frequency for a given damping factor, the faster the response time of the system.

For higher-order systems, the gain and phase margin and the crossover frequency of the frequency response of the open-loop transfer function do correlate to the transient response of the system. However, as can be imagined from the simple example of the second-order system, the exact correlation between time domain and frequency domain characteristics cannot generally be expressed in a mathematical equation.

As a result, the designer who wants to verify the transient response or time domain response of his design must either perform the root-

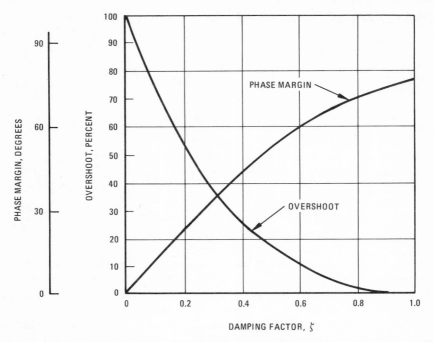

Figure 7-32. A Plot of Phase Margin and Overshoot as a Function of
Damping Factor for a Second-Order System

locus analysis or a dynamic simulation of the system. In many cases
the simulation is the best and easiest method because with the simulation other system characteristics can be studied, such as the effect of
nonlinearities and random disturbances.

Although Bode methods are primarily applicable to linear time-invariant systems, they can be used for certain simple nonlinearities as
discussed in References 1, 4, and 6. Similarly, the Bode method is
useful in the consideration of the probable effect of random disturbances if the spectral density of the disturbance is known as a function of frequency. If the particular disturbance has a spectral density
that is limited to a particular frequency band, then the designer may
be able to reduce the gain of the control system in the band of disturbance frequencies. In that way the control system does not respond
to the unwanted disturbance inputs, i.e., it filters the disturbances.

However, the exact manner in which the control system responds to random disturbance as a function of frequency is determined by the closed-loop frequency response of the system, as shown by Equation 7.78. Therefore, the designer can only achieve an educated estimate of the effect of his design on the response of the control system to random disturbances from the open-loop frequency response of the system.

To more accurately evaluate the response of his design to random disturbances using the Bode method, the designer can program a computer to perform those operations shown in Equation 7.78. This is an especially useful tool if the same program computes the gain and phase curves of the open-loop system as a function of frequency.

Finally, Bode methods can be effectively applied to sampled-data systems. For those systems expressed in z-transforms the bilinear w-transform is substituted for each z-transform to obtain an open-loop transfer function in w-transforms only. Then, the fictitious frequency, jv, is substituted for each w in the same manner as $j\omega$ is substituted for the Laplace operator for a continuous-data system.

Once the open-loop transfer function is expressed in terms of the fictitious frequency, the Bode methods of plotting gain and phase curves using asymptotic approximations are performed to determine the gain and phase margin of the system. If the system exhibits insufficient positive gain and phase margin, the designer performs the necessary modifications to the system parameters or adds a compensation network to achieve the desired results.

If he adds a compensation network, he can reverse the transformation process once the design is complete to determine the compensation network in z-transform form. Once in z-transform form, the designer can convert the compensation network to difference equation form for implementation in a digital computer, if that is the means of implementation of the compensator.

If the open-loop transfer function of the sampled-data system is a combination of z-transforms and sampled Laplace transforms, then, in some cases, the simplest approach to using Bode methods of analysis is to convert the z-transform portion of the transfer function to a w-transform representation and leave the sampled Laplace portion as an infinite series like Equation 7–31. This is especially true if the sampled Laplace portion of the transfer function can be approximated by one term of the series.

Then, in the combination open-loop transfer function, the w-transforms are replaced by jv and the Laplace transforms by $j\omega$. Next, the gain and phase curves are computed as a function of the real frequency, computing the fictitious frequency using $v = \tan(\omega T_s/2)$. This technique is not practical except in very simple systems or if the computations are performed by a digital computer. If a digital computer program is used to compute the gain and phase curves as a function of the real frequency, ω, then the effect of random disturbances on the control system can be predicted using Equation 7.78.

REFERENCES

1. Katsuhiko Ogata, *Modern Control Engineering,* Prentice-Hall, Inc., 1970.

2. Julius T. Tou, *Digital and Sampled-data Control Systems,* McGraw-Hill Book Company, 1959.

3. D. G. Schultz and J. L. Melsa. *State Functions and Linear Control Systems,* McGraw-Hill Book Company, 1967.

4. D. Graham and D. McRuer, *Analysis of Nonlinear Control Systems,* John Wiley and Sons, Inc., 1961.

5. J. J. D'Azzo and C. H. Houpis, *Feedback Control System Analysis and Synthesis,* McGraw-Hill Book Company, 1966.

6. John E. Gibson, *Nonlinear Automatic Control,* McGraw-Hill Book Company, 1963.

7. P. M. De Russo, R. J. Roy, and C. M. Close, *State Variables for Engineers,* John Wiley and Sons, 1965.

8. J. T. Tou, *Modern Control Theory,* McGraw-Hill Book Company, 1964.

9. John G. Truxal, Ed., *Control Engineer's Handbook,* McGraw-Hill Book Company, 1958.

10. W. B. Davenport and W. L. Root, *An Introduction to the Theory of Random Signals and Noise,* McGraw-Hill Book Company, 1958.

11. John G. Truxal, *Control System Synthesis,* McGraw-Hill Book Company, 1955.

8

Dynamic Models and Simulation Methods

8.1 Introduction

The major elements of a dynamic positioning control system are illustrated in a closed-loop form in Fig. 8–1. To analyze this control system according to the methods given in Chapter 7 requires a dynamic model for each element. In this chapter such models are given.

The basis for the thruster and sensor models given in this chapter is contained in previous chapters. Thus, only the models themselves are described in this chapter. Likewise, the vessel model is based on other reference material which is cited at the appropriate time.

The control system analysis methods described in the previous chapter are limited to linear systems and a small set of nonlinear systems. To analyze more complex systems and verify the designs obtained using linear analysis techniques usually requires a simulation. These simulations use a computer to solve the mathematical equations which model the system.

The solution given by the computer is an estimate of the real-world time response of the systems to various simulated inputs. Several types of computers can be used to simulate control systems. In this chapter, brief descriptions of analog, digital, and hybrid computers are given.

8.2 Mathematical modeling

The dynamic characteristics of many physical systems can be described by mathematical equations. These systems include electrical,

371

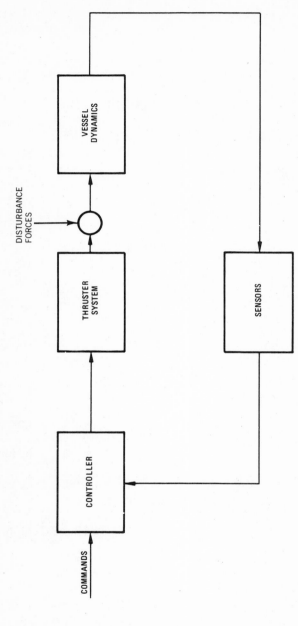

Figure 8-1. The Major Elements of the Dynamic Positioning System

mechanical, thermal, hydraulic, biological, etc. Usually the mathematical descriptions are obtained from physical laws governing a particular system.

For example, Newton's laws can be used to describe mechanical systems and Kirchoff's laws can be used for electrical systems. These mathematical representations of the dynamic characteristics of a physical system are models of the system and, before control system analysis and design can begin, each system element in the control system must be modeled in some manner.

It must be recognized that a model is an approximation which characterizes the dynamic behavior of a physical system. As it is an approximation, the degree to which a model is accurate varies. Generally speaking, the more accurate the model, the more complex it is. Model complexity usually makes the task of analysis more difficult. There is a tendency for the users of the model to lose touch with the physics of the device. This can endanger the user's ability to judge the reasonableness of the final design.

Creating a more accurate and complex model also requires more effort and, therefore, more cost. Analysis costs are also escalated by more complex models because the designer must utilize more complex analysis tools, such as computer simulators. Thus, a key consideration in math modeling is the tradeoff between accuracy and return on investment.

Modeling usually begins during the preliminary design. At that time the models should be simple, describing only the major dynamic characteristics of the system. Then, as the design progresses, the models are made more accurate in order to meet the needs and budget of the design. Naturally, there is a limit to the simplicity that can be used in modeling without completely misrepresenting the system.

It is generally the designer's responsibility to determine the degree of simplicity appropriate to the model based on his experience and the preferred methods of analysis and design.

As explained in the previous chapter, the characteristics of physical systems are nonlinear, at least to some degree. Furthermore, the methods of analysis for nonlinear systems are limited and frequently can only be realized by computer simulations. Thus, the most frequently used modeling simplification is the linearization of the characteristics of a physical device.

Fortunately, many control system devices display nearly linear characteristics over sufficiently large regions of operation to make linearization meaningful. In fact, many control devices are designed to operate in the most linear region of their operation.

The simple example shown in Fig. 8–2 illustrates many of these points. The modeling equations for the spring-mass system shown in Fig. 8–2 are derived using Newton's laws and result in the following equation:

$$m \frac{d^2 x}{dt^2} + B\left(\frac{dx}{dt}\right) + K(x) = f(t) \tag{8.1}$$

where:

$m =$ the mass

$x =$ the displacement of the mass

$B\left(\dfrac{dx}{dt}\right) =$ the damper force which is a function of velocity

$K(x) =$ the spring force which is a function of displacement

$F(t) =$ the forcing function

If the spring or the damping force are nonlinear functions of their indicated arguments, then Equation 8.1 is a nonlinear second-order differential equation which may or may not have a closed-form solution. Fig. 8–3 illustrates potential nonlinear characteristics for the

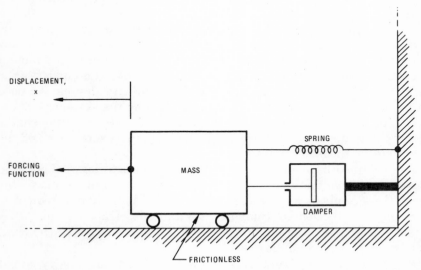

Figure 8-2. A Simple Spring-Mass System with Damping

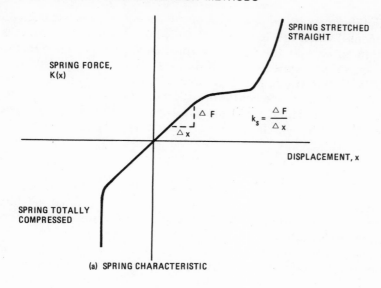

SPRING STRETCHED
STRAIGHT

SPRING FORCE,
K(x)

$\triangle$ F

$k_s = \dfrac{\triangle F}{\triangle x}$

$\triangle$ x

DISPLACEMENT, x

SPRING TOTALLY
COMPRESSED

(a) SPRING CHARACTERISTIC

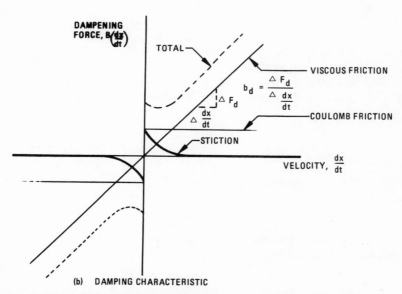

DAMPENING
FORCE, $B\left(\dfrac{dx}{dt}\right)$

TOTAL

VISCOUS FRICTION

$b_d = \dfrac{\triangle F_d}{\triangle \dfrac{dx}{dt}}$

$\triangle F_d$

COULOMB FRICTION

$\triangle \dfrac{dx}{dt}$

STICTION

VELOCITY, $\dfrac{dx}{dt}$

(b) DAMPING CHARACTERISTIC

Figure 8-3. Potential Characteristics for a Spring and Damper

spring and damper. If the operation of the spring mass is restricted to displacements which are less than those where the spring characteristics begin to exhibit the effects of total compression or extension, then the spring performs as a linear device.

Similarly if the stiction and coulomb friction components of the damper characteristics are negligible, then the damper can be modeled as a linear device. Thus, Equation 8.1 becomes the following linear differential equation which can be solved in a standard manner:

$$m \frac{d^2 x}{dt^2} + b_d \frac{dx}{dt} + k_s x = f(t) \qquad (8.2)$$

where b_d is the damping coefficient and k_s is the spring constant.

Equation 8.2 is the same second-order equation as studied in Chapter 7 except the parameters are expressed in terms of the physical parameters. If Equation 8.2 is expressed in terms of a natural frequency and a damping factor from which transient response characteristics can be computed using the equations given in the previous chapter, then the following equivalence of parameters are required:

$$\text{Natural frequency, } \omega_n = \sqrt{\frac{k_s}{m}} \qquad (8.3)$$

$$\text{Damping factor, } \zeta = \frac{b_d}{2\sqrt{mk_s}} \qquad (8.4)$$

The hydraulic control valve system shown in Fig. 8-4 is considered a second example of modeling and the process of linearization.[4] The flow characteristics of the control valve are nonlinear as illustrated in Fig. 8-5, and can be expressed by the following functional relationship:

$$Q = f(x, \Delta P) \qquad (8.5)$$

where:

$$Q = \text{the flow from the supply}$$
$$x = \text{the displacement of the control valve}$$
$$\Delta P = \text{the pressure difference across the piston}$$

Around a given operating point, the valve flow can be linearized as expressed by the following equation:

$$Q - Q_o = k_x (x - x_o) - k_{\Delta p} (\Delta P - \Delta P_o) \qquad (8.6)$$

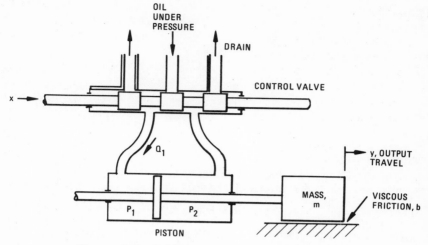

Figure 8-4. Hydraulic Control Valve System

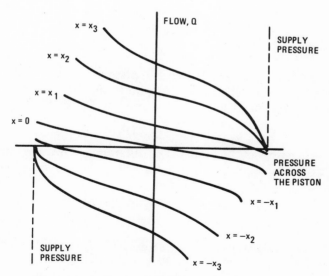

Figure 8-5. Control Valve Characteristics

where:

$$Q_o = f(x_o, \Delta P_o)$$

$$k_x = \frac{\partial Q}{\partial x} \bigg|\; x = x_o, \Delta P = \Delta P_o$$

$$k_{\Delta P} = \frac{\partial Q}{\partial \Delta P} \bigg|\; x = x_o, \Delta P = \Delta P_o$$

$x_o = x$ operating point value

$\Delta P_o =$ the pressure difference operating value

Normally, for a control valve system like the one shown in Fig. 8–4, the operating point is for zero x-displacement and zero difference of pressure across the piston, which reduces Equation 8.6 to the following:

$$Q = k_x x - k_{\Delta P} \, \Delta P \qquad (8.7)$$

This linearization is only valid over a limited portion of the total operating range of the control valve system. Thus, if the system is operated in another region which is sufficiently removed from where k_X and $k_{\Delta P}$ are computed for $x = 0$ and $\Delta P = 0$, then k_X and $k_{\Delta P}$ must be recomputed.

Like the spring-mass system, the hydraulic control valve system can be modeled using Newton's laws which yields:

$$m \frac{d^2 y}{dt^2} = -b \frac{dy}{dt} + F_{piston} \qquad (8.8)$$

where F_{piston} is the force exerted on the piston.

The force exerted on the piston is equal to the difference in pressure across the piston times the area of the piston, i.e.,

$$F_{piston} = A_p \, \Delta P \qquad (8.9)$$

where A_p is the area of the piston. The difference in pressure across the piston is given by Equation 8.7, which yields when substituted into Equation 8.9, the following equation:

$$F_{piston} = \frac{A_P}{k_{\Delta P}} [k_x \, x - Q] \qquad (8.10)$$

The flow, Q, at any instant can be derived from the fact the piston moves a differential distance, dy, determined by the amount of input oil over a differential amount of time or in equation form:

$$Q \, dt = A_p \, \rho dy \qquad (8.11)$$

where ρ is the density of the oil. When Q from Equation 8.11 is substituted into Equation 8.10, i.e.:

$$F_{\text{piston}} = \frac{A_P}{k_{\Delta P}}\left[k_x\, x - A_p\, \rho\, \frac{dy}{dt}\right] \tag{8.12}$$

then the overall system differential equation from Equation 8.8 becomes:

$$m\,\frac{d^2y}{dt^2} + \left(b + \frac{A^2\rho}{k_{\Delta P}}\right)\frac{dy}{dt} = \frac{A_p}{k_{\Delta P}}\,k_x\, x \tag{8.13}$$

which yields the following transfer function in Laplace transform notation:

$$\frac{Y(s)}{X(s)} = \frac{\dfrac{A_p k_x}{k_{\Delta P}}}{s\left[ms + \left(b + \dfrac{A_p^{\,2}\rho}{k_{\Delta P}}\right)\right]} \tag{8.14}$$

which reduces to the following:

$$\frac{Y(s)}{X(s)} = \frac{K}{s(\tau s + 1)} \tag{8.15}$$

where:

$$K = \frac{A_p k_x}{b k_{\Delta P} + A_p^{\,2}\rho}$$

$$\tau = \frac{m k_{\Delta P}}{b k_{\Delta P} + A_p^{\,2}\rho}$$

Using the system classification presented in the previous chapter, the hydraulic control valve system is a Type 1 system which can be used in a position feedback control system to give zero error to a step function input. Such a position control system is shown in block diagram form in Fig. 8–6a. A physical implementation of the feedback and error sensor mechanism using mechanical components is illustrated in Fig. 8–6b. The transfer functions corresponding to the mechanical couplings in Fig. 8–6b are given as follows:

$$H(s) = \frac{\ell_1}{\ell_1 + \ell_2} \tag{8.16}$$

$$B(s) = \frac{\ell_2}{\ell_1 + \ell_2} \tag{8.17}$$

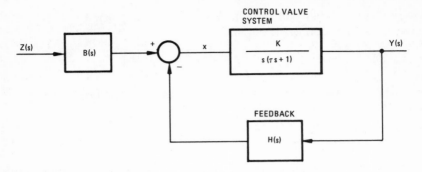

(a) BLOCK DIAGRAM REPRESENTATION

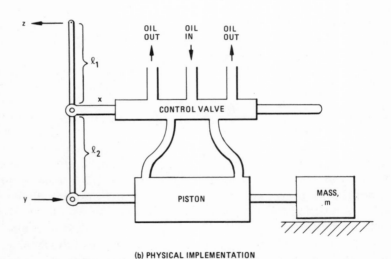

(b) PHYSICAL IMPLEMENTATION

Figure 8-6. A Position Control System Using a Hydraulic Control
Valve

where the lever arm lengths are defined in Fig. 8–6b. An electrical
feedback and error-sensing scheme can also be used.[4]

This has been a brief introduction to the modeling of physical systems as modeling relates to control systems. Little has been said about

the specific techniques of obtaining the differential equations because this book is not devoted to subject of deriving models. The interested reader is referred to References 2, 4, 5, and 7 for more details on methods of deriving the model equations.

Also, the Lagrangian and Hamiltonian approaches to deriving model equations and the benefit of such approaches to designs based on the modern control are given in Reference 6.

8.3   Thruster system models

In Chapter 4, the thruster system is presented by means of functional block diagrams and equations describing various elements of the thruster system. This section describes models for two commonly used thruster systems which are suitable for the control system analysis of the dynamic positioning systems of the vessels on which these thrusters might be installed.

The two thruster systems are the controllable speed and the controllable pitch systems, both of which are discussed in Chapter 4. This section does not include the derivation of the models. However, references are given from which the models can be derived.

The thruster models given in this section contain the primary effects, such as inflow degradation, dynamic response, and limiting features. More detailed models can be created and for certain analysis are needed. However, too much detail, especially in the beginning, can be harmful because it obscures the basic operation of the system.

The major elements of the thruster system to be modeled are the thrust-producing mechanism, i.e., the propeller, and the thrust control system. The propeller is assumed to have a fixed axis. In the controllable speed case, the propeller has a fixed pitch and the thrust control system controls either the speed of rotation or the armature current of the dc motor which rotates the propeller. Such a control system is illustrated in Fig. 4-19.

For the controllable-pitch propeller, the propeller is driven by an ac induction motor at an approximately constant speed of rotation. The thrust control system sets the pitch of the propeller as a function of the command from the dynamic positioning controller. Fig. 4-18 illustrates a controllable-pitch thruster system.

In both systems, the propeller produces thrust as a function of the thrust-command variable, and as a function of the inflow conditions to the propeller. Other parameters influence the output thrust of the propeller, such as draft and turbulence, or flow, caused by other propellers, or surface wave action. These effects are not necessarily secondary, but they are not included in the models in this section.

The thrust produced by a propeller is a nonlinear function of the thrust-command variable and the inflow-current velocity as discussed in Chapter 4. Under normal conditions a thruster is producing thrust around an average operating point. The degree of variation in the thrust around the operating point depends on the disturbances acting on the vessel and any noise entering the dynamic positioning control system.

Usually the variations are of such a magnitude to permit linearization of the thrust versus command variable function around the operating point, at least for the preliminary design where controller parameters are selected to achieve closed-loop system stability, transient response characteristics, and noise immunity.

In the final design analysis, where a computer simulation is used, the thrust versus command variable functions for each thruster can be stored in the computer memory in as accurate a representation of their nonlinear nature as desired or as is economically feasible.

The same preliminary modeling philosophy is used for the effect of the inflow current, i.e., linearization: Then, in the more complete design evaluations, nonlinear functions of the effect of inflow current on the thrust output of the propellers can be included. As a result, in the preliminary design phases the propeller is modeled as shown in Fig. 8–7a, and in the detailed design phase the propeller model is as shown in Fig. 8–7b.

In the model of the effect of thrust degradation caused by the inflow current shown in Fig. 8–7, the current velocity is the component of the inflow current along the axis of rotation of the propeller. Even more important than the inflow current velocity is the relative velocity of the propeller to the water surrounding the vessel.

As such, this velocity is the sum of the component of the sea current velocity along the axis of the propeller plus the velocity of the propeller through the water with zero sea current. The propeller velocity is composed of the translational velocity of the vessel along the axis of the propeller plus the tangential velocity of propeller as the vessel rotates about the center of rotation. The tangential velocity is directly proportional to the distance that the propeller is from the center of rotation of the vessel.

The propeller model including the effect of vessel motion, is illustrated in Fig. 8–8.

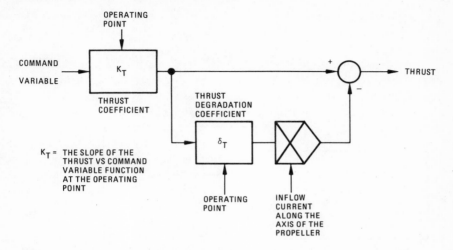

(a) PRELIMINARY MODEL

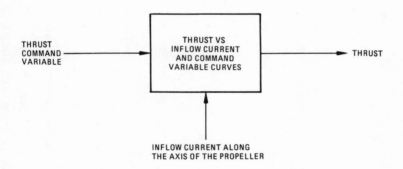

(b) DETAILED MODEL

Figure 8-7. Thrust Producing Mechanism Model

Fig. 8–8 also shows that the thrust of each thruster acts on the vessel model to produce a moment and translation force. In Fig. 8–8 the assumption is made that the axis of the thruster is parallel to one of the major axes of the vessel. If not, the thrust from the propeller must be properly resolved to coincide with the axes of the vessel.

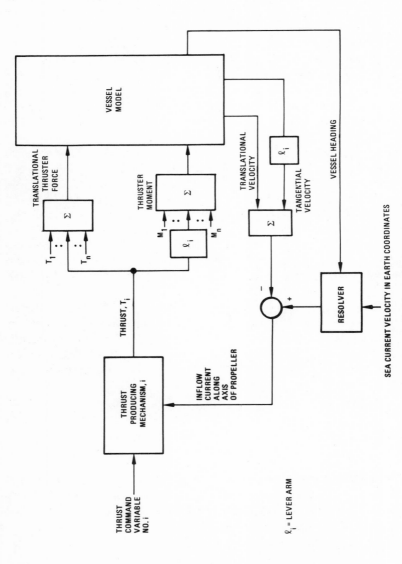

Figure 8-8. A Complete Model of the Thrust Producing
Mechanism for a Thruster System

A further assumption of the model of the thrust-producing mechanism shown in Fig. 8–8 is that the output thrust is directed along the axis propeller. This is a simplification and is not necessarily true. If the necessary data is available to model the direction of the thrust from the propeller, then the thrust can be resolved into its separate components to be applied to the vessel dynamics. However, this is another obvious complication to the thruster model which is not yet complete.

Another conflicting effect not shown in Fig. 8–8 is interaction among thruster units. This interaction occurs when two thrusters are arranged in close proximity in such a way that the flow conditions from one thruster influence the flow from the other unit. Frequently, in thruster installations where there are stern lateral thrusters near the main screws, there is interaction between the main screws and the lateral thrusters.

To model this interaction, a transfer function which couples the two interacting thrusters is required. If the interaction is between thrusters in different axes of vessel, like in the case of the main screws and the lateral thrusters, the interaction functions act as cross-coupling terms between control axes. If the cross-coupling interaction is significant, it influences the design of the control system.

This complication is not considered in this book. Data regarding such interactions are generally not readily available except from a model basin testing where a physical model of the hull of the vessel is built with the thrusters appropriately installed.

To complete the model of the thruster system, the model of the thrust control system must be added to the thrust-producing mechanism model. The thrust control system for the controllable speed system is shown in a more complete block diagram form in Fig. 8–9. The dc motor with a constant field can be modeled by the following transfer function:[2][3]

$$\frac{\dot{\theta}\,(s)}{E_a\,(s)} = \frac{K_m}{L_a J s^2 + (L_a b + R_a J)s + R_a b + K_m K_e} \tag{8.18}$$

where:

K_m = the motor torque constant, $\dfrac{\text{ft-lb}}{\text{amp}}$

L_a = the inductance of the armature, henrys

R_a = the resistance of the armature, ohms

K_e = the back emf constant, $\dfrac{\text{volt-sec}}{\text{radian}}$

J = the equivalent moment of inertia referred to the motor shaft

$$= J_m + \left(\frac{N_1}{N_2}\right)^2 J_1$$

J_m = the moment of inertia of the motor
J_1 = the moment of inertia of the load
b = the equivalent viscous friction referred to the motor shaft

$$= b_m + \left(\frac{N_1}{N_2}\right)^2 b_1$$

b_m = the viscous friction of the motor
b_1 = the viscous friction of the load

The moment of inertia of the load is the sum of the moments of inertia of the propeller, its drive shaft, hub, and a term which is a function of the water by the propeller. Likewise, the viscous friction of the load includes a term for the viscous friction in the propeller bearings plus a term which is a function of the viscous friction caused by the entrained water in the propeller.

The terms, caused by the entrained water, model to some degree the lag of the thrust-producing mechanism in moving the water through it when the thrust-command variable changes suddenly.

Frequently, the inductance of the armature, L_a, is negligible and the motor model can be simplified to the following:

$$\frac{\dot{\theta}\ (s)}{E_a\ (s)} = \frac{K_M}{(\tau_M\ s + 1)} \tag{8.19}$$

$$\tau_M = \frac{R_a J}{R_a b + K_m K_e}$$

where:

$$K_M = \frac{K_m}{R_a b + K_m K_e}$$

The speed of the rotation sensor or tachometer is generally modeled by an output voltage which is linearly proportional to the measured speed of rotation or in equation form:

$$e_T = K_T\ \dot{\theta} \tag{8.20}$$

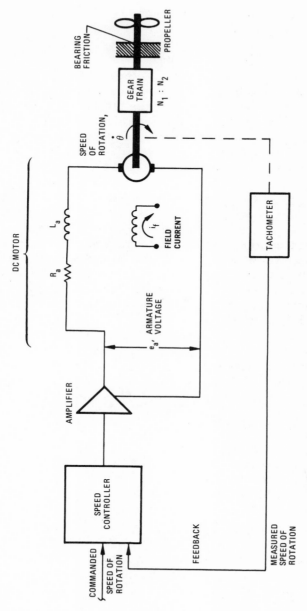

Figure 8-9. A Controllable Speed Thrust Control System

where K_T is the constant of proportionality. This model does not include any of the dynamics of the device. However, in comparison to the motor and its load, the tachometer responds essentially in an instantaneous manner.

The speed controller shown in Fig. 8–9 computes the difference between the commanded and measured speeds of rotation and computes an armature voltage which produces the proper armature current to rotate the motor at the commanded speed of rotation. For the motor to rotate the propeller at a non-zero speed, the armature voltage must be non-zero. At the same time, for the speed of rotation of the motor to be equal to the commanded speed of rotation, the difference in the two speeds of rotation must equal zero.

As a result, the controller must include some means of computing an output command with zero input. Such a computational means is achieved by an integration. With the addition of the integration and error-signal calculation, the block diagram of a controllable speed propeller is as shown in Fig. 8–10 and the overall transfer function of the system is given as follows:

$$\frac{\dot{\theta}_p\,(s)}{\dot{\theta}_c\,(s)} = \frac{\dfrac{N_1}{N_2}\,K_I K_M}{\tau_M s^2 + s + K_I K_M K_T} \tag{8.21}$$

where:

$\dot{\theta}_p$ = the speed of rotation of the propeller

$\dot{\theta}_c$ = the commanded speed of rotation of the propeller

K_I = the integration gain

The time response characteristics of the speed controller can be adjusted by varying the integration gain, K_I. However, with only the integration gain, the designer is limited to the degree of adjustment he can actually achieve. As a result, usually the controller includes in addition to the integrator, a proportional gain, K_P, making the controller a proportional plus integral type. With the addition of the proportional term, the overall speed controller transfer function becomes:

$$\frac{\dot{\theta}_p\,(s)}{\dot{\theta}_c\,(s)} = \frac{\dfrac{N_1}{N_2}\,(K_I + K_P s)\,K_M}{\tau_M s^2 + (1 + K_M K_T K_P)\,s + K_M K_T K_I} \tag{8.22}$$

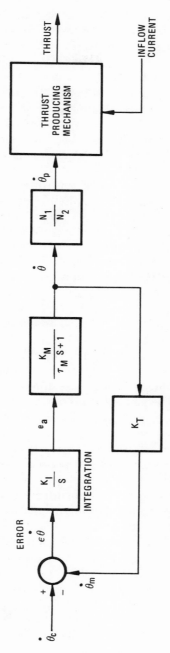

Figure 8-10. The Block Diagram of a Controllable Speed Thruster

whose poles and zero can be adjusted by varying the parameters of the controller to achieve the desired time response more easily than in the case of system with integration only.

The thruster system contains nonlinearities, some of which are by design and some of which are caused by physical effects. For example, the gear train may exhibit a backlash or hystersis-type of nonlinearity. Also, the bearings may exhibit coulomb (dry friction) and stiction in addition to the viscous friction. The amplifier has a maximum output. The motor has a power limit so that for a particular inflow condition the propeller can rotate only so fast, producing a limited amount of thrust. These nonlinearities are natural features of the system.

Another nonlinearity that a speed controller contains is a limit on the rate that armature voltage is applied to the motor. This limit is included by design and is used to prevent overloading the power system and drawing excessive current through the armature windings and commutator, causing a power blackout or damage to the motor.

The pitch control system for a controllable pitch thruster is illustrated in Fig. 8–11. As Fig. 8–11 shows, a hydraulic piston rotates the propeller blades through a mechanical coupling network. The pitch of the blade, which is related to the angle of rotation of the propeller blade, is generally not a linear function of the displacement of the hydraulic piston (see Chapter 4).

As a result, the feedback path or the input command path contains an electronic device which can compensate for the nonlinearity in the mechanical coupling device. Then, the difference between the commanded and measured variables can be computed accurately to form the command to the electrohydraulic control valve, which opens proportionally to the difference to move the piston until the difference is zero.

A block diagram of a pitch control system is shown in Fig. 8–12. It is like the one shown in Fig. 8–11, except the inverse of the nonlinearity of the mechanical coupling to the propeller blades is used to convert the commanded pitch to an equivalent piston position command, y_c. By converting the pitch command to a piston position command, the feedback loop is a more linear control loop than a pitch command system. The overall transfer function for the piston position command system is given as follows:

$$\frac{\lambda_p\,(s)}{\lambda_c\,(s)} = \frac{K_A\,K}{\tau s^2 + s + K_A\,K\,K_{ps}\,K_f} \tag{8.23}$$

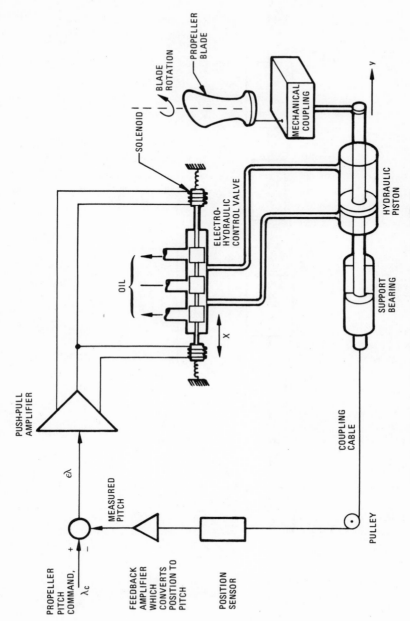

Figure 8-11. A Pitch Control System for a CP Thruster

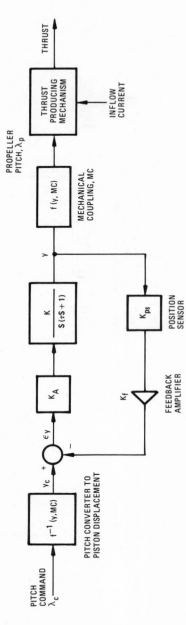

Figure 8-12. The Block Diagram of the Pitch Control System of a Controllable Pitch Propeller

where:

$$K_A = \text{the gain of the push-pull amplifier}$$
$$K_{ps} = \text{the gain of the position sensor}$$
$$K_f = \text{the gain of the feedback amplifier}$$

Unlike the speed control system, the pitch control system does not require an integration in the controller. However, the push-pull amplifier shown in the pitch control system generally contains certain filter networks to improve the performance of the system. This additional capability may be necessary so that the desired time response or noise immunity of the system can be achieved.

Not shown in Fig. 8–11 is the ac induction motor which rotates the propeller. To prevent overloading this motor, the current in each phase is sensed. When the sensed current is greater than a pre-established level, the pitch command to the system is either limited or reduced, depending on the nature of the overload.

Also, the rate at which the pitch of the propeller can be changed is also limited by limiting the flow to the hydraulic piston or the speed of the response of the push-pull amplifier to a sudden increase in error.

So far no examples of the parameters of the thruster systems are given. These parameters are not generally published data from the thruster system manufacturers. In fact, usually their published data is limited to the thrust produced by the thruster under bollard conditions as a function of the thrust command variable. Also, they provide the time required to change the pitch or speed of the thruster from full forward to full reverse.

However, this reversal time is usually dominated by the rate limits on the thrust command variable, imposed by the manufacturer to prevent overloads.

When the thruster manufacturers supply time-response data for their thrust control systems, the responses to a step-function command input are like the responses shown in Fig. 8–13. In the first part of the response the thrust control system is rate limited and only as the response approaches the commanded value is the control system operating as a linear system.

However, the final approach of the output of the thruster control system to the commanded input does not typically reveal any overshoot, implying that the control system is overdamped when operating linearly. Likewise, the linear recovery portion of the step-function

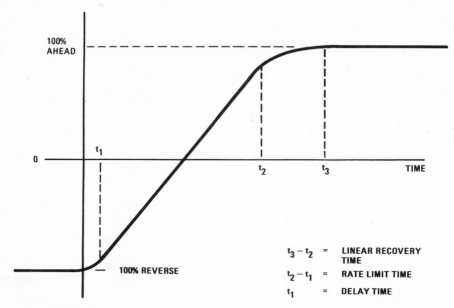

Figure 8-13. Typical Response Characteristic of a Thruster Control
System

response implies that when the thruster is operating in a small
region around a steady-state operating point, it is performing as a
linear overdamped system.

  Typically, for a thruster used on a dynamically positioned vessel,
the full ahead to full reverse time is 10 to 20 seconds, the delay time
is 1 or 2 seconds, and recovery time is 3 or 4 seconds. As a result, the
thruster system is frequently modeled, at least for initial studies, as a
first-order lag with a time constant equal to 1 or 2 seconds. For a more
detailed analysis which includes nonlinear effects, the rate limit is
included with the simple lag, using the block diagram shown in
Fig. 8–14.

8.4 Vessel model

  The dynamically positioned vessel is a mass floating in a fluid, i.e.,
water with wind and waves acting on it, as it performs a given task
such as station keeping or track following. In the general case, the

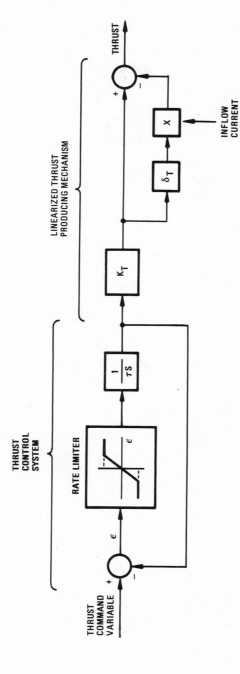

Figure 8-14. A Simplified Thruster Model

fluid is moving relative to a fixed-earth reference, imposing another force on the vessel along with other hydrodynamic forces caused by the vessel's own motion in the fluid.

Countering the various forces acting on the vessel which are tending to drive the vessel away from the commanded position are the thrusters of the dynamic positioning system. The equations of motion or modeling equations of the vessel when derived in their most complete form are quite complex not only because of the sheer number of equations and forces, but also because the hydrodynamic forces are complex. See References 1, 7, 9, and 10 for further details.

In this section a simplified vessel model is presented for only the three primary control axes of the dynamic positioning system. The three axes are x-y positioning and heading. The primary simplifications are the linearization of the hydrodynamic forces. Also, the equations of motion for the vessel are separated into two parts.

The first part involves the motion of the vessel caused by the higher-frequency oscillatory wave forces. The second part involves the remaining forces which are generally at lower frequency.

The coordinate system to be used is shown in Fig. 8–15. The positive x-direction is directed toward the bow of the vessel and is usually referred to as the surge axis. The positive y-direction is directed toward the starboard side of the vessel and is usually referred to as the sway axis. The yaw angle of the vessel is positive in a clockwise direction around the z-axis, or heave axis of the vessel, which is directed downward.

The reference coordinate system shown in Fig. 8–15 is usually the earth coordinate system with X_R pointing north and Y_R pointing east. In such a system the heading of the vessel is the angle clockwise from north to the x-axis of the vessel.

The equations of motion of the vessel in the reference coordinate system are expressed by the following differential equations:

$$
\begin{aligned}
m(\dot{u} - v\dot{\psi}) &= \Sigma F_x \\
m(\dot{v} + u\dot{\psi}) &= \Sigma F_y \\
I_{zz}\,\ddot{\psi} &= \Sigma M_z
\end{aligned}
\qquad (8.24)
$$

where:

ΣF_x, ΣF_y = the summation of x-axis and y-axis forces respectively

ΣM_z = the summation of moments about the z-axis

m = the mass of the vessel

I_{zz} = the moment of inertia of the vessel about the z-axis
u, v, $\dot{\psi}$ = the velocities for the x-axis, y-axis, and yaw axis

The mass of the vessel is equal to the displacement of the vessel times the density of the fluid in which the vessel is immersed. The moment of inertia about the z-axis is equal to the mass of the vessel times the square of the radius of gyration of the vessel about the z-axis.

Forces acting on the vessel include hydrodynamic, aerodynamic, thruster, and any force related to the task that the vessel is performing such as tensioning a pipe as it is being laid. The aerodynamic and thruster forces are quite straightforward.

For example, the aerodynamic forces can be expressed as follows since the velocity of the vessel is generally small compared to the velocity of the wind:

$$F_{XA} = C_{XA} (\alpha_A) v_W^2$$
$$F_{YA} = C_{YA} (\alpha_A) v_W^2$$
$$M_{ZA} = C_{ZA} (\alpha_A) v_W^2 \qquad (8.25)$$

where the $C (\alpha_A)$'s are the aerodynamic drag coefficients in units of force or moment per wind speed units squared which are a function of the angle of attack of the wind to the vessel and v_W is the magnitude of the wind velocity. The aerodynamic forces are referred to the vessel coordinate system because the wind is usually measured relative to vessel coordinate system.

However, in a time-domain simulation the wind is defined in terms of the earth coordinate system, so that as the vessel changes heading, the wind direction does not shift. With the wind defined in earth coordinates the magnitude of the wind velocity does not change, but the angle of attack of the wind does change so that $\alpha_A = \psi_A - \psi$ where ψ_A is the wind azimuth with respect to north.

Typically, the aerodynamic drag coefficients for an offshore vessel, especially ship-shaped hulls, resemble sinusoidal functions of the angle of attack of the wind to the vessel, α_A. Thus, quite frequently, reasonable approximations to the aerodynamic drag forces on these vessels can be expressed as follows:

$$F_{XA} = K_{XA} v_W^2 \cos \alpha_A$$
$$F_{YA} = K_{YA} v_W^2 \sin \alpha_A \qquad (8.26)$$
$$M_{ZA} = K_{ZA} v_W^2 \sin 2\alpha_A$$

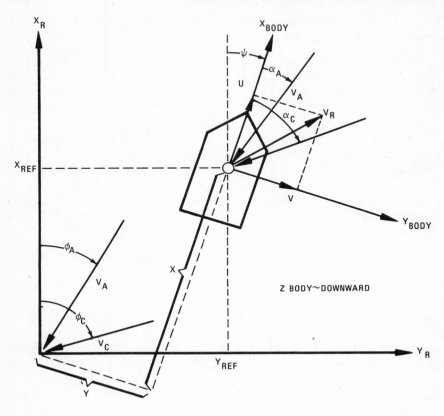

X_R, Y_R	REFERENCE AXES
$X_{BODY}, Y_{BODY}, Z_{BODY}$	BODY AXES
$X_{REF}, Y_{REF},$	VESSEL POSITION MEASURED IN REFERENCE COORDINATES
X, Y	VESSEL POSITION MEASURED IN BODY COORDINATES
ψ	VESSEL HEADING RELATIVE TO REFERENCE AXES
U, V	VESSEL VELOCITY COMPONENTS IN BODY COORDINATES
V_C	SEA CURRENT VELOCITY
ϕ_C	ANGLE OF SEA CURRENT RELATIVE TO REFERENCE AXES
V_A	WIND VELOCITY
ϕ_A	ANGLE OF WIND RELATIVE TO REFERENCE AXES
V_R	VESSEL VELOCITY RELATIVE TO THE WATER
α_A	ANGLE OF WIND RELATIVE TO BODY AXES
α_C	ANGLE OF RELATIVE VELOCITY IN BODY COORDINATES

Figure 8-15. Coordinate System and Nomenclature

where the K's are coefficients which give the amplitudes of the drag forces and moments.

The wind speed and direction which generate the aerodynamic forces on the vessel vary randomly in the real world. Generally speaking, wind speed and direction can be characterized with a mean value around which the speed and direction vary randomly. The variations in wind speed about its mean value can be modeled by a spectral density in Chapter 1.

The randomness of the wind direction can also be modeled by a spectral density. However, the author is not aware of the existence of a published wind direction spectral density.

To simulate the variable portions of the wind, i.e., its gustiness, the spectral densities of wind speed and direction can be approximated by a finite number of sinusoids whose frequencies and amplitudes are determined by partitioning the densities. The same process is used to simulate the wave spectrum. This is described in more detail later in this section and in Appendix B.

Once the variable portions of the wind are generated, they are added to the appropriate mean values and used in Equation 8.25 to generate the wind forces to be applied to the vessel dynamics.

When simulating the random effects of the wind for the purpose of frequency domain analysis as discussed in Chapter 7, the spectral density used to model the wind speed must be modified to a force spectral density. Once the random variations of the wind are modeled as a force spectrum, then the wind can be entered into the control system as shown in Fig. 7–2 as a disturbance force.

Then the spectral density of some variable in the control system can be computed using Equation 7.77 from which the variance of the wind variable can be computed by integration of the spectral density (see Equation 7.78). For example, the spectral density of the output in Fig. 7–2 is given as follows:

$$S_c\ (f) = \left| \frac{G\ (j\omega)}{1 + G\ (j\omega)\ H\ (j\omega)\ D\ (j\omega)} \right|^2 S_F\ (f) \qquad (8.27)$$

where $S_F\ (f)$ is the spectral density of wind force.

The thruster forces and moment are even more straightforward than the aerodynamic quantities if the axes of the thruster are fixed. In equation form the thruster forces and moment for fixed-axis thrusters are as follows:

$$F_{XT} = \sum_{i=1}^{M} F_{Ti}$$

$$F_{YT} = \sum_{j=1}^{N} F_{Tj} \tag{8.28}$$

$$M_{ZT} = \sum_{i=1}^{M+N} \ell_i F_{Ti}$$

where ℓ_i is the distance that the thrust axis of the ith thruster is from the center of rotation of the vessel, M is the number of thrusters with their thrust axis aligned with the x-axis of the vessel, and N is the number of y-axis thrusters.

If there is interaction between thrusters, additional terms must be added appropriately to the equation. It should be remembered that in the thruster model the effect of thrust degradation caused by inflow current is already included. Thus, velocity outputs from the vessel model must be fed back to the thruster models.

Modeling the forces and moments caused by the work tasks may be quite straightforward. If they depend on vessel motions and hydrodynamic effects, however, they may be quite complex. For example, modeling the forces and moments caused by the pipe on a pipe-laying vessel is complex unless the pipe is modeled as a simple catenary, which, of course, is an oversimplification, except, perhaps, in very deep water.

Simplified hydrodynamic forces are modeled for the x-y and yaw axes to include terms which act as added mass and damping. Both of these terms are generally modeled as a function of frequency. The data can be computed or measured in model basins and relies on the assumption that the forcing function is purely oscillatory and that its amplitude is sufficiently small so that the process is linear.[7,9] Thus, these hydrodynamic terms are applicable in determining the motion of the vessel to the higher-frequency sea spectrum as modeled by the Pierson-Moskowitz or Bretschneider wave spectra given in Chapter 1.

Added mass decreases and damping increases as a function of frequency as shown in Fig. 8–16, whereas added inertia varies as a function of frequency as shown in Fig. 8–16. For a ship-shaped hull the added mass for the x-axis is much smaller than the added mass for the y-axis, reflecting the difference in hydrodynamic resistance between the two axes for a ship-shaped hull. For example, the x-axis added

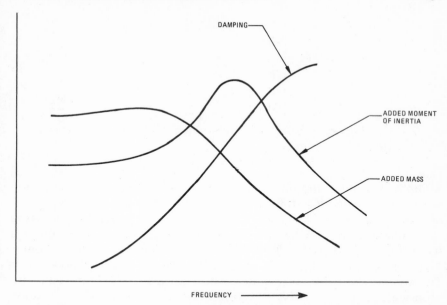

Figure 8-16. Hydrodynamic Parameters as a Function of Frequency

mass is typically 10 to 20% of the mass while the y-axis added mass is typically 80 to 100% of the mass of the vessel for a ship-shaped hull.

The hydrodynamic forces modeled by the added mass and damping are 180 degrees out-of-phase with the inertial acceleration of the vessel. As a result, the equations of motion for the vessel when these forces are included appear as follows:

$$(M + A\ (\omega))\ \bar{\ddot{q}} + B\ (\omega)\ \bar{\dot{q}} = \bar{F}\ (t) \qquad (8.29)$$

where:

$$\bar{q} = [X,\ Y,\ \psi]^{-1}$$

$$M = \begin{bmatrix} \rho\nabla & 0 & 0 \\ 0 & \rho\nabla & 0 \\ 0 & 0 & \rho k^2_{zz}\nabla \end{bmatrix}$$

$$A\ (\omega) = \begin{bmatrix} A_{11}\ (\omega) & A_{12}\ (\omega) & A_{13}\ (\omega) \\ A_{21}\ (\omega) & A_{22}\ (\omega) & A_{23}\ (\omega) \\ A_{31}\ (\omega) & A_{32}\ (\omega) & A_{33}\ (\omega) \end{bmatrix}$$

$$B\ (\omega) = \begin{bmatrix} B_{11}\ (\omega) & B_{12}\ (\omega) & B_{13}\ (\omega) \\ B_{21}\ (\omega) & B_{22}\ (\omega) & B_{23}\ (\omega) \\ B_{31}\ (\omega) & B_{32}\ (\omega) & B_{33}\ (\omega) \end{bmatrix}$$

$$F(t) = \begin{bmatrix} F_x\ (t) \\ F_y\ (t) \\ M_z\ (t) \end{bmatrix}$$

$\rho =$ the density of the water in which the vessel is immersed
$\nabla =$ the displacement of the vessel
$\omega =$ the radian frequency

The added mass and damping terms in the A and B matrices vary not only as a function of frequency, but also as a function of the geometry of the vessel. The subscripting used in the A and B matrices is not standard for the yaw axis. Normally the yaw axis subscript is 6 instead of 3. In the case of a ship-shaped hull, for instance, $A_{12}\ (\omega) = A_{13}\ (\omega) = A_{21}\ (\omega) = A_{31}\ (\omega) \cong 0$ and $B_{12}\ (\omega) = B_{13}\ (\omega) = B_{21}\ (\omega) = B_{31}\ (\omega) \cong 0$. The other off-diagonal terms, i.e., A_{23}, A_{32}, B_{23}, and B_{32}, represent the coupling of motion between the yaw and sway axes of the ship-shaped vessel. Generally, the off-diagonal terms are small compared to the sum of the diagonal terms and the mass matrix terms and are, therefore, neglected in preliminary system investigations.

The forces and moment given in Equation 8.29 are oscillatory with the following form:

$$F_i(t) = F_{oi} \cos\ (\omega t + \phi_i) \qquad\qquad (8.30)$$

where F_{oi} is the amplitude of the ith sinusoid and ϕ_i is the phase angle of the ith element of the force vector, $F\ (t)$. These forces and moment can be caused by anything. However, for an offshore vessel, the forces and moment which are oscillatory at a sufficiently high frequency are caused by the wind-generated waves.

Using Equation 8.29 is unwieldy because of the dependence of the damping factors and added masses on frequency. As a result, Equation 8.29 is solved to determine the transfer functions between the wave-induced vessel motions and the waves which cause the motion.[9][11] With these transfer functions, the dynamic model of the vessel can be partitioned into a higher-frequency wave-induced motion section and a section which generates the vessel motions due to other forces. The total motions from the two sections is shown in Fig. 8–17.

Usually, the non-wave section of the vessel model generates the

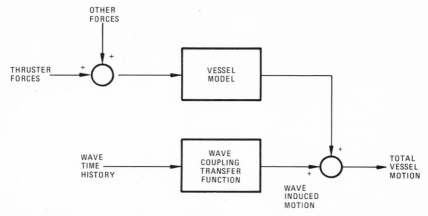

Figure 8-17. Vessel Model with the Separation of Wave Motion

responses with lower frequency content than the wave portion and is, therefore, often referred to as the low-frequency portion of the vessel model.

These transfer functions are referred to as response amplitude operators (RAOs) and indicate the motion characteristics of the vessel in waves. The RAOs supplied with the vessel often indicate the vessel coordinates which most affect the work task for which the vessel is designed.

In the case of a drillship, for instance, the owner of the vessel gives the roll, pitch, and heave RAOs to show how well his vessel performs in rough weather. However, none of these are directly applicable to the design of a dynamic positioning system for the drillship.

For a typical drillship, whose length is 400–500 ft and displacement is 10–15,000 tons, the RAOs for x (surge), y (sway), and yaw are as shown in Fig. 8–18. Fig. 8–18 also shows the spectral density of the higher-frequency waves. The overlap of the RAOs and the wave spectrum is a measure of the motion of the vessel in relation to the particular wave spectrum, i.e., the greater the overlap, the greater the vessel motion.

Thus, for a smaller vessel whose RAOs are shifted to higher frequencies, the RAOs indicate that the smaller vessel experiences larger motions in waves than a drillship.

The two-section vessel model shown in Fig. 8–17 is helpful in performing a frequency analysis of the dynamic positioning control

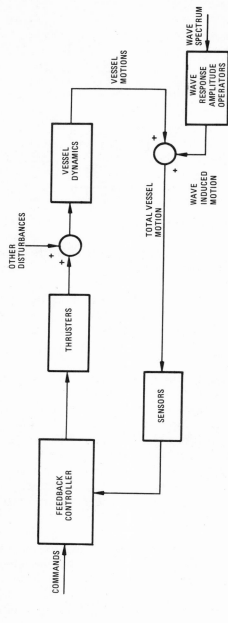

Figure 8-18. Wave Induced Motion Input to the DP System

system. Then the wave-induced motion portion of the total system can be treated as an input to the control system as illustrated in Fig. 8–19. The spectral density of the wave-induced motion is computed by multiplying the spectral density of the waves by the square of the RAO as shown in Equation 7.76 in the previous chapter. Then the spectral density of any signal within the control loop can be determined using Equation 7.77.

Generating the wave-induced motions of the vessel in a time domain simulation is slightly more complicated than frequency domain simulation. First the wave profile must be generated by approximating the wave spectrum with a finite number of discrete frequency sinusoids. The frequencies of the sinusoids should be nonharmonically related and are selected by partitioning the wave spectral density into a finite number of areas as shown in Fig. 8–20.

There are various methods of partitioning the spectrum. For example the spectrum can be divided according to equal energy or areas. The frequency of each partition is the center frequency of the area and the amplitude is proportional to the square root of the area. For equal-area partitioning, all the sinusoids have the same amplitude. In equation form, the approximate wave profile becomes as follows:

$$h_{wave}(t) = \sum_{i=1}^{N} a_{wi} \sin (\omega_i t + \epsilon_i) \qquad (8.31)$$

where:

$h_{wave}(t)$ = the wave amplitude about the mean water level as a function of time

N = the number of discrete frequencies used to simulate wave spectrum

a_{wi} = the amplitude of the ith sinusoid

ω_i = the frequency of the ith sinusoid

ϵ_i = the random phase of the ith sinusoid

The number of sinusoids used to approximate the wave spectrum varies depending on the desired accuracy in simulating the spectral density function selected to model the real-world waves.

Since the selected spectral density is, in itself, a model with only a given degree of accuracy, using too many sinusoids to simulate the wave spectrum is of questionable value. This is especially true if the simulation is being performed with a computer of limited capabilities.

After the wave profile is generated by means of Equation 8.31, the

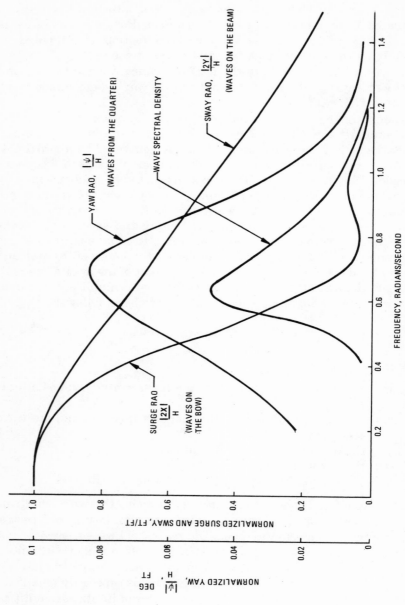

Figure 8-19. Typical Drillship RAOs in Surge, Sway, and Yaw

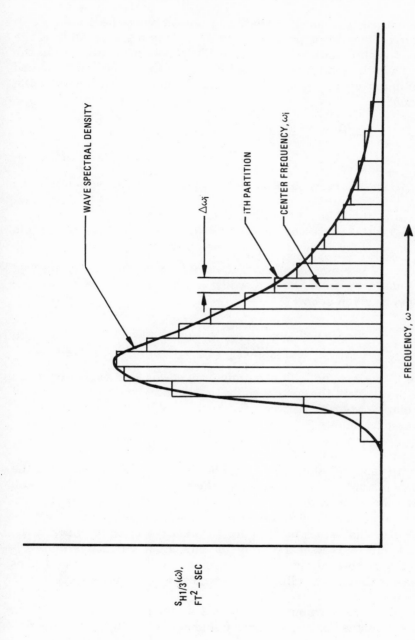

Figure 8-20. The Partitioned Wave Spectral Density

wave-induced motion is computed by multiplying the amplitude of each sinusoid by the appropriate factor from the RAO and shifting the phase of each sinusoid according to the phase information from the RAO. Then the total wave-induced motion of the vessel is determined by summing up the modified sinusoids as indicated in the following equation:

$$q_{wave}(t) = \sum_{i=1}^{N} a_{wi} RAO(\omega_i) \sin(\omega_i t + \epsilon_i + \phi_{RAO}(\omega_i) \tag{8.32}$$

where:

$q_{wave}(t) =$ the vessel coordinate for which the wave motion is to be computed

$RAO(\omega_i) =$ the value of the RAO at ω_i

$\phi_{RAO}(\omega_i) =$ the phase shift caused by vessel dynamics for a sinusoid of frequency ω_i

There are two remaining hydrodynamic forces. The first is the drag force caused by ocean current which is predominantly a square-law function as given below:

$$\begin{aligned} F_{xc} &= C_{xc}(\alpha_c) V_R^2 \\ F_{yc} &= C_{yc}(\alpha_c) V_R^2 \\ M_{zc} &= C_{zc}(\alpha_c) V_R^2 \end{aligned} \tag{8.33}$$

where $C(\alpha_c)$ are the hydrodynamic drag coefficients expressed as an angle of attack of the current to the vessel and V_R is the magnitude of the relative velocity between the current and the vessel, which is expressed as follows:

$$V_R^2 = (u - u_c)^2 + (v - v_c)^2 \tag{8.34}$$

where the speed of the current equals $\sqrt{u_c^2 + v_c^2}$ and u_c and v_c are the components of the current speed in the x-y coordinates of the vessel.

If the current is defined in earth coordinates, then the current velocity will necessarily be resolved into vessel coordinates through the heading of the vessel to compute the relative velocity between the current and the vessel. Defining the current in earth coordinates is generally performed for a time-domain simulation of the vessel so as the vessel changes heading, the current does not shift direction with the vessel.

The final hydrodynamic force is another force caused by the waves. However, instead of being higher frequency, as in the case of the previously discussed wave forces, this force is low frequency and models the slow drift of the vessel in waves. The apparent mechanism which creates this slowly varying hydrodynamic force is the presence of the hull of the vessel in the waves affecting the mass transfer by reflecting waves back into the on-coming waves.[1][12][13]

This interaction leads to the wave-drift forces which have not yet been as accurately modeled as the other forces acting on the vessel. In some cases these wave drift forces are modeled as a constant force and in other cases a low-frequency model is used as discussed in Reference 1. The force used in the model discussed in Reference 1 is derived from an expression for the lateral drift force given as follows[13]:

$$F_{YWD} = \frac{1}{2}\rho g A_w^2 R^2 L \sin^2\alpha \qquad (8.35)$$

where:

ρ = the density of water
g = the gravitational constant
A_w = the amplitude of the incident wave
R = the drift force coefficient
L = the reference length
α = the angle of incidence of the wave to the vessel

For a free-floating vessel, the drift force coefficient for a vertical plate appears to match some measurements made by Laangas on a Series 60 ship as shown in Fig. 8–21.

Over the frequency range of the variations in the wind, current and wave drift forces and moment, the hydrodynamic forces are essentially those represented by the added mass terms of the vessel. This is true of any other slowly varying (low-frequency) forces. As a result, the equations of motion or model of vessel for low-frequency effects can be expressed as follows:

$$\rho\nabla(\dot{u} - v\dot{\psi}) + A_{11}\dot{u} = F_{XC} + F_{XA} + F_{XWD} + F_{XD} + F_{XT}$$
$$\rho\nabla(\dot{v} + u\dot{\psi}) + A_{22}\dot{v} + A_{23}\ddot{\psi} = F_{YC} + F_{YA} + F_{YWD} + F_{YD} + F_{YT}$$
$$(k_{zz}^2\rho\nabla + A_{33})\ddot{\psi} + A_{32}\dot{v} = M_{ZC} + M_{YA} + M_{ZWD} + M_{ZD} + M_{ZT} \qquad (8.36)$$

where F_{XWD}, F_{YWD}, and M_{ZWD} are the wave-drift forces and F_{XD}, F_{YD}, and M_{ZD} are any other slowly varying disturbance forces.

The low-frequency model of the vessel represented by Equation 8.36 also includes the forces and moment generated by the thrusters. The reason for this is that the dynamic positioning controller purposely

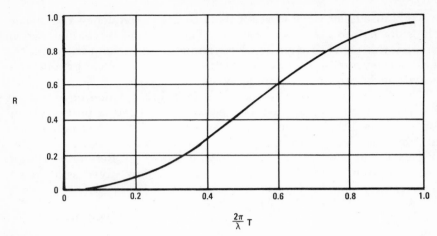

Figure 8-21. The Wave Drift Force Coefficient for a Vertical Plate in Beam Waves [13]

filters the sensor data to minimize the response of the thrusters to the higher-frequency wave-induced motions.

This same criteria of a lower-frequency range than the waves, must be applied to any force or moment included in the low-frequency model of the vessel because only then can the hydrodynamic forces be modeled as a constant-added mass term.

Wave-induced motion and the low-frequency vessel motion comprise the total vessel motion. Computation of the total vessel motion can be performed by adding the position variables of Equation 8.36 and wave-induced position variables of the low-frequency model together with the summation of the wave-induced motions to obtain the total vessel motion.

This is shown in block diagram form in Fig. 8–22. The contents of the various blocks are expressed by the appropriate equations given earlier in this section. As Fig. 8–22 shows, the outputs of the total vessel model are fed to the models of the sensors. Since in Fig. 8–22 only x-y position and yaw are fed to the sensor, it is assumed that the sensors measure only x-y position and yaw angle. However, if velocity or acceleration sensors are used in the actual system being modeled, then velocity and acceleration must be output from the vessel model.

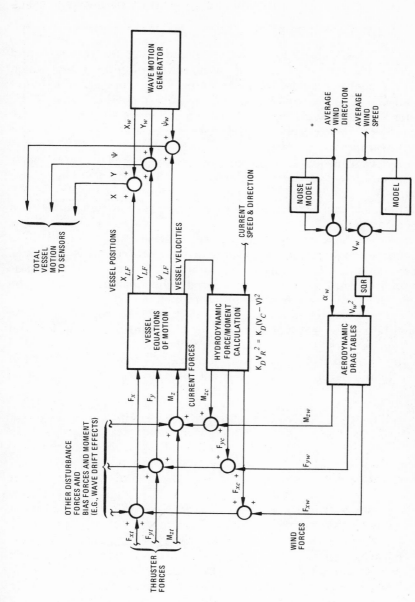

Figure 8-22. The Total Model for the Vessel

8.5   Sensor models

Thus far this chapter has dealt with models for the thrusters and the vessel. The final element of the dynamic positioning system which requires modeling to close the loop on the DP control system is that of the sensors. The controller does not require modeling because the transfer functions which make up the controller are known equations. If modeling of the controller is performed, it is usually a simplification of the known transfer functions.

Three sensor models are required to close the three DP control loops: (1) a heading sensor, (2) x-position sensor, (3) and y-position sensor. Usually, a model of a wind sensor is also required because most DP controllers include wind feed-forward compensation commands as a standard item.

Very often the x-position and y-position sensors are provided by a dual-channel position sensor. There are several position reference systems (Chapter 2). In this section, only the short baseline acoustic position reference system is considered. The heading sensor to be considered in this section is the gyrocompass.

If the short baseline acoustic position reference is properly installed and calibrated, a reasonable model for the resulting x-position and y-position sensors is one of the vertical reference sensor in parallel with the position measurement as shown in Fig. 8–23. The model of the vertical reference sensor depends on the type of sensor being modeled.

For this analysis the vertical reference sensor is assumed to be a vertical gyro whose transfer function is given in Chapter 3. The com-

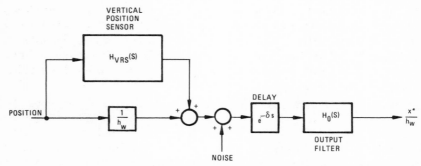

Figure 8-23. A Model of a Single Axis of an Acoustic Position Reference Sensor

bined transfer function of the vertical gyro and the acoustic position measurement is approximated for small position offsets where the acoustic system is assumed to be a pinger system as follows:

$$\Theta_M = \left[1 - s^2 \frac{h_w}{g} \frac{\omega_n^2}{s^2 + 2\zeta\omega_n s + \omega_n^2} \right] \tag{8.37}$$

where:

$$\Theta_M \cong \frac{\text{position offset}}{\text{water depth}}$$

$h_w = $ the water depth
$g = $ the gravitational constant
$s = $ the Laplace operator

To complete the model of the acoustic position reference system, noise is added to the measured position and vertical reference sensor output (Fig. 8–23). Then a pure computational delay is included in series with the noise, the measured position, and vertical reference sensor output in order to model position measurement made in a periodic discrete fashion as acoustic energy is received from the subsea pinger.

Finally, the position reference sensor system may have a filter to smooth the output position measurement data.

The noise injected into the position measurement model represents a random uncertainty in the position measurements. It is usually modeled as white noise with zero mean and a given variance. The variance of the noise depends on the particular sensor.

Although a gyrocompass has certain dynamics, if it has settled after its initial alignment cycle, the gyrocompass responds to the changes in heading of a ship sufficiently fast so that it can be modeled as a perfect sensor with unity transfer function. Likewise, the measurement of heading does not exhibit any significant sources of noise. Thus, a reasonable model of the heading sensor is a block containing a unity transfer function.

The wind sensor does exhibit a delay or lag to changes in wind speed and direction. These effects can be successfully modeled as simple lags, as shown in Fig. 8–24. Noise can be added to the measurement of the wind speed and direction if such data is available. The sensor may be in error in its measurement of the speed and direction by a constant factor because of shading or height above a free surface.

One possible model of the constant errors in speed and direction in-

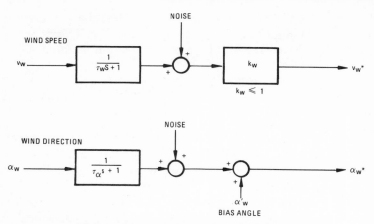

Figure 8-24. A Wind Sensor Model

cludes a gain factor for speed and an offset angle in the direction measurement.

8.6   Simulation methods

Chapter 7 contains various analysis techniques for closed-loop control systems. They use the mathematical models of the various elements of the control system and mathematical methods to estimate the performance of the closed-loop system. The techniques given in Chapter 7 are primarily for linear representations of the system and give indirect measures of system performance from quantities derived in the frequency domain.

For example, if the control system requirements call for a transient response of a given percent overshoot to a step-function input, then the designer adjusts the frequency response characteristics of the system to achieve a given amount of gain and phase margin.

The amount of gain and phase margin required to achieve the desired transient characteristics is usually determined by the experience of the designer in previous designs where he has had the opportunity to evaluate the transient response of a system for a given amount of phase and gain margin.

As mentioned in Chapter 7, when the mathematical analysis techniques are not applicable because of nonlinearities or multiple variables, then other methods of estimating the performance of the control system are required. Even when a system can be successfully modeled in a linear manner about various operating points or in various operating regions, another method of estimating the system performance over the entire region of operation is necessary to insure that there are no undesirable effects in transitioning from one region to another.

In fact, for the linear system, some method of verifying that the design achieves the required transient response is often needed or desirable. Most commonly, this method of estimating total system performance is the time-domain solution of the mathematical equations which represent the closed-loop control system. This method is referred to as a simulation in which, except in the simplest system, a computer performs the solution of the equations.

There are two major types of computers used for simulations. The first, the analog computer, was more commonly used 10 or 15 years ago. The second is the high-speed digital computer. Simulations are also implemented on a combination analog and digital computer referred to as a hybrid computer. In this section methods are given for implementing in the computers the equations of motion for a closed-loop control system.

The computer is used to solve the mathematical equations which represent the control system. In a control system which includes dynamic elements the computer solves the system of differential equations or first-order multidimensional differential equations (state equations). In the case of linear elements the transfer functions are often represented in Laplace transform format. These three equation forms are illustrated in Fig. 8–25.

The primary element of an analog computer upon which the equations of a system are implemented is the operational amplifier. The operational amplifier is composed of a high-gain, high-input impedance amplifier around which various impedances are arranged as shown in Fig. 8–26. The high-input impedance allows the transfer function of the operational amplifier to be written as follows:

$$\frac{e_0}{e_i} = \frac{-\dfrac{1}{Z_i}}{\dfrac{1}{Z_f} + \dfrac{1}{K}\left(\dfrac{1}{Z_i} + \dfrac{1}{Z_f}\right)} \qquad (8.38)$$

where:

e_0 = the output voltage of the amplifier
e_i = the input voltage of the network

Z_i = the impedance of the input element
Z_f = the impedance of the feedback element
K = the voltage gain of the amplifier

If K is very large, say 10^6 or 10^8, then Equation 8.38 can be approximated as follows:

$$\frac{e_0}{e_i} = -\frac{Z_f}{Z_i} \qquad (8.39)$$

For example, if the feedback element is a capacitor of capacitance C and the input element is a resistor of resistance R, then the transfer function of the analog computer amplifier is given as follows:

$$\frac{e_0}{e_i} = -\left(\frac{1}{Cs}\right)\left(\frac{1}{R}\right) = -\frac{1}{RCs} \qquad (8.40)$$

where s is the Laplace transform operator. Equation 8.40 is the transfer function of an integrator with a gain of 1/RC. Thus, if R equals one megohm and C equals one microfarad, then the transfer function of the analog element is that of pure integration with a sign inversion (see Chapter 7).

When an integration is performed, the output of the integration is equal to the sum of the integral of the input function plus the initial value of the output variable. To provide the analog integrator with the initial value of the output variable, the feedback capacitor is initially charged to a voltage equal to the initial value of the output variable. Then, as the input voltage is applied to the integrator, the output voltage starts at the initial value.

In a more complex example where the feedback element is a combination of a capacitor in parallel with a resistor and the input element is a resistor, the resulting transfer function of the analog computer element is:

$$\frac{e_0}{e_i} = \frac{\left(\dfrac{R_f}{R_fCs + 1}\right)}{R_i} = \frac{\dfrac{R_f}{R_i}}{R_fCs + 1} = \frac{K}{\tau s + 1} \qquad (8.41)$$

where $K = \dfrac{R_f}{R_i}$ and $\tau = R_fC$. The analog computer element simulated by Equation 8.41 is a first-order lag.

If the control system is represented in block diagram form as shown in the second part of Fig. 8–25, then an analog computer circuit or

DIFFERENTIAL EQUATION FORMAT

$$\frac{d^n x}{dt^n} + a_1 \frac{d^{n-1} x}{dt^{n-1}} + \ldots + a_{n-1} \frac{dx}{dt} + a_n x = u(t)$$

TRANSFER FUNCTION FORMAT WITH LAPLACE TRANSFORMS

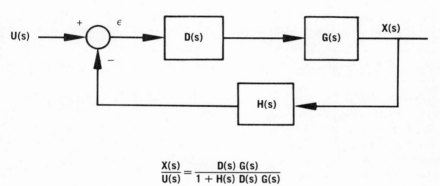

$$\frac{X(s)}{U(s)} = \frac{D(s)\ G(s)}{1 + H(s)\ D(s)\ G(s)}$$

STATE SPACE FORMAT

$$\frac{d}{dt}\ \bar{x}(t) = \underline{A}(t)\ \bar{x}(t) + \underline{B}(t)\ \bar{u}(t)$$

$$\bar{y}(t) = \underline{C}(t)\ \bar{x}(t) + \underline{D}(t)\ \bar{u}(t)$$

Figure 8-25. Three Possible Control System Representations

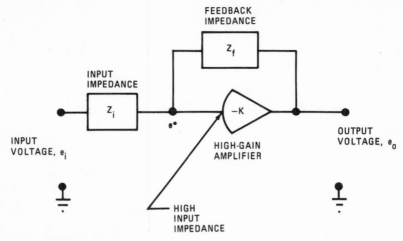

Figure 8-26. The Basic Analog Computer Operational Amplifier

combination of circuits can be implemented to represent each block of the system. In each case the output of one amplifier or transfer function becomes the input to the next amplifier. As a result, the structure or interconnection of the analog computer is exactly the same as the block diagram.

This can be useful to the designer if he has modeled the system so that the outputs of the analog computer elements are physical variables of the system. If they are, he can record or measure the output voltages as a function of time to various system inputs to determine the time response characteristics of various physical elements of the system other than the output. This can be especially important in a system which has been linearized so that the assumption of linearization can be verified.

Most commercially available analog computers do not provide the user with a generalized input and feedback impedance from which he can synthesize his desired transfer function. Instead they provide four basis elements: (1) multiple input integrator, (2) a multiple input summer, (3) an inverter, and (4) a gain. These elements are illustrated in Fig. 8–27 with their corresponding equation.

To synthesize more complex transfer functions from the basic analog computer elements, the transfer functions or differential equations can be rewritten in a form which is composed of first-order de-

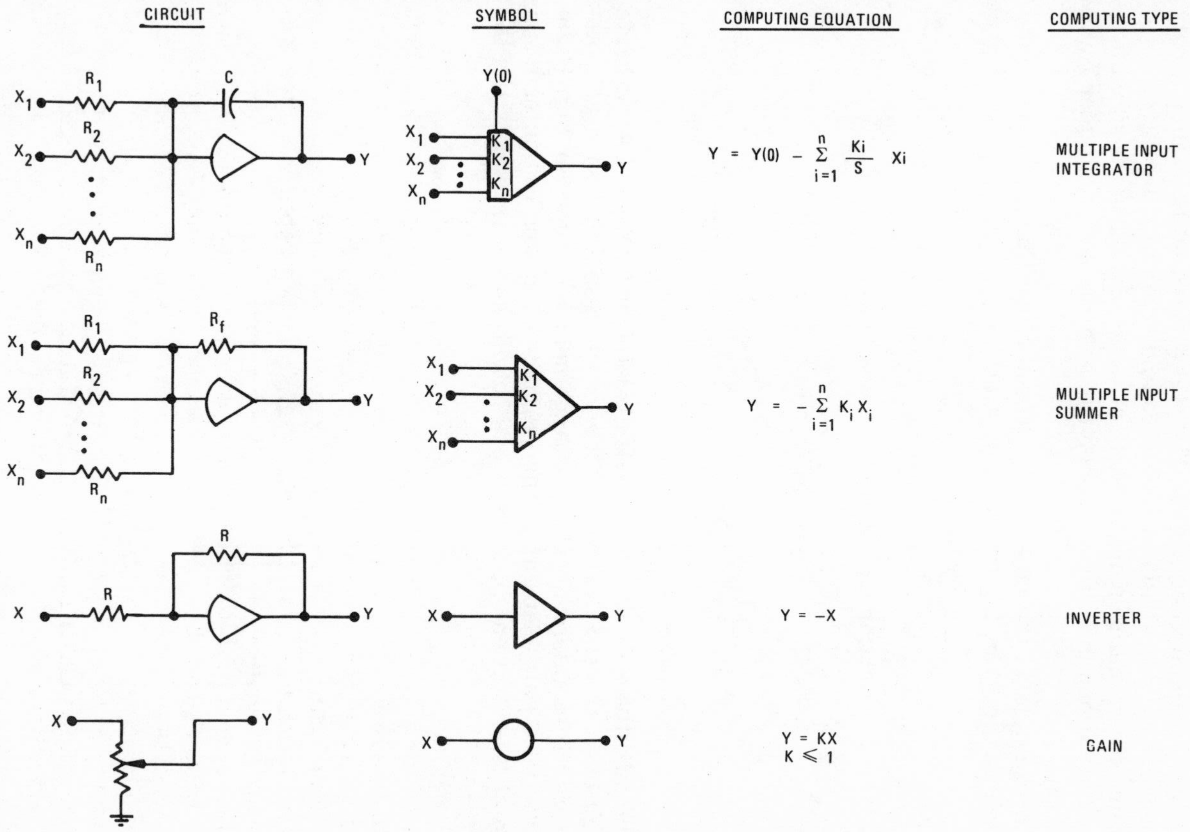
CIRCUIT
SYMBOL
COMPUTING EQUATION
COMPUTING TYPE
X1 R1 X2 R2 Xn Rn C Y
Y(0) X1 X2 Xn K1 K2 Kn Y
Y = Y(0) − Σ (i=1 to n) Ki/S Xi
MULTIPLE INPUT INTEGRATOR
X1 R1 X2 R2 Xn Rn Rf Y
X1 X2 Xn K1 K2 Kn Y
Y = − Σ (i=1 to n) Ki Xi
MULTIPLE INPUT SUMMER
X R R Y
X Y
Y = −X
INVERTER
X Y
X Y
Y = KX
K ≤ 1
GAIN

Figure 8-27. Common Analog Computer Elements

rivatives or single integrations. The resulting equations can be easily implemented with single integrations and summing amplifiers. Such decomposition methods are equivalent to those used to rewrite differential equations in state variable form, as discussed in Chapter 7 and Reference 6.

An example is the case of the second-order linear system represented by the following transfer function:

$$\frac{X(s)}{U(s)} = \frac{\omega_n^2}{s^2 + 2\zeta\omega_n s + \omega_n^2} \tag{8.42}$$

The transfer function can be rewritten as follows:

$$X_2(s) = \frac{1}{s}\left[-2\zeta\omega_n X_2(s) - \omega_n^2 X_1(s) + \omega_n^2 U(s)\right]$$

$$X_1(s) = \frac{1}{s}\left[X_2(s)\right] \tag{8.43}$$

$$X_1(s) = X(s)$$

If the quantities in the brackets are treated as inputs to two separate integrators, the second-order system can be simulated with two integrators in an interconnection arrangement as shown in Fig. 8–28.

The basic gain element of the commercial analog computer has a gain of less than one because it is a potentiometer (Fig. 8–27). If the

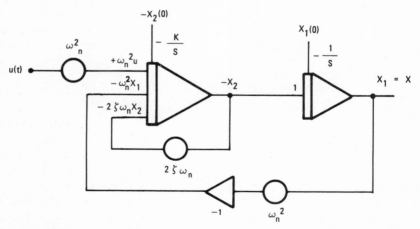

Figure 8-28. A Simulator Diagram for a Second-Order Linear System

coefficients of a transfer function are greater than one, then the potentiometer must be combined with an active computer element which has an input gain that is greater than one. Most commercial analog computers provide their summers and integrators with multiple inputs with gains of one and ten. Then with a combination of a pot and gain of ten, a transfer function coefficient with a value between zero and ten can be achieved.

If larger numbers are required for coefficients, then additional stages of amplification are required. However, there are limits to large coefficients and to small coefficients because of practical considerations of the analog computer elements. In the case of the summers and the integrators, their output voltages are linear functions of their inputs up to a given value. Beyond that value the linearity decreases until the output voltage no longer increases for an increase in the input to the device.

Thus, when large transfer function coefficients are being simulated, output voltages may necessarily be amplified until they are larger than the linearity limit which tends to invalidate the resulting solution of system equations.

The solution to this problem requires the proper scaling of the simulation equations not only in the magnitude of the equation variables but also in terms of time.

Magnitude scaling is required not only to limit the computer variables to linear regions of computer operation, but also to scale the units of the physical variables to the units of the computer, i.e., volts. To limit the computer variables to the linear regions of the computer operation requires at least an initial estimate of the maximum expected magnitude of the physical variables. Then the magnitude scale factors can be computed as follows:

$$\alpha_M = \frac{|\text{Maximum expected value of the physical variable}|}{|\text{Maximum linear voltage of the computer}|} \quad (8.44)$$

where the physical variable is equal to α_M times the computer variable in volts.

Once the computer simulation is programmed, test runs must be performed to verify that proper magnitude scaling has been assumed. If the scaling is not correct, then an adjustment must be made. It is important that all computer variables are within the linear limits. It is equally important that the computer variables are of sufficient amplitude for accurate computation and display.

Time scaling is often necessary to improve computing accuracy or to adjust the length of time required to perform the simulation or a combination of both. If the simulation represents a system which operates beyond the frequency capabilities of the computer, then the simulation will not be accurate.

Thus, to perform an accurate simulation, the simulation must operate slower than the actual system does. Likewise, if the physical system has a very low natural frequency, then the simulation requires a considerable amount of time to perform and may be inaccurate because of electronic drifts in the amplifiers of the computer.

Time scaling is performed by replacing physical or real-world time with a scaled computer time, or in equation form:

$$\tau_c = \alpha_T t \tag{8.45}$$

where τ_c is computer time and t is real time. In a time-invariant system the elements of the modeling equations where time appears are the derivatives, or in terms of the Laplace operators, appear as follows:

$$s = \frac{d}{dt} = \alpha_T \frac{d}{d\tau_c} = \alpha_T S \tag{8.46}$$

Magnitude and time scaling change the equation coefficients of the modeling equations of a system. This is best seen in the second-order system given earlier. By defining the computer variables

$$\begin{aligned} X_1 &= \alpha_{M1} E_1 \\ X_2 &= \alpha_{M2} E_2 \\ U &= \alpha_{M3} E_3 \end{aligned} \tag{8.47}$$

where X_1, X_2, and U are the physical variables and E_1, E_2, and E_3 are the computer variables, then Equation 8.43 becomes:

$$E_2(S) = \frac{1}{S} \left[-\frac{2\zeta\omega_n}{\alpha_T} E_2(S) - \frac{\omega_n^2 \alpha_{M1}}{\alpha_T \alpha_{M2}} E_1(S) + \frac{\omega_n^2 \alpha_{M3}}{\alpha_T \alpha_{M2}} E_3(S) \right] \tag{8.48}$$

$$E_1(S) = \frac{1}{S} \left[\frac{\alpha_{M2}}{\alpha_{M1} \alpha_T} \right] E_2(S)$$

Thus, the elements of the computer which establish the simulated equations' coefficients, i.e., the gains of the amplifiers and the potentiometers, contain the scale factors. Adjustment to scaling factors can be made after performing test runs by changing the gain factors of the simulator.

Once the computer equations are programmed and scaled, then the elements of the computer are interconnected to implement the programmed equations and the pots are set to the proper values to achieve the desired equation coefficients. Then the initial conditions are set for the system variables and the test input is applied.

The resulting time response from the applied input is usually recorded on some form of a permanent recording device, such as a strip chart recorder or x-y plotter.

Nonlinearities can also be implemented on analog computers. Nonlinear transfer functions which can be represented either exactly or approximately by a finite set of piecewise linear segments can be mechanized with diode function generators. Diodes can also be connected around the electronic amplifiers to simulate various forms of amplitude limiting. Multiplier circuits are available for creating variables raised to integer powers or products of variables.

Other special circuits of frequently encountered nonlinear operations are supplied as a part of a standard analog computer. These circuits include divider, sine, cosine, and square root. Reference 14 gives a complete discussion of analog computers and their use in simulating physical systems.

Logical and digital devices can be combined with the analog circuits to execute logical decisions and control timing. For example, a comparator which has a two-state logical output can be used to sense when an analog voltage exceeds a pre-set threshold. The logical output of the comparator can then be used to stop an integrator, holding the output of the integrator constant.

The same output from the comparator may change the state of a flip-flop or start a one-shot whose outputs control some other logic which is controlling the output of other analog devices. In fact, with the additions of logical control and timing electronics, the analog computer can be operated in a repetitive solution mode.

In the repetitive solution mode, the analog computer is automatically transferred from one of the three operating modes to another. The time duration of each mode can be set by the computer operator. The three modes are initialization, standard computation, and holding the final solution. Once the pre-set time has expired for the three modes, the computer will automatically start the three-mode cycle again.

With proper time scaling, the solution time of the computer can be increased to the point that several solutions can be performed in one second. This capability can be used with advantage to study the effect of parameter variations, especially if the pertinent parameter is set by a single pot.

With this capability the operator can place the computer in the repetitive solution mode and adjust the pot, observing the resulting effect on the output of the computer.

When digital computers are used to simulate control systems, they solve the same mathematical equations that the analog computers solve in performing a simulation. Instead of using a voltage which is proportional to a physical variable, the digital computer uses a number.

Thus, in performing the basic mathematical operations involved in the simulation equations, such as addition, subtraction, multiplication, and division, the digital computer does a very accurate, time-invariant job. Magnitude scaling is not a problem for a digital computer with floating point arithmetic. Nonlinearities and logical operations can be implemented in digital computers with ease. The one simulation task which can be more difficult for the digital computer than the analog computer is the integration of the differential equations.

Various numerical integration routines are available for performing integration in a digital computer. In each routine an incremental procedure is used to compute the solution to the differential equation. With each step another increment is computed which is related to the function being integrated and the step size. The increment is, then, added to an accumulated sum of past incremental computations. This step-wise process is illustrated in Fig. 8–29.

The very simplest numerical integration routine, the Euler method, uses the first two terms of the Taylor series expansion to determine the integrated variable, that is:

$$y(n\Delta T) = y(n\Delta T - \Delta T) + \frac{d}{dt}\left[y(n\Delta T - \Delta T)\right]\Delta T \qquad (8.49)$$

where y (nΔT) is the output of the numerical integrator, ΔT is the integration interval, and dy/dt is the input to the integrator.

Although Equation 8.49 is very easy to implement on a digital computer, it is not as accurate as other methods, and if ΔT is not small enough, the solution exhibits oscillations or instabilities.

An improved version of the Euler method uses an average value of the input to the integrator as expressed in the following expression:

$$y \, (n \, \Delta \, T) = y \, (n \, \Delta \, T - \Delta \, T)$$

$$+ \left\{ \frac{d}{dt} \left[y(n\Delta T - \Delta T) \right] + \frac{d}{dt} \left[y(n\Delta T - 2\Delta T) \right] \right\} \frac{\Delta T}{2} \quad (8.50)$$

However, the integrator stepsize, ΔT, must be sufficiently small to avoid oscillations and achieve satisfactory accuracy. Usually, if ΔT is smaller than the shortest significant periods in the system, oscillations or instability does not occur. A few trial runs with differing values of ΔT may be necessary, however, to determine the proper integration stepsize.

In the case where successive integrations are required, such as in higher-order differential equations, they are most conveniently rewritten in a series of first-order derivatives. Then, each first derivative can be integrated using either Equation 8.49 or Equation 8.50.

For example, in the case of the previously given second-order system, there are two integration expressions as follows:

$$x_2(n \, \Delta \, T) = x_2 \, (n \, \Delta \, T - \Delta \, T) + [-2\zeta\omega_n \, x_2 \, (n \, \Delta \, T - \Delta \, T)$$
$$-\omega_n^2 x_1 \, (n \, \Delta \, T - \Delta \, T) + \omega_n^2 \, U(n \, \Delta \, T - \Delta \, T)] \, \Delta \, T \quad (8.51)$$
$$x_1(n \, \Delta \, T) = x_1 \, (n \, \Delta \, T - \Delta \, T) + x_2 \, (n \, \Delta \, T - \Delta \, T) \, \Delta \, T$$

Complex numerical integration techniques exist which are more accurate but require more computational time.[15] These methods include Runge-Kutta and various predictor-corrector techniques. The Runge-Kutta is generally noted over the predictor-corrector techniques because it is self-starting, but the predictor-corrector methods run faster than Runge-Kutta.

They are of comparable accuracy and, therefore, Runge-Kutta is often used to start the predictor-corrector techniques. Most computer services have standard numerical integration routines of either variety.

Unlike the analog computer, unless the digital computer is under some form of clock, the computer time is only coincidently equal to real time. The time required for each cycle is not only a function of the computations required to integrate the equations but also the time required to output data to various computer peripherals.

If these peripherals are high speed, such as a vector-graphic display or a high-speed line printer, then the integration process is as rapid as the progress of the integration process. However, if the integration calculations are quite involved either because of the dimension of the system or because of the smallness of the integration stepsize, the

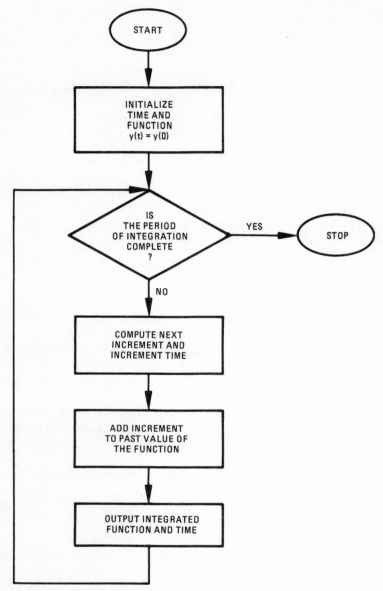

Figure 8-29. Basic Numerical Integration Process

simulation can take longer than real time. This often happens if the system contains significant high-frequency elements, such as structural bending modes.

In those cases where certain parts of a system are more efficiently simulated on an analog computer and other parts are more efficiently simulated on a digital computer, the hybrid computer is useful. The hybrid computer is a combination of a high-speed analog computer with mode and timing control plus a high-speed digital computer. The two computers are linked by analog-to-digital, digital-to-analog converters, and various control lines.

The digital computer, under its program control, controls the overall process, automatically cycling the analog computer through its operating modes, changing parameters as prescribed by its program, and recording solution data.

However, events which are detected on a predetermined criterion on the analog computer can interrupt the digital computer program, causing a preprogrammed course of action.

REFERENCES

1. J. N. Langfeldt and F. L. Galtung, "A Computer Simulation Program for Evaluation of Station Keeping Systems," Automation in Offshore Oilfield Operation, IFAC/IFIP Symposium, Bergen, Norway, June 1976.

2. V. Gourishanker and D. H. Kelly, *Electromechanical Energy Conversion,* Intext Educational Publishers, 1973.

3. K. Ogata, *Modern Control Engineering,* Prentice-Hall, Inc., 1970.

4. J. G. Truxal, Editor, *Control Engineer's Handbook,* McGraw-Hill Book Co., 1958.

5. Herbert Goldstein, *Classical Mechanics,* Addison-Wesley Publishing Co., 1959.

6. D. G. Schultz and J. L. Melsa, *State Functions and Linear Control Systems,* McGraw-Hill Book Co., 1967.

7. S. N. Blagoveshchensky, *Theory of Ship Motions,* Volume 1, Dover Publications, Inc., 1962.

8. R. G. Schieman, E. A. Wilkes, and H. E. Jordan, "Solid-State Control of Electric Motors," Proceeding of the IEEE, Volume 62, No. 12, November 1974, pp. 1643–1660.

9. B. V. Korvin-Kroukovsky, *Theory of Seakeeping,* The Society of Naval Architects and Marine Engineers, 1961.

10. M. A. Abkowitz, *Stability and Motion Control of Ocean Vehicles,* MIT Press, 1969.

11. J. P. Comstock, *Principles of Naval Architecture,* The Society of Naval Architects and Marine Engineers, 1967.

12. J. N. Newman, "The Drift Force and Moment on Ships in Waves," *Journal of Ship Research,* March, 1967), p. 51.

13. G. F. M. Remery and G. Van Oortmerssen, "The Mean Wave, Wind, and Current

Forces on Offshore Structures and Their Role in the Design of Mooring Systems," OTC 1741, 1973.

14. G. A. Korn and T. M. Korn, *Electronic Analog and Hybrid Computers*, McGraw-Hill Book Co., 1964.

15. Anthony Ralston, *A First Course in Numerical Analysis*, McGraw-Hill Book Co., 1965.

9

Procurement and Operational Considerations

9.1 Introduction

At some point in an offshore operation the requirement for positioning the vessel or vessels must be considered. In many cases, if the vessel is to be at a fixed location and the water depth is not unusually deep, the positioning method that is selected is the conventional multipoint mooring system.

In an ever-increasing number of applications, however, dynamic positioning is the means selected for positioning the vessel. This is especially true when mobility, quick response time, and the ability to operate in deep water are important.

A decision between the two types of positioning systems may not be made until a trade-off study is performed. This process frequently occurs in applications where more than one method will meet the positioning requirements and it is a question of which one is the best.

In fact, there are several examples where more than one positioning method is installed on a vessel in order to optimize the method of positioning to the particular operational situation. For example, the *Discoverer 534* and the *Saipem Due* both have a multipoint mooring system and a full dual-redundant dynamic positioning system. In shallower water they use the mooring system; in deep water and during mooring system deployment they use the dynamic positioning system.

There are many considerations in making the decision that a vessel should use dynamic positioning. Once the DP decision is made there are several steps to the process of successfully procuring, installing, and placing in operation a dynamic positioning system.

During this discussion, no particular application for the vessel is assumed except to illustrate certain points. The process described in this chapter is by no means the only way to approach this decision process. It does represent, however, a method that has been used quite successfully on several operational dynamically positioned vessels and it is most heavily weighted toward the aspects of the process covered in previous chapters.

9.2   Initial steps

A requirement develops to perform a task or mission in an offshore area. To perform the task, certain equipment is required. The equipment is not usually self-supporting and, therefore, requires a platform suitable to support the equipment, the personnel to operate the equipment, and the supplies required to operate and maintain the equipment.

The platform may also need to furnish hotel accommodations for the personnel as well as the features required of a seaworthy offshore vessel with proper registry to operate in a given area.

A good example of this process is offshore exploratory drilling for petroleum. The probable existence of offshore petroleum is established by careful studies of geophysical surveys of the ocean floor in some offshore area of the world. The group which holds the lease on the area then contracts a drilling company to drill a number of exploratory wells at pre-established points in the lease. The drilling company, then, moves one of their drilling rigs mounted on an appropriate platform to the site of the first well.

If the water depth is sufficiently shallow, and the bottom conditions are proper, the drilling company may choose to employ a fixed platform, such as a jack-up rig. For deeper water, the drilling company uses a floating platform. The type and size of the platform are determined by the requirements necessary to properly perform the drilling operation. This is generally true for any offshore activity such as mining, pipe laying, and support.

The general requirements for a floating offshore vessel to properly perform a given task include:

- The vessel shall provide sufficient space and displacement to carry the necessary equipment and to allow for safe, efficient operation of the equipment.
- The equipment shall be so positioned on the vessel as to provide the best operating environment over the range of platform motions for the operating area.
- Space and displacement shall be provided for the necessary support equipment.
- Space and facilities shall be provided for the personnel to operate all the equipment on the vessel and to manage the vessel.
- The vessel shall provide sufficient space for the positioning equipment and the equipment required to move the vessel to and from the site of the work task.

In the case of the offshore drilling operation, these general requirements create a vessel layout for a ship-shaped hull similar to that shown in Fig. 9-1. The equipment required for drilling is located amidship in order to provide sufficient space for all drilling equipment, the riser, and drill pipe, and to minimize vessel motions in more significant weather.

With the derrick and other drilling equipment amidship, the vessel can more freely change its heading without changing its position over the wellhead.

The support equipment, such as the power system and the cranes, is located where it can most efficiently support the drilling equipment.

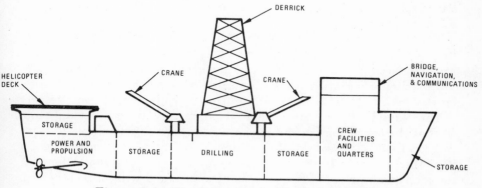

Figure 9-1. The Layout of a Typical Drillship

Since the ship is only wide enough for the drilling equipment, and because the power system must also supply power to the main propulsion system of the drillship, the power system is located aft of the drilling equipment and below, near the main propulsion.

Even in a dynamically positioned drillship, the power system is located aft because the most demand for propulsion power is aft with the main screws and the aft lateral thrusters.

Cranes on the drillship are located near the drilling equipment to support the drilling operation by lifting the riser casing, drill pipe, and drilling supplies from smaller vessels which come alongside the drillship. The area aft and above the power generating system is used for storage of drilling supplies and a helicopter landing deck. The helicopter deck is required to rotate the crew, to carry special supplies, and evacuate injured or very sick personnel. Regulations even establish the minimum size and clearance of the helicopter deck.

Personnel quarters are located in the forward part of the drillship, ahead of the drilling equipment and equipment storage (Fig. 9–1). With crew sizes numbering around one hundred, a significant area is required for quarters and messing facilities. Since the crew works two 12-hour shifts, the crew members not working need a certain amount of area for recreation and relaxation in the two to four weeks they are on board the ship. Storage areas are likewise required for supplies for the crew.

An advantage of the drillship over other offshore drilling platforms is that it is self-propelled and capable of navigation at a top speed of 12 to 14 knots all around the world. During the transit time between locations the ship is operated as an ocean-going vessel and requires a sailing crew.

This same crew remains on the drillship during drilling to maintain the ship, stand watch, and assist in the drilling operation. They are quartered with the drilling crew and operate the ship from a standard marine bridge with standard marine nagivation aids. The marine crew usually operates the communication equipment on the vessel.

This equipment is usually more sophisticated than a normal ship so that the operating personnel can reach shore supply points in very remote areas of the world. If the ship is dynamically positioned, space near the bridge is allocated for the dynamic positioning controller and its support equipment. The crew is also enlarged with operators and maintenance personnel for the DP system.

If the drillship is dynamically positioned, lateral thrusters are required in the forward part of the ship. If the drillship is to be moored with a multipoint mooring system, then space forward and aft must be provided for the anchor line winches, the fairleaders, the storage racks for the anchors, and the anchors. Since several thousand feet of mooring line or chain are required for each anchor, and since the anchors weigh 30,000 to 45,000 lb, the multipoint mooring system adds a significant space and displacement requirement.

A vessel meeting these requirements is approximately 500 ft long and 70 ft wide with a draft of 25 ft which displaces 15,000 tons. A drillship of this size has sufficient storage to operate far from land for long periods at a time, and to survive in weather with 60 to 80 knot winds and 30 to 40-ft seas.

The area of the world where the task must be performed also has a significant impact on the vessel system requirements, because it determines the range of water depths, the weather, and the sea conditions. Supply is also determined by the area of operation. The more removed the operating area is, the more difficult it is to supply, resulting in the need for more storage to be self-sufficient.

The vessel requirements establish many of the requirements for the dynamic positioning system, some directly and some indirectly. The vessel requirements which directly establish DP requirements are:

- Positioning and maneuvering requirements plus work task description
- The accuracy with which positioning must be maintained to ensure performance of the vessel's given task
- The environmental limits in which the vessel must perform the given task, plus the limits that the vessel must survive without performing the task
- The range of operating water depths
- The number of consecutive days of uninterrupted operation

A typical set of these requirements for a drilling vessel is shown in Fig. 9–2.

The next design step is sizing the power and thruster systems. If these two systems are not properly sized for the vessel, then the positioning accuracy requirement for the design operating environment will not be met. This step is of such importance to a successful dynamic positioning system that the next section of this chapter is devoted to a discussion of it.

For the remainder of this section, however, the assumption is made

BASIC POSITIONING TASK: HOLDING POSITION DIRECTLY OVER A WELLHEAD

HOLDING POSITION ACCURACY: A WATCH CIRCLE CENTERED ON THE WELLHEAD WITH
                            A RADIUS OF 6% OF WATER DEPTH

RANGE OF OPERATING WATER DEPTHS: 200 TO 6000 FEET

MAXIMUM OPERATING ENVIRONMENTAL CONDITIONS:	STEADY WIND VELOCITY:	40 KNOTS
	WIND GUST VELOCITY:	50 KNOTS
	SEA CURRENT:	2 KNOTS
	SIGNIFICANT WAVE HEIGHT:	16 FEET

MAXIMUM SURVIVAL ENVIRONMENTAL CONDITIONS:	STEADY WIND VELOCITY:	65 KNOTS
	WIND GUST VELOCITY:	85 KNOTS
	SEA CURRENT:	2 KNOTS
	SIGNIFICANT WAVE HEIGHT:	30 FEET

DAYS OF CONSECUTIVE OPERATION: 150 DAYS

Figure 9-2. Typical DP Requirements for a Drilling Vessel

that the power and thruster systems have been properly sized or the requirements changed to fit the power and thruster systems that can be installed on the given vessel. As might be expected, a compromise is generally necessary between the requirements and the size of the power and thruster systems.

During the process of sizing the power and thruster systems, the vessel may change in size and shape from its initial design (usually larger) which must be factored in the thruster and power system sizing effort. Thus, thruster and power system sizing and vessel design become an interrelated process.

Another effort parallel to the thruster and power system sizing is the prediction of the vessel's motion characteristics and performance in various weather conditions. The prediction of the motion characteristics are very often performed in a model basin. At the same time, vessel model parameters can be determined to refine the power and thruster sizing estimates and for use in dynamic simulations of the vessel.

The first of these dynamic simulations can be used to estimate the vessel's dynamic positioning system performance and to size the power and thruster system under dynamic conditions. More is said about the dynamic simulation for thruster and power system sizing in the next section.

In the process of sizing the power and the thruster system, the number and size of power generators and thrusters is determined. The number and size of the power generators and thruster may be affected by certain redundancy requirements.

For example, it is common to require that the thruster and power system both be capable of performing in the operating environment with one unit out-of-service. The approximate locations of the thrusters are also determined during the thruster sizing effort by the turning moment loading imposed on the vessel in the operating environment. The type of thrusters is also usually determined during the thruster sizing exercise because of the parameters that are required to determine the total thrust that is needed.

However, there may still be a choice available after sizing between various types of thrusters which have equivalent thrust producing efficiencies. For example, if the sizing study is performed with fixed-axis thrusters, then either controllable-pitch or controllable-speed thrusters can be used.

With the sizing and selection of the power and thruster system, the next step in placing a dynamic positioning system in operation on an offshore vessel is the preparation of a specification for the elements of the dynamic positioning system, i.e., the power system, the thrusters, the controller, and the sensors.

Presently, with the number of manufacturers who supply dynamic positioning system controllers and sensors, the specification need not be very detailed because most manufacturers have "off-the-shelf" system configurations which directly meet the requirements for a wide range of applications, such as drilling, diver support, mining, etc. The situation is very likely the same with the power system and the thrusters.

When there are special or unique requirements, then a specification reflecting these requirements is needed. Or if certain equipment preferences or redundancy arrangements are desired, such as an acoustic position reference system as opposed to a taut wire system, then a more detailed specification is required.

A typical beginning specification or system description from which a supplier can prepare a quotation is given in Fig. 9–3. The DP sup-

```
1.   BASIC DESCRIPTION OF VESSEL'S MISSION OR WORK TASK
2.   PERFORMANCE SPECIFICATION (SEE FIGURE 9.2)
3.   THRUSTER TYPE, INTERFACE, AND CONFIGURATION
4.   BASIC VESSEL CHARACTERISTICS (SIZE, DISPLACEMENT, ETC.)
5.   REDUNDANCY REQUIREMENTS
6.   MODES OF CONTROL (JOYSTICK, AUTOMATIC, MANUAL , ETC.)
7.   SENSOR:  PREFERRED TYPES AND BACK-UP REQUIREMENTS
8.   COMPUTING SYSTEM REQUIREMENTS
     a. PROGRAM STORAGE REQUIREMENTS (FIXED OR MODIFIABLE)
     b. RESTART REQUIREMENTS
     c. MEMORY PROTECT
     d. HARD COPY REQUIREMENTS
     e. ADDITIONAL COMPUTING CAPABILITIES:  PRESENT AND FUTURE
9.   DISPLAYS:  NORMAL AND SPECIAL PURPOSE
10.  AUXILIARY INSTRUMENTATION
     a. ENVIRONMENTAL
     b. VESSEL MOTION
     c. WORK RELATED
11.  DATA RECORDING REQUIREMENTS
     a. MAGNETIC TAPE
     b. STRIP CHART
12.  SPACE ALLOCATION AND LOCATION OF EQUIPMENT
13.  DOCUMENTATION REQUIREMENTS
14.  TRAINING REQUIREMENTS
15.  TESTING AND SEA TRIAL DEMONSTRATION REQUIREMENTS
16.  SPARE PARTS
17.  SERVICE REQUIREMENTS
18.  REGULATORY AGENCY UNDER WHICH VESSEL TO BE BUILT
```

Figure 9-3. An Example of the Items for an Initial Specification of a Dynamic Positioning System

plier usually asks for certain data about the vessel, some of which he uses to estimate the cost of his design and development of the system, some he uses to judge the feasibility of the project as specified, and some he uses to better understand the project. Included in this list of data are the items shown in Fig. 9-4.

Once the specification and information are completed and released to the various suppliers, the next task is selection of the supplier. The selection process depends on business as well as technical considerations which are a function of each individual operation and the people involved.

So far the steps in placing a dynamic positioning system in operation have been covered to the point of selecting the suppliers of the elements of the dynamic positioning system. These steps are outlined in flow chart form in Fig. 9-5. Further discussions of initial DP system planning and design considerations are given in References 1, 2, and 3 at the end of this chapter.

PHYSICAL CHARACTERISTICS OF VESSEL (IF NOT COMPLETE IN SPECIFICATION)
    DRAFT AND ITS VARIATIONS WITH THE OPERATION
    GENERAL VESSEL ARRANGEMENT
    DISPLACEMENT AS A FUNCTION OF DRAFT
    CENTER OF ROTATION AND RADIUS OF GYRATION AS A FUNCTION OF DRAFT

VESSEL MODELING CHARACTERISTICS (REQUIRED IN SURGE, SWAY, AND YAW)
    ADDED MASS
    WIND DRAG FORCES
    CURRENT DRAG FORCES
    WAVE DRIFT FORCES
    VESSEL MOTIONS

THRUSTER SYSTEM CHARACTERISTICS
    THRUSTER TYPE, SIZE, AND NUMBER
    THRUSTER CONTROL AND INTERFACE
    THRUSTER LOCATIONS
    TIME RESPONSE CHARACTERISTICS
    ACOUSTIC PROPERTIES

SPACE LOCATION FOR DP EQUIPMENT
    DP ROOM
    WIND SENSOR
    ACOUSTIC HYDROPHONE
    VERTICAL REFERENCE SENSOR

Figure 9-4. Example of Type of Initial Data that is Useful to the Potential DP Supplier of the Control System

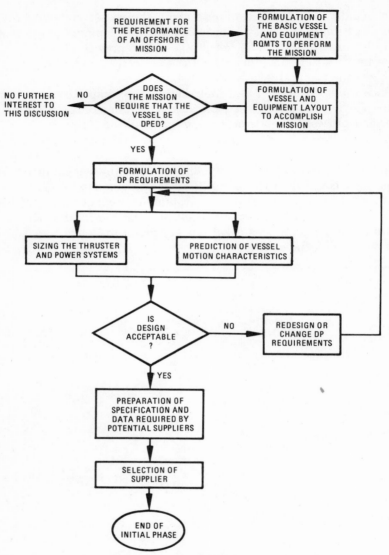

Figure 9-5. A Flow Chart of the Initial Steps to Obtaining a Dynamic
Positioning System

9.3   Power and thruster systems sizing

Perhaps the most important initial step in installing a dynamic positioning system on a vessel is the sizing of the power and the thruster systems. Without sufficient thrust and power the dynamic positioning system cannot maintain the required positioning accuracy in the design environment.

Furthermore, without the knowledge of the number and size of the thrusters and the power generators, the vessel design cannot be performed. In fact, as can easily happen, the original estimate of required thrust and horsepower is optimistic and, therefore, the vessel sizing is underestimated.

As a result, the true power and thrust requirements are established by a rigorous sizing analysis, and the operation planners are faced with the following alternatives:

- Increase the size of the vessel to accommodate the increases in thrust and power systems
- Reduce the previously planned spaces in size to make room for the increases in the power and thruster systems
- Reduce the positioning accuracy or design environment specification
- Place the thrusters and power generating units in spaces not normally allocated for them or mount the thrusters in unconventional appendages on the hull

Obviously, none of these alternatives are desirable. All of them, except the reduction in specifications, represent increased cost either in the vessel, the thruster, or the power system equipment. The reduction in the specifications may also be expensive because of the loss of work contracts.

Sizing the power and thruster systems involves determining the performance of the dynamic positioning system for limiting conditions. Two approaches are commonly used for determining the limits. One uses the specifications of the environment and the positioning accuracy to determine the required thrust and power to satisfy the specification.

The other uses the given thrust and power capacity planned for the vessel in combination with the positioning accuracy specification to determine the environment in which the vessel can operate and maintain the given positioning accuracy. Thus, the first synthesizes the thrust and power requirements and the second analyzes a given thruster and power system configuration.

In the first approach, the results may exceed the originally planned size of the thruster and power system. Then, one of the previously mentioned alternatives must be adopted. Usually a compromise solution results. The compromise includes a partial reduction in the specifications as well as an increase in the installed thrust and power.

The reduction in the specification may be the result of a logical interpretation of the original specification in the face of reality. For example, for a ship-shaped vessel, the environment that can be held on the beam is less than the environment that can be held on the bow or the stern because of the difference in drag characteristics between the two axes of the vessel.

Thus, in the environmental specification, as given in Fig. 9-2, the conditions should be given as a function of the angle-of-attack of the environmental elements to the vessel. Likewise, it should be decided whether or not the sizing is to be performed with coincident environmental elements, which is very often the most demanding combination.

The second approach of analyzing the thruster and power system can also yield unfavorable results when the given thruster and power system does not satisfy the environmental specification. Just as with the first approach, one or more of the previously mentioned alternatives must be adopted in either a strict or compromising manner.

To determine either the thrust limits or the environmental limits requires a method of predicting the performance of the dynamic positioning system under a given set of conditions. Since the environmental conditions and some work-task related conditions are time-varying, an accurate determination involves the dynamics of the vessel and the DP system.

Furthermore, since the thruster and power system perform at their limits during sizing, the DP system is operating in a nonlinear region, so linear analysis techniques are not valid. As a result, one of the methods of dynamic simulation discussed in Chapter 8 becomes a primary candidate to perform the sizing exercise.

Performing a dynamic simulation requires the proper modeling information for the vessel, the environment, and any work-related loads which require thrust. However, the thruster and sensor modeling information and the exact nature of the controller are of less importance because the thrusters are performing at their limits when the limits of the DP system performance are reached.

When the thrusters are producing their maximum outputs, the feedback through the sensors to the controller and on to the thrusters to close the loop has no effect on the thruster outputs because they are already at their limits. Consequently, the DP system is no longer a feedback control system. Instead it is an open-loop control system.

The situation within the DP system when the thruster limit is reached is better demonstrated by a simplified example of a DP system. If the motion of the vessel is restricted to a single translational axis and only the forces caused by the wind and current are considered, then the model of the DP system is as shown in Fig. 9–6. Forces due to wave-drift effects and work-related tasks can be considered as zero or included in the steady-wind budget.

If the wind and current are assumed to be coincidental and acting along the assumed translational axis of the vessel, with the wind composed of a steady component plus a gust component as shown in Fig. 9–7, then the equation of motion for the vessel is approximately given as follows:

$$M \frac{d^2q}{dt^2} = - C_{qA}(v_{AO} + v_G)^2 - C_{qc}(v_c + \dot{q})^2 + F_{qT} \qquad (9.1)$$

where:

$q =$ the translational axis of the vessel, either x or y
$M = \rho \nabla +$ added mass in the q-axis
$C_{qA} =$ the aerodynamic drag coefficient in the q-axis when the angle-of-attack is equal to the direction of the q-axis.
$C_{qc} =$ the hydrodynamic drag coefficient in the q-axis when the angle-of-attack is equal to the direction of the q-axis
$v_{AO} =$ the steady wind speed
$v_G =$ the gust speed
$v_c =$ the current speed
$F_{qT} =$ the thrust in the q-axis

In addition, if the q-coordinate sensor is assumed to be perfect and the controller is assumed to be a PID-type, then the commands to the q-axis thrusters are computed as follows:

$$F_{qTc} = K_P \, \epsilon_q + K_D \, \frac{d\epsilon_q}{dt} + K_I \int_0^t \epsilon_q d\tau \qquad (9.2)$$

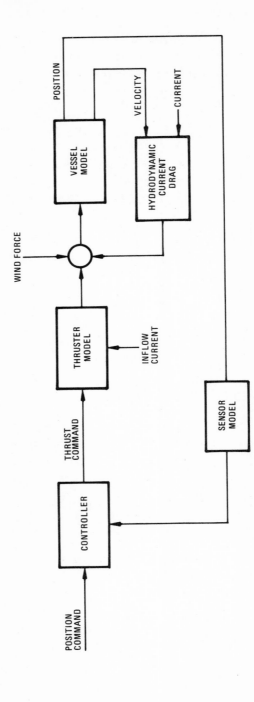

Figure 9-6. A Single-Axis Position Model of a DP System

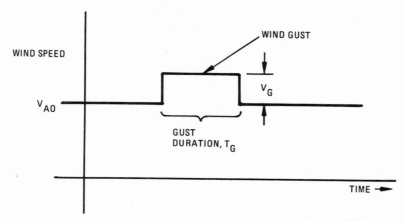

Figure 9-7. The Assumed Model of the Wind for the Simplified DP
Example

where:

F_{qTc} = the thrust command to the thrusters
$\epsilon_q = q_c - q$
q_c = commanded value of the q-coordinate
q = the measured value of the q-coordinate with a perfect sensor
K_P = the proportional gain
K_D = the derivative gain
K_I = the integral gain

The control loop is closed with the thruster model (see Fig. 9.6) which for this example is assumed to have current inflow degradation, no dynamics, and a maximum output of F_{QTmax}. The transfer function of the thrusters is given as:

$$F_{qT} = F_{qTc} \, (1 - \epsilon_{qT} \, (v_c + \dot{q}))$$ (9.3)

where:

ϵ_{qT} = the thrust degradation factor,
$F_{qTc} = F_{QTC}$ if $|F_{qTc}| < |F_{QTmax}|$
$= F_{QTmax}$ otherwise

If the gust magnitude is expressed as a fraction of the steady-state wind speed and if the constant forces in the vessel equation of motion are collected, Equation 9.1 becomes:

$$M \frac{d^2q}{dt^2} = F_{qss} - C_{qA}v_{AO}^2(2\delta_G + \delta_G^2) - C_{qc}(2v_c + \dot{q})\dot{q}$$

$$+ F_{qTc}(1 - \epsilon_{Tq}(v_c + \dot{q})) \tag{9.4}$$

where:

$$\delta_G = \text{gust factor} = \frac{v_G}{v_{AO}}$$

$$F_{qss} = -C_{qA}v_{AO}^2 - C_{qc}v_c^2$$

The modeling of the wind gust as pulse as shown in Fig. 9–7 with the gust magnitude as a fraction of the steady-state wind speed is not a real model of the wind. However, the average gust magnitude has been shown to be a fraction of the average or steady-state wind speed over a wide range of average wind speeds.[4]

The gust factor has also been shown to vary with respect to the periods over which the ratios are determined according to the following equation:[4]

$$\delta_G = \frac{v_G}{v_{AO}} = 0.45 - 0.07 \ln T_G \tag{9.5}$$

where T_G is the period over which the average gust magnitude is determined and v_{AO} is determined from a 10-minute average of the wind speed. Thus, it is not totally unreasonable to select the gust factor magnitude based on Equation 9.6 for a given wind gust duration or averaging period.

When the output of the q-axis thrusters is less than their maximum value, i.e., unsaturated, and the system is at steady-state prior to the application of the gust, the closed-loop action of the control system adjusts the thrust to balance the steady-state wind and current forces.

This countering of steady-state thrust can be determined from Equation 9.4 by setting all the derivatives and the gust factor to zero and solving for thrust. The result is:

$$F_{qTc} = \frac{-F_{qss}}{1 - v_c\epsilon_{qT}} = \frac{C_{qA}v_{AO}^2 + C_{qc}v_c^2}{1 - v_c\epsilon_{qT}} \tag{9.6}$$

For the controller to maintain the steady thrust command output given by Equation 9.6, the integral term given in Equation 9.2 must build until its output is equal to Equation 9.6. Otherwise, the error between the command and the feedback is not zero, and the integrator output continues to change.

So long as F_{qTc} is less than the maximum output of the q-axis thruster system, Equation 9.6 is valid. If F_{qTc} is greater than F_{qTmax}, then the thruster system can never balance the wind and the current forces, and steady-state is never reached. If steady-state is never reached, then the translational position of the vessel increases without bound. When $F_{qTc} = F_{QTmax}$, then Equation 9.6 yields the maximum steady-state environment that the q-axis thrusters can balance or hold the vessel against.

If the duration of the wind gust is short compared to the time required for the integrator of the controller to respond, the proportional and derivative portions of the controller react to the wind gust and the integrator generates the thruster command to counter the steady forces. If v_c is significantly larger than the maximum velocity of the vessel, Equation 9.4 can be written as follows when the thrusters are not saturated:

$$M \frac{d^2q}{dt^2} = F_{qss} + \Delta F_w - 2C_{qc}v_c \frac{dq}{dt}$$

$$- \left(K_P q + K_D \frac{dq}{dt} + \frac{F_{qss}}{1 - \epsilon_{qT}v_c}\right)\left(1 - \epsilon_{qT}\left(v_c + \frac{dq}{dt}\right)\right) \quad (9.7)$$

or when terms are collected and products eliminated:

$$\frac{d^2q}{dt^2} + \frac{1}{M}\left[2C_{qc}v_c + K_D(1 - \epsilon_{qT}v_c) - \frac{\epsilon_{qT}F_{qss}}{1 - \epsilon_{qT}v_c}\right]\frac{dq}{dt}$$

$$+ \frac{K_P}{M}(1 - \epsilon_{qT}v_c)q = \frac{\Delta F_w}{M} \quad (9.8)$$

where $\Delta F_w = -C_{qA}v_{AO}^2(\delta_G^2 + 2\delta_G)$

Equation 9.8 is a linear second-order differential equation which approximately represents the response of the DP system when a wind gust is applied to the vessel and when the thrusters are not saturated. The natural frequency and damping factor of this system are:

$$\omega_n = \sqrt{\frac{K_P(1 - \epsilon_{qT}v_c)}{M}}$$

$$\zeta = \frac{1}{2\sqrt{K_P M(1 - \epsilon_{qT}v_c)}}\left[2C_{qc}v_c + K_D(1 - \epsilon_{qT}v_c) - \frac{\epsilon_{qT}F_{qss}}{1 - \epsilon_{qT}v_c}\right] \quad (9.9)$$

where ω_n is the natural frequency and ζ is the damping factor.

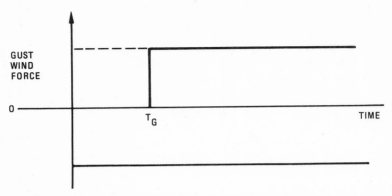

Figure 9-8. Modeling the Wind Gust for Easy Laplace Transformation

It is interesting to note that the damping factor of the system is a function of the steady sea current. Thus, as the sea current speed increases, the vessel response to wind gust is more damped. The term in the damping factor caused by the steady-state thruster force and thruster degradation factor reduces the damping, but it is much smaller than the other terms because of the size of ϵ_{qT}, and, therefore, is not very significant.

The transient response of the system for the unsaturated case can be determined from Equation 9.8 using Laplace transforms. The Laplace transform for the wind gust is determined by modeling the gust as the sum of a negative-going step function at time zero and a positive-going step function of the same amplitude at time equal to T_G as shown in Fig. 9.8.* This model yields the following equation:

$$\Delta F_w(s) = \Delta F_w \left(\frac{1 - \epsilon^{-T_G s}}{s} \right) \tag{9.10}$$

Although the transient response can be determined from Equations 9.8 and 9.10, the solution which is of interest to this discussion is the maximum excursion of the vessel caused by the wind gust. The peak excursion can be determined from the solution of the transient response by differentiation and setting the derivative equal to zero. Another way is to consider the response of a second-order system to the two step functions shown in Fig. 9–8.

* The wind gust is an increase in wind speed acting along the q-axis. Thus, the force on the vessel is negative driving the vessel in the negative q-direction. The negative sign is in the drag coefficient. Thus, Equation 9.8 is still positive.

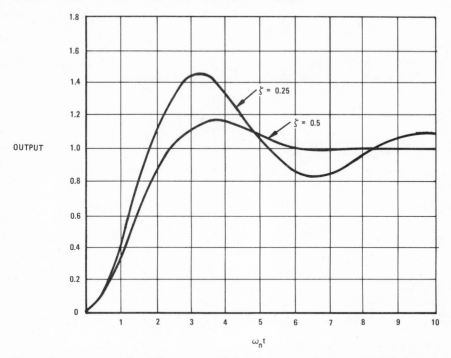

Figure 9-9. Unity Step Function Response of a Second-Order System

The response of a second-order system to unity step function is shown in Fig. 9–9 for several values of damping. The composite response to two step functions of opposite sign depends on when the second step occurs relative to the first.

Typically, the gust duration used for such analysis of dynamic positioning systems is 60 seconds for a control system with a natural frequency equal to 0.0628 radians per second. Since the peak of the first negative-going response occurs at $\omega_n t = \pi / \sqrt{1 - \zeta^2}$ (see Equation 7.58) and ζ is between 0.25 and 0.5, depending on the damping caused by the current, the peak occurs between 52 and 58 seconds which is before the positive-going step function occurs.

As a result, the peak excursion caused by the wind gust is equal to the peak overshoot caused by the first step and the overall response is a composite as shown in Figure 9–10. For a second-order system the peak overshoot is given as follows:

$$q_{MAX} = \left\{ 1 + \exp\left[\frac{-\zeta\pi}{\sqrt{1-\zeta^2}}\right]\right\} \frac{\Delta F_w}{K_p(1 - \epsilon_{qT}v_c)} \tag{9.11}$$

which, when the expressions for ΔF_w and K_p are substituted into Equation 9.11 and the offset is converted to percent of water depth, becomes:

$$q_{MAX}, \% \text{ water depth} = \frac{100}{h_w}\left\{ 1 + \exp\left[\frac{-\zeta\pi}{\sqrt{1-\zeta^2}}\right]\right\} \frac{C_{qA}(2\delta_G + \delta_G^2)}{\omega_n^2 M} v_{AO}^2 \tag{9.12}$$

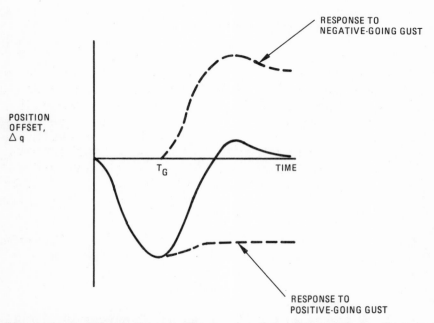

Figure 9-10. The Composite Response of the Second-Order System to the Wind Gust Pulse

where h_w is the water depth. With this result, it can be seen that the maximum excursion resulting from the wind gust increases as the square of the steady-state wind velocity as long as the thrusters do not saturate during the gust.

When the thrusters are saturated, Equation 9.4 is expressed as follows:

$$M \frac{d^2q}{dt^2} = F_{qss} + \Delta F_w - 2\, C_{qc}\, v_c \frac{dq}{dt} + F_{QTmax}\, (1 - \epsilon_{qT}v_c - \epsilon_{qT} \frac{dq}{dt})$$

(9.13)

or collecting all q-coordinate terms and defining a new thruster force term as δF_T:

$$\frac{dq^2}{dt^2} + \frac{(2\, C_{qc}v_c + F_{QTmax}\, \epsilon_{qT})}{M} \frac{dq}{dt} = \frac{\Delta F_w}{M} + \frac{\delta F_T}{M}$$

(9.14)

where:

$$\Delta F_w = -C_{qA}\, v_{AO}{}^2\, (\delta_G{}^2 + 2\delta_G)$$
$$\delta F_T = F_{QTmax} - C_{qA}\, v_{AO}{}^2 - v_c\, (\epsilon_{qt}\, F_{QTmax} + C_{qc}\, v_c)$$

Equation 9.14 is valid only if δF_T is greater than zero. Otherwise the steady-state forces are greater than the thruster capacity, and as mentioned earlier, the vessel never reaches a steady-state position.

The transient response when the thrusters are saturated can be determined from Equation 9.14 by using Laplace transforms and the previous definition for ΔF_w given in Equation 9.10, and by treating δF_T as a step function starting at the same time as the wind gust. The result is as follows:

$$q(t) = c(t - \tau_q\, (1 - \epsilon^{-t/\tau_q}))\, u(t)$$

$$+ d(t - T_G - \tau_q \left(1 - \epsilon^{-\frac{t - T_G}{\tau_q}}\right)) u(t - T_G) \quad (9.15)$$

where:

$$c = \frac{\Delta F_w + \delta F_T}{2C_{qc}v_c + \epsilon_{qT}\, F_{QTmax}}$$

$$d = \frac{-\Delta F_w}{2C_{qc}v_c + \epsilon_{qT}\, F_{QTmax}}$$

$$\tau_q = \frac{M}{2C_{qc}v_c + \epsilon_{qT}\, F_{QTmax}}$$

The maximum value of Equation 9.15 occurs at the following time:

$$t_{max} = -\tau_q\, \ln\left[\frac{c + d}{c + d\epsilon^{T_G/\tau_q}}\right] = \tau_q\, \ln\left[1 - \frac{\Delta F_w}{\delta F_{qT}}(\epsilon^{T_G/\tau_q} - 1)\right] \quad (9.16)$$

When the time at which the maximum occurs is substituted into Equation 9.15, the resulting maximum value of excursion due to the wind gust with the thrusters saturated is:

$$q_{max} = ct_{max} + d(t_{max} - T_G) \tag{9.17}$$

or in terms of the forces:

$$q_{max} = \tau_q \left[\frac{\delta F_{qT}}{2C_{qc}v_c + \epsilon_{qT} F_{QTmax}} \right] \ln \left[1 - \frac{\Delta F_w}{\delta F_{qT}}(\epsilon^{T_G/\tau_q} - 1) \right]$$

$$+ \frac{\Delta F_w T_G}{2C_{qc}v_c + \epsilon_{qT} F_{QTmax}} \tag{9.18}$$

To show the difference in control system response when the thrusters are saturated and unsaturated, Equations 9.11 and 9.18 are plotted in Fig. 9.11 for the equation variables given in Table 9-1. As Fig. 9-1 illustrates, the maximum offset of the vessel increases sharply once the output of the thrusters reaches maximum during their response to the wind gust.

In fact, if the position holding accuracy specification in 6% of water depth, the effect of thruster saturation determines the average wind speed that the example vessel can hold against in all water depths greater than 200 ft. Actually, the effect of thruster saturation extends to even shallower depths because the thrusters are responding in a saturated mode as well as a linear mode in the transition region between Equations 9.11 and 9.18.

If active wind compensation is used as a part of the controller, as it is in most dynamic positioning systems, then the thrusters are driven into saturation more quickly because of the feed-forward effect of the compensation (see Chapter 5). Thus, the result of the feed-forward effect is a more abrupt change from the linear response curve given by Equation 9.11 to the saturated response curve given by Equation 9.18 and a reduction of the amplitude of maximum excursion caused by wind gust in the linear region.

As previously noted, the maximum limit of the thruster holding capacity is reached when the force caused by the average wind speed and sea current without a disturbance is equal to the maximum output of the thruster system as expressed by Equation 9.6.

For this example, the predicted limit of thruster output with the assumed sea current of two knots occurs for an average wind speed equal to 36.5 knots. In Fig. 9-11 this limiting wind speed appears to

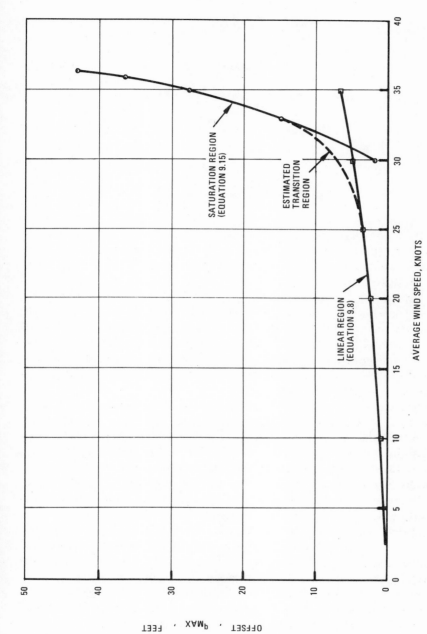

Figure 9-11. Vessel Offset as a Function of Average Wind Speed

TABLE 9–1
PARAMETER VALUES USED IN THE
EXAMPLE CALCULATIONS

M	= 2,000,000 slugs
C_{qc}	= 20,000 lb/kt^2
C_{qA}	= 60 lb/kt^2
F_{QTmax}	= 200,000 lb
ϵ_{qT}	= 0.1/knot
ω_n	= 2 π/100
T_G	= 60 second
v_c	= 2 knots
$\delta_{i.}$	= 0.24 knots
ζ	= 0.4 or peak overshoot of 25%

be an asymptote for the saturation response curve. However, the value of maximum offset given by Equation 9.18 for a wind speed equal to 36.5 knots is 43.6 ft.

This offset is the distance that the vessel moves because of the wind gust and remains without returning to its original position because there is no restoring thruster force. Thus, although the maximum position offset caused by wind gust as predicted by Equation 9.18 is not infinity, it might as well be because the vessel never returns to its original position.

Furthermore, if additional gusts are applied to the vessel, the vessel continues to move further from its original position.

At the other extreme of static holding, the steady wind speed is equal to the average wind speed plus the gust. Using Equation 9.6 with the wind speed equal to the average wind speed plus the gust to determine the limiting wind speed gives a maximum average wind speed of 29.45 knots. As shown in Fig. 9–11, this value of average wind speed is approximately equal to the value where the maximum offset predicted by Equation 9.18 is zero which is the beginning of the saturation region.

Therefore, the maximum wind speed at which the maximum specified position offset occurs is bounded by those values predicted by Equation 9.6 with the average wind speed and the average wind speed plus the gust. The exact limiting wind speed, however, must be computed using Equation 9.18 or by a computer simulation. The same procedure can be used to determine the limiting value of current if the average wind speed and the gust magnitude are given.

The simplified example demonstrates several important results regarding the sizing of the thrusters:

- Unless very precise position control is specified, the thrusters are operating in a region of maximum thrust output during the time when the specified limiting positioning performance of the vessel is reached.
- Since the thrusters are operating at maximum thrust output, the feedback control loop is essentially open-loop and, therefore, the models of the sensors, the controller, and the thruster system dynamics are not as critical as the vessel and environmental force models.
- The actual colinear environment that the thrusters and control system can counter in order to satisfy the position-holding specification is bounded by the steady-state environment and the steady-state plus disturbance environment.

The example also demonstrates that without simplifying assumptions the determination of maximum offset of the vessel caused by an environmental disturbance is complicated, justifying a computer simulation. The use of a computer becomes a necessity if more than one axis of motion is simulated at the same time wave-induced motions are simulated or if the effect of additional nonlinearities are to be investigated.

The alternative to the dynamic simulation using a computer is to calculate the bounds of the actual environment that the dynamic positioning system can hold. As mentioned above, these bounds are determined by the steady-state environment and the steady-state-plus-disturbance environment.

The computations of the bounds are considerably simpler because they involve the solution of algebraic equations which represent a static balance of the thruster forces and steady-state environmental forces. In the example, Equation 9.6 determines the environmental bounds for the steady-state case in one axis of the vessel.

If the thruster system uses independent units for each control axis, then a separate algebraic equation is involved to determine the bounds in that axis. Generally, however, thruster units supply force in more than one control axis and nearly always supply moment for the heading control axis. As a result, the algebraic equations are coupled through the thruster forces.

Furthermore, the thruster forces and moment used in the algebraic equations must obey the rules of thruster allocation logic. Thus, for the very common thruster configuration of fixed-axis thrusters shown in Fig. 9–12, the algebraic force equations take the following form:

$$F_{XT} = F_{XC} + F_{XA} + F_{XWD} + F_{XTR}$$

$$F_{FT} + F_{AT} = F_{YC} + F_{YA} + F_{YWD} + F_{YTR} \qquad (9.19)$$

$$F_{FT}\ell_F + F_{AT}\ell_A = M_{ZC} + M_{ZA} + M_{ZWD} + M_{ZTR}$$

where:

F_{XC}, F_{YC}, M_{ZC} = the force and moment caused by the current
F_{XA}, F_{YA}, M_{ZA} = the force and moment caused by the wind
$F_{XWD}, F_{YWD}, M_{ZWD}$ = the x-y force and moment caused by wave drift effects
F_{XT}, F_{FT}, F_{AT} = the thrust from the x-axis thruster, the forward thruster, and the aft thruster

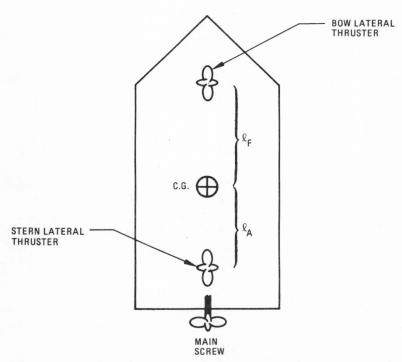

Figure 9-12. A Common Fixed-Axis Thruster Configuration

The set of algebraic equations can be solved for the cases where the environmental forces and moment represent the steady-state values of each element and the steady-state values plus disturbance values. As used in the example the disturbance can be caused by a wind gust.

Additionally, the equations can be solved with either the environmental forces and moment given or the maximum thrust output of the thruster system given. In either case, since the environmental forces and moment are a function of the angle-of-attack of the environmental elements, the equations must be solved at each angle-of-attack, describing a continuous-holding boundary as a function of the angle-of-attack of the environment.

If the limiting values of the environment or thrust as a function of the angle-of-attack are plotted on a polar plot, they form closed curves, often referred to as holding rosettes because of their shape. Typical examples of holding rosettes are shown in Fig. 9–13 for a ship-shaped drilling vessel. Only half of the rosettes are shown in Fig. 9–13 because the drillship is symmetrical about its longitudinal centerline. The elongation of the holding rosettes in Fig. 9–13 is caused by the environmental drag being larger in the lateral direction than in the longitudinal direction.

Solving Equation 9.19 when the environmental forces are given in straightforward even if thrust degradation resulting from inflow current is included because the equations are linear with a sufficient number of equations and unknowns, and the thrust is not limited. The situation is more complicated when the thruster system is given with its maximum thrust outputs and locations on the vessel.

Then, Equation 9.19 becomes a set of nonlinear algebraic equations because the environmental forces are generally square-law functions. Likewise, there is no unique answer unless three of the four environmental elements are assumed to be constant. Normally, the wave drift and work-related environmental elements are held constant. Then, either wind or current is held constant and the other solved for using Equation 9.19.

In solving Equation 9.19 for either the steady wind or current, the limiting value of the variable parameter is determined when at least one group of thrusters reaches its maximum output. For example, with the fixed-axis thruster configuration shown in Fig. 9–12 when the environment is on the bow or the stern, the x-axis thrusters reach their maximum output first and, therefore, determine the limiting value of current or wind in that region of environmental angle-of-attack.

The other two equations determine the amount of lateral thrust used when the x-axis thrusters are at their maximum output. With this information the total power demand can be computed at the same time that the limiting environment is computed. As the environment shifts toward the beam of the ship, the y-axis thrusters reach their maximum output first and the x-axis thruster output decreases.

Because the set of algebraic equations are being solved under several constraint equations, i.e., the thrust-limiting equations and thrust-allocation rules, it is often much easier to use a digital computer program to solve Equation 9.19. The program is especially useful if thrust degradation resulting from inflow current is included or a complex thruster configuration is specified. A convenient program uses an iterative technique where the environmental parameter being solved for is iteratively increased until a thruster limit is reached.

The real value of the static thruster sizing study can be seen in Fig. 9–13. First of all, the outside rosette shows that for a given thruster configuration that the dynamic positioning system holds the specified steady wind and so many knots of current as a function of environmental angle-of-attack. In the example shown in Figure 9–13, the environmental elements are colinear or at a constant given angle-of-attack.

The inside rosette of Fig. 9–13 shows the current that the thruster system counterbalances with a specified steady wind plus a gust as a function of the angle-of-attack of the environmental elements. Thus, if the values of current bounded by the two rosettes are less than the value given in the performance specification, the proposed thruster system has no chance of satisfying the performance specification.

If the specified current is not achieved, then either the specification can be changed as indicated in Fig. 9–14 or the thruster system reconfigured. If the specification is to be changed, then the static rosettes are very helpful in its redefinition. If lesser values of specified wind over limited angles-of-attack, such as on the beam, or if increased thrust output are defined, then a new static thruster sizing gives fast and easy-to-understand results. Generally, these are sufficiently accurate results to make initial design decisions, such as what adequate thruster system size and location will be. The predicted thrust usage can be used to compute estimates of power usage for sizing the power system.

Static thruster sizing is also less expensive than the dynamic computer simulation. First, much less time is required to compute solutions to the static equations than the differential equations of the

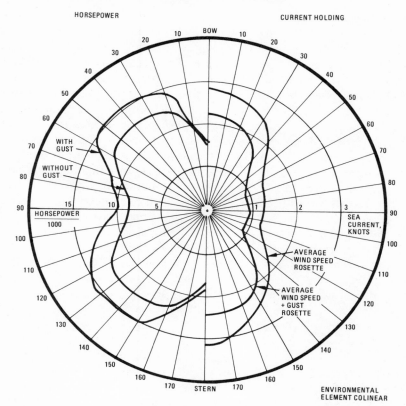

Figure 9-13. A Typical Current Holding and the Corresponding
Horsepower Rosettes for a Ship-Shaped Hull

dynamic simulation. Secondly, the only data required for the static
solution are the static environmental drag forces and the static
thruster forces, all of which are much easier to generate than the data
required for a dynamic simulation as Table 9–2 illustrates.

The value of the static thruster sizing study must not be exaggerated. When either more precise positioning information is required
or the effect of more complex characteristics is to be evaluated, the
dynamic simulation is required. However, the static analysis is very
useful in supporting dynamic simulations which are being used to
determine limiting thruster and power system conditions.

In this support role, the static analysis defines areas of environ-

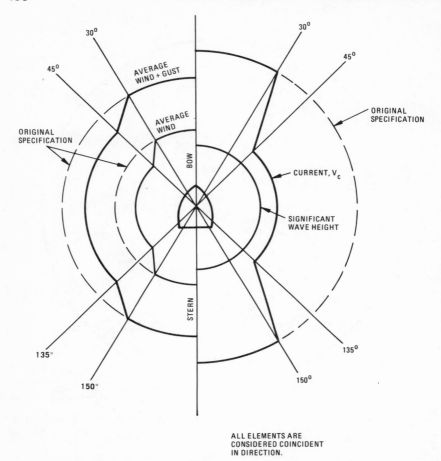

Figure 9-14. Adjusted Environmental Performance Specification

mental parameters where limiting occurs. Often bounding the range of the parameters to be tried in the dynamic simulation can save significant computational time.

In the early stages of a project the decision to use the dynamic simulation or the static analysis is difficult to make. It is often best to start with the static analysis to establish project costs and feasibility and, then, turn to the dynamic simulation to refine the information required to sell the project. In the next section the real importance of the dynamic simulation is shown.

TABLE 9–2

DATA REQUIREMENTS FOR INITIAL THRUSTER SIZING STUDIES

	Study type	
	Static	*Dynamic*
Vessel		
Displacement at operating draft		X
Center of rotation in the horizontal plane	X	X
Yaw radius of gyration		X
Thruster system		
Number of thrusters and their maximum thrust output	X	X
Location of thruster relative to the center of rotation	X	X
Thrust degradation as a function of current inflow conditions	X	X
Thruster dynamic modeling information (response characteristics and nonlinearities)		X
Thrust and power as a function of the thrust command variable		X
Hydrodynamic		
Definition of coordinates and coordinate center of hydrodynamic model	X	X
Added mass in surge, sway, and yaw		X
Added mass coupling among vessel axes		X
Current drag coefficients in x-y and yaw axes as a function of the angle-of-attack of current	X	X
Aerodynamic		
Definition of coordinates and coordinate center for aerodynamic model	X	X
Wind drag coefficients in the x-y and yaw axes as a function of the angle-of-attack of the wind	X	X
Wave		
Definition of coordinates and coordinate center for wave model	X	X
Response amplitude operators in the surge, sway, and yaw axes as a function of the angle-of-encounter to the waves		X
Mean wave drift coefficients in the x-y and yaw axes as a function of the angle-of-encounter to the waves	X	X
Work Task		
Definition of coordinates and coordinate center of the work-task model	X	X
Average forces and moments in the x-y and yaw axes	X	X
Dynamic forces and moments in the x-y and yaw axes		X

Note: The data listed in this table assumes that the study is an initial investigation where complete data is not available. Some of the data listed in the table can often be estimated accurately enough that special model basin and wind tunnel tests are unnecessary.

9.4   Controller design and fabrication

Following the award of the contract to produce a controller for a dynamic positioning system, the selected supplier starts the design steps which lead to the fabrication and delivery of the system. Production of a controller can be divided into the following phases:

• Design
• Fabrication
• System test
• Delivery and dockside checkout
• Sea trial

An important aspect in the ownership of a dynamic positioning control system is training the operators. Training starts before the system is operational and is performed parallel with the last three phases given above. Training will be discussed in detail in Section 9.5.

The design of the dynamic positioning control system is like other engineering design efforts except that it involves additional considerations created by real-time feedback control aspects of the system. If, in addition, the controller is implemented in a digital computer, then the production of software as well as hardware is necessary.

In fact, once the design specifications are prepared, the software and hardware efforts proceed, more or less, on their own until the system test phase.

The first key step in the design process is the preparation of the final system requirements from which the design specification can be written. These requirements are usually finalized in a meeting between the future owner and other suppliers of elements of the dynamic positioning system. These suppliers usually included the thruster, power system, and ship yard representatives.

Another important topic of discussion in this supplier coordination meeting is the interfaces between suppliers. The control system supplier is concerned with the thruster interface and the installation of the control system on the vessel. Consequently, the control system supplier, thruster manufacturers, and ship-yard representatives exchange preliminary information regarding necessary data for design.

For example, the control system and thruster system suppliers describe their respective equipment to find a mutually acceptable interchange of information between their two systems. Usually, the interface between the control system and the thruster system is electrical or electrohydraulic. Thus, the two suppliers agree on signal and impedance levels, the scaling of the signals, and the definitions of signal polarities.

For example, if the thrusters are controllable-speed propellers and the thruster supplier is also supplying the propeller speed controller and the propeller prime mover, then the interface between the control system and the thruster system is usually electrical. The thrust command variable is usually propeller rpm's which are transmitted from the control system to the thruster controller as a proportional analog voltage as illustrated in Fig. 9–15.

The scaling of the analog voltage is a given number of rpm's per volt and positive voltage produces thrust in the positive direction. The importance of the definition of the polarity of the command signal cannot be overemphasized because the incorrect polarity of the command signal produces an unstable control system.

Besides the command signal, the thruster interface includes a feedback signal for the thrust command variable or a measure of the thrust command variable. Usually the feedback signal is scaled the same as the command signal. However, different scaling is acceptable because the computer can make the necessary conversion before it makes a comparison of the command to the feedback for the purpose of alerting the operator to off-nominal performance.

As previously mentioned for the thrust command variable, the definition of polarity meaning is important. It is, however, not absolutely critical because the feedback signal is only used for monitoring purposes.

As shown in Fig. 9–15, the thruster interface sometimes includes signal lines to request that a thruster be made ready. Then, once the thruster is ready for use in the control system, a signal from the thruster control system is activated to notify the DP system. These signals are discrete or switch closure-type information and, therefore, can be implemented by relay or optical isolator-type devices so that the thruster and control system can be electrically isolated.

Electrical isolation is desirable to avoid ground loops and noisy control signals. For this reason, the agreement between the thruster and control system suppliers regarding grounding and wiring practices is critical and should be specified in the first interface meeting. Then, as interface drawings are prepared for installation purposes, very clear instructions regarding cabling and wiring can be included on the drawings. This practice saves many hours of checkout.

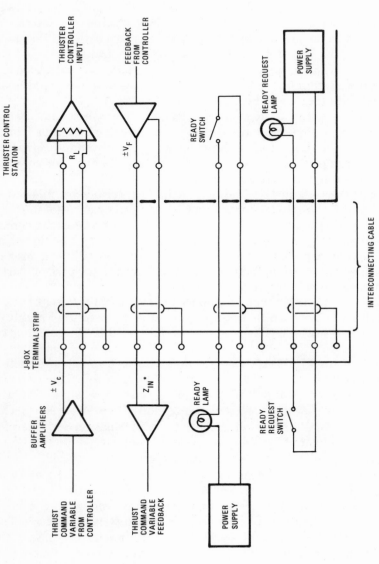

Figure 9-15. A Typical Thruster Interface

The control system supplier in this first meeting also identifies the necessary modeling data for control system design and simulation. The list of parameters includes:

- Thrust versus thrust command variable over the entire range of thrust
- Maximum thrust output
- Thrust degradation as a function of inflow current
- Time-response characteristics or a dynamic model of the thruster controller including any nonlinearities such as rate limiting
- Special command sequences or delays before thruster accepts commands from the DP controller
- Command limiting in the thruster controller or prime mover to protect the thruster system from damaging overloads
- Acoustic noise data if available

If any of the modeling data appears to be unacceptable based on the experience of the control system supplier, then a potential problem can be identified early in the project and either corrective action can be considered or an analysis to verify that there is a problem can be initiated.

As an example, the thruster supplier may state that the response time of his thruster is much longer than the control system supplier knows to be acceptable for a responsive control system design.

A similar exchange of data between the control system supplier and the owner regarding the vessel and environmental modeling should be performed in this early interface meeting. If this data has been already received by the control system supplier during the thruster sizing effort, then only a recheck of the data is necessary to ensure that no changes have occurred.

The control system and vessel suppliers require an exchange of interfacing information. This information generally is concerned with the installation requirements of the DP controller and sensors. Usually, the control system supplier furnishes the sensors in addition to the controller. However, another supplier can be involved, requiring another attendant at the interface meeting.

Layouts and interface drawings for the DP control room equipment, sensor locations, and installation techniques are required. Cabling requirements between the DP control room and the thruster system and between the DP control room and the sensors must be defined. Also, cable routing with respect to potential sources of interference is important. These sources include not only power lines but also radio and radar transmitters and their antennas.

There are other miscellaneous topics that should be on the agenda for the first interface meeting:

- Delivery schedules for the various procurements and equipment installations
- Interface drawing completion schedules so that designs can proceed smoothly and on schedule
- Data exchange schedules so that designs and simulations can proceed smoothly and on schedule
- Destination of all deliveries
- Training and testing schedules

The first interface meeting is very crucial to the success of the entire project. From it comes the information needed by the various suppliers to begin their designs. The interface meeting begins the process of defining the interconnection and transfer of information among the various elements of the system. Without nearly perfect interconnection and information transfer the system will not achieve its best performance. So the sooner the definition and design of these interfaces can begin, the better.

Following the first interface meeting, the control system supplier prepares the necessary design specifications so that the detailed design can begin. The design specifications allow the hardware and software efforts to proceed simultaneously with only sufficient coordination to ensure that the two elements come together properly at the end of the fabrication phase.

The hardware effort follows standard patterns for electronic equipment (Fig. 9–16). In fact, since most control system suppliers have manufactured more than one unit, a significant share of the hardware design may be a repeat of a previous design. Naturally, design improvements and upgrading of various hardware devices occur so that the quality of the control system continues to keep pace with the state of the art. Presently, with the rapid improvements in digital technology, keeping pace involves nearly constant change.

Another hardware area that requires special attention with nearly each DP system is the design of the control console. Since the operator cannot see most of the DP system or, in many cases, where the vessel is positioned relative to the desired position, he must depend, like a

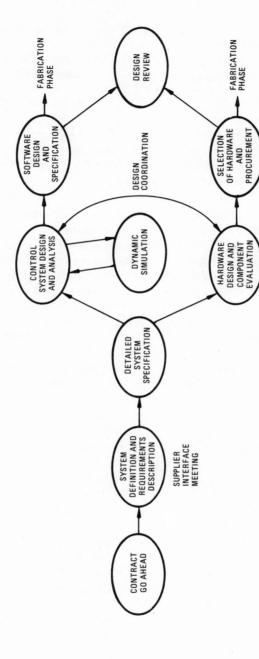

Figure 9-16. The Steps of the Design Phase [5]

pilot of an aircraft at night, on the displays and controls mounted in the control console.

In addition, there are usually some differences from one control system to another and these differences are generally apparent at the control console. As a result of the system differences and the importance of the control console, special attention is given in its design with each system order.

The other path of design effort is the control system and software design (Fig. 9–16). The control system design involves feedback control analysis to establish the necessary control laws and parameters which yield a system with sufficient responsiveness and noise rejection so that the dynamic positioning system will meet the performance specification. In support of the feedback analysis, dynamic simulations are performed to verify the control system design when system nonlinearities and cross-coupling among control axes are included.

The control system design furnishes the necessary descriptions and specifications for the next step of the design phase, i.e., the software design and specification. The software specification is in the form of engineering equations and flow charts plus the necessary data base. The software specification also includes the requirements for operator displays, program priorities, and other support software such as performance monitoring and data logging.

Software is fabricated during the next phase. However, before fabrication begins, a thorough design review is recommended. This review should not only address the hardware and software design and their compatibility, but also the interface designs. When the review is complete and the resulting deficiencies corrected, the fabrication phase can begin and the interface designs can be transmitted to the other suppliers and the owner for their final inspection and use.

Hardware fabrication for a DP control system is the same as for most instrumentation and computer-type systems. The subassemblies, such as computers and displays, are purchased or built-up from piece parts and then checked out independently as much as possible. The checkout also serves to "burn-in" the electronics.

In fact, the subassemblies are often turned on for extended periods of time for "burn-in." "Burn-in" has been found very beneficial with solid-state electronics not only in stabilizing the characteristics of the circuit elements, but also in allowing marginal components to fail before the equipment is placed in service.

Once the subassemblies are fabricated they are integrated into the equipment enclosures and the cabling among subassemblies is installed. The next level of hardware checkout includes open-loop type tests, aided by special diagnostics; it begins once the proper continuity and power supply checks are completed.

For example, the various sensors to be used with the control panel switches can be operated in a simulated manner to see that their inputs are read into the computer memory with the proper polarity and scaled magnitude. Likewise, special diagnostics in the computer can output preset messages to the various peripherals and signal outputs, such as the thruster commands, to verify that the output interfaces are properly operating.

When this testing is complete, a great deal of confidence can be placed in the hardware and its inter-connection. Without this thorough, end-to-end testing any problems between hardware and software are extremely difficult to isolate.

A similar fabrication procedure is used for the software. The only differences are the exact methods of assembly and testing. The assembly is accomplished by the control system designers, working with computer programmers to transform the software specification into computer code. Then, the various subassemblies, or subroutines, of the overall software program are tested with special preset test conditions.

Following the complete subroutine checkout, the integration of the subroutines with the main controlling program is performed. Checkout or software verification of the overall program comes next.

The most convenient and realistic checkout of the control system program is the dynamic simulation. With the additions of some switches and analog inputs to the simulator, all of the inputs and outputs of the actual system can be simulated. Then, each mode of control can be carefully verified and the response of the system recorded. As problems are identified they can be corrected.

The software verification on the dynamic simulator can be performed in a very realistic manner if the simulator is in one computer set-up and the system software program in another. This method is illustrated in Fig. 9–17. Even greater realism can be achieved if the computer system and its peripherals in which the DP control software is operating is the same as the operational computer system. Then the results from the software verification can be used as benchmarks for the next phase in the manufacture of the DP controller, i.e., system test.

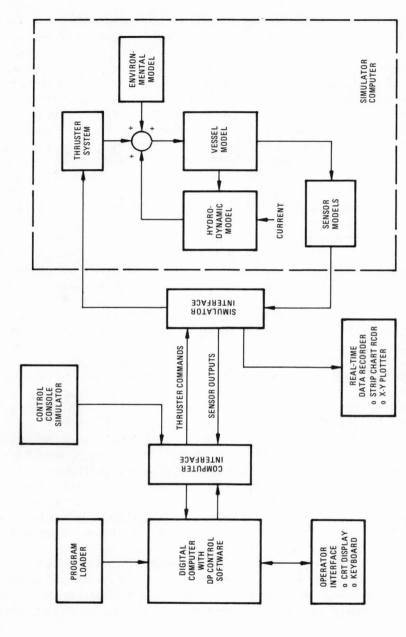

Figure 9-17. Software Verification Set-up Using a Dynamic Simulation

Another use of this set-up to verify the control system software program with the dynamic simulator is to estimate the performance of the overall dynamic positioning system. Since the software at the completion of the verification process is very nearly the same as will be used in the actual system and the dynamic simulation is a model of the remainder of the dynamic positioning system, any performance results from this combination are as good an estimate of expected performance that is available until the system goes into full operation.

In fact, performance test cases should be performed and the resulting system responses recorded for purposes of comparison with future operational responses. If there are any significant differences between predicted and actual responses which cannot be explained by the tolerances in the models, there is data to enable the differences to be traced and explained.

Very often this step is never performed because the systems do perform within the original specification even though the performance is not quite what was predicted. Therefore, since such investigations involve additional expense which is not warranted because the system is operating, they are never funded.

By the time that system test begins, the software and hardware have received a considerable amount of independent checkout. During the system test, these two elements are combined by the software program being loaded into the computers and final testing begins. The first stage of system test includes end-to-end or open-loop tests to verify that signals pass through the entire system as they should. This procedure verifies polarity, magnitude, proper signal routing, and signal control.

For example, the gyrocompass can be turned to a heading which is a few degrees from the value commanded at the control console. Then the output command from the controller outputs to the thruster interface can be verified to be equal to the correct static amount for the given difference between the commanded and measured heading. Such a test, especially if intermediate signal points are also checked, verifies that many system elements, both hardware and software, are performing within the tolerances of the system.

Following the open-loop tests, the combined hardware-software system is connected to the dynamic simulator (Fig. 9–18) and closed-loop testing begins. During this time, the simulator performs in real-time, simulating actual operation as closely as possible. Again, closed-loop tests, like those performed for the software verification, are

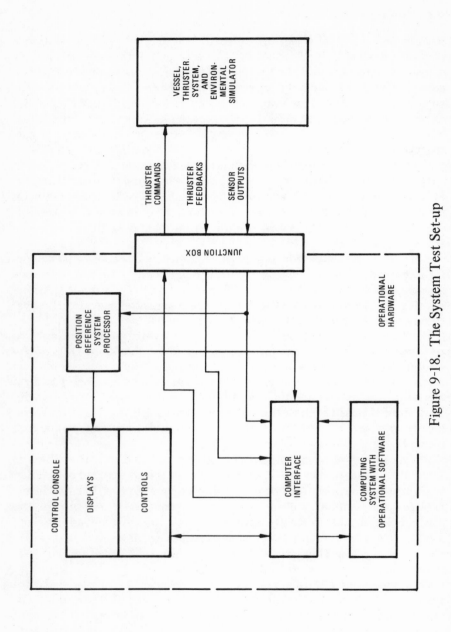

Figure 9-18. The System Test Set-up

performed to verify all system modes and responses to various nominal and off-nominal situations for which the system is designed. Recordings of the responses in the form of hard-copy print-outs and time histories should be collected as benchmark information. The overall system set-up should also be operated for extended periods of time for additional burn-in hours under the normal sequence of control system cycling.

Before system testing is complete, a test plan which has been agreed to by the DP control system supplier and the owner is executed in the presence of the owner or his designated representative. (If the vessel is being registered under a given regulatory agency, a representative of the regulatory agency is usually required to be in attendance during the factory acceptance tests.) These tests are commonly referred to as factory acceptance tests. They serve at least two purposes: first, to demonstrate to the owner that the system meets the design specification as well as can be determined by the dynamic simulator and the combined system hardware and software set-up, and second, to train maintenance and operating personnel.

The training aspect of the factory acceptance test set-up is quite important because the system performs as a closed-loop feedback control system, teaching on-board personnel to identify nominal from off-nominal performance.

Next, the equipment is packed for shipment to where it will be installed on the vessel. Usually the place of installation is the shipyard where the vessel is being built or modified. Installation takes place before the vessel is ready to go to sea.

The installation of the DP control system equipment is usually performed by shipyard personnel under the supervision of the DP supplier's technicians. The first step of the installation is simply mounting the equipment in the spaces previously allocated in the interface designs. The primary part of this equipment mounting effort takes place in the dynamic positioning control room which is usually located near the bridge and navigation area of the vessel. The other part of the installation effort is the mounting of the sensors in their assigned locations.

Next, the cabling to and from the DP control room is installed. Generally, the cabling inside the DP control room interfaces to the outside through a common junction box to which the external cables are routed and connected as illustrated in Fig. 9–19. The junction box is very often one of the items which everyone forgets to include in their

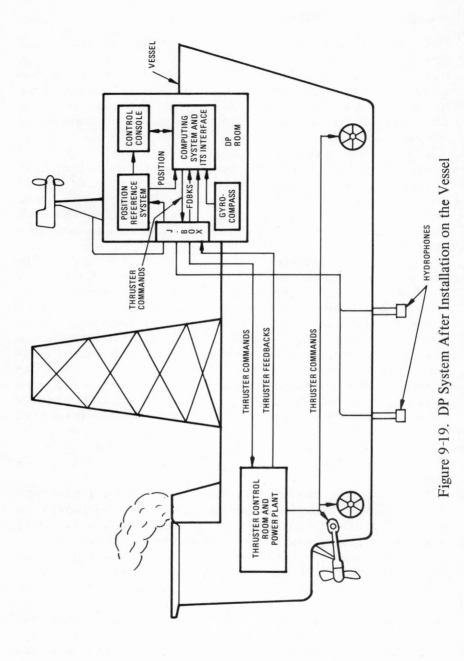

Figure 9-19. DP System After Installation on the Vessel

delivery. Similarly, normal room furnishings such as chairs and tools are often overlooked.

The power for the DP control system is also supplied through the junction box. In the junction box there are numerous electrical terminal strips across which electrical circuits are completed from the control system to external system elements. Since there is no assurance that the cables coming to the junction box from the external world are correctly connected, a detailed checkout of external cabling is performed before any circuits are completed.

Once the cabling checkout is complete, then the final interface of the control room equipment to the remainder of the system is completed. At this time, another series of tests is performed. Since the vessel is usually still located at a dock, these tests are referred to as dockside tests.

During the dockside testing, only a limited amount of the system can be checked out because the vessel is tied to a dock and the water is generally very shallow. For example, the thrusters cannot be safely operated at full output. At best, the thruster actuator systems can be tested with the control system equipment if they are independent of the propeller rotating as in the case of a controllable-pitch thruster. Usually, the position reference systems cannot be tested either, especially if they are acoustical, because of the shallowness of the water.

However, the dockside tests can verify most of the system operation prior to the vessel sailing. Additional training can be performed, especially in the area of maintenance. In a dual computer system, operator training, using a special training arrangement of the two computers, is also possible if the training feature has been included as a part of the system delivery. More is said of this feature in the next section.

With the completion of the vessel and the installation of the dynamic positioning system, the final phase of making a DP system operational is ready to begin. In this phase the vessel goes to sea where all the major systems are tested. Some of the testing is monitored by regulatory agencies to certify the classification of the vessel. In the near future the dynamic positioning system will undergo a certification, which in most cases will not be unlike the sea-trial testing now performed by every DP system supplier.[7]

Actually, the sea trials for the dynamic positioning system should be more of a demonstration than a test because of all the prior testing under simulated dynamic conditions. However, the sea-trial effort is usually performed as a test with at least a semiformal test plan document. Specific tests are performed and the responses evaluated against specified responses given in the test plan document. The witnesses either decide that corrections are necessary, that variances are allowable within the test conditions or tolerances of the measurements, or that the test results are acceptable. The present state-of-the-art is such that the sea-trial tests go smoothly and there are few post sea-trial corrections needed.

After the completion of the sea trials or during the sea trials, another training opportunity for operating and maintenance personnel is available. The sea trial is particularly good because the operational system under nearly all operational conditions is available for training. The operational equipment used to perform the intended mission is generally all that is missing during the sea trial.

On a drillship, for example, the BOP stack and riser are not in place when the sea trial is performed. Thus, operator training does not cause the stress that is present in the operational situation, allowing the operator the opportunity to learn under less pressure.

A goal which is usually set for the sea trials is performance evaluation of the dynamic positioning system. However, a major objective of the owner is to limit the sea trial to the minimum number of days so that the vessel can start earning a return on his investment.

Therefore, if the weather does not cooperate, that is remains less than the design environment, the performance evaluations are limited to whatever occurs and plans are set to evaluate the performance once the vessel is operational. More about this is discussed in the next section.

9.5   Operational considerations

Although the operation of the system does not begin until after the sea-trial period and final outfitting, planning for the operational period on the vessel and its dynamic positioning system begins months earlier. This planning for the DP system includes the following items

- Selection of operational personnel
- Training of operational personnel
- Preparation of operational procedures
- Outfitting the vessel with the equipment required to maintain the the DP system

Following the planning phase and the sea trials, the operation begins. The author has had limited experience with the operation of the DP system, therefore, only the more obvious considerations are given in this section. For more information, see References 8 through 12.

In the selection of operating personnel two types of knowledge appear to be useful. First, the prospective operator should have a basic understanding of electronics. This understanding can be acquired as an electronics technician or through a training school. However, practical experience is very important.

The second type of knowledge is marine experience. The value of marine experience is that it gives the operator a feel for the natural responses of a vessel and the procedure used to operate a seagoing vessel. The feel for vessel responses is useful in shaping the operator's judgment regarding normal versus abnormal conditions.

Similarly, the appreciation for standard marine practices gives the operator the ability to make the greatest use of the marine crew of the vessel. An experienced operator understands what actions the crew will take under certain conditions and what shipboard life is like.

As a result, some of the most qualified DP operator candidates are electronics mates who have maintained and operated electronic equipment on ocean-going ships. With the electronics capability, the operator may be able to diagnose and repair equipment or perform preventive maintenance as he stands watch on the DP system. Standing watch can be many hours of very boring, sedentary work for the DP operator because the control system is automatic. Some useful activity is very welcome.

Furthermore, the computer is performing an untiring, continuous watch of the system performance for the operator. So when an off-nominal response is detected, the DP controller sounds an audible and visual alert, calling the attention of the operator to a specific system element. If he is near the console, he can very quickly stop whatever he is doing and respond to the alert.

If the prospective operator has some knowledge of the work that the vessel is performing, he is better equipped to perform his duties and use the DP system to support the work. As control methods become more sophisticated in the offshore industry and more people accumulate experience in the offshore industry, more personnel with this kind of experience will be available.

Naturally, even for the best equipped candidate, training is required. As mentioned in the previous section, training can start as early as during system test, which generally occurs one or two months before the system is delivered. Training can continue through sea trials and beyond if desired.

Once the system is in operation, training of new personnel as the turnover of shipboard personnel occurs is usually performed by existing operating personnel. The new operator may visit the DP supplier to obtain some training on the separate parts of the system, such as the acoustic position reference system or the computers. However, the only realistic place to obtain training on the system that he is ultimately going to operate is on board the operating vessel.

An alternative exists for DP systems with dual computers: one computer can be used to simulate the vessel and the other can be used in its normal fashion. This setup is very similar to the one used in the system test shown in Fig. 9–18. Naturally, the redundancy of the system is lost when one computer is used as a simulator. Consequently, this type of on-board training is only possible when redundancy is not required, such as in port or when traveling to a new location.

The DP control system supplier furnishes a full set of operations and maintenance manuals when he delivers his system. These manuals are designed to assist operating personnel in the operation and maintenance of the DP control system and usually contain procedures for troubleshooting equipment problems and for reacting to various equipment and operational difficulties.

However, the DP control system supplier may not understand the work task to be performed by the vessel as well as the owner or the future service organization that will actually operate the vessel. In certain cases, the supplier-furnished procedures may not fit the manner in which the owner plans to operate his vessel or the operational procedures required to perform the work.

As a result, the owner or his operating service organization may develop their own procedures or work with the DP control system supplier to develop the necessary operational procedures. As the operating personnel accumulate experience, the procedures can be rewritten or revised to handle nearly any conceivable situation. However, prior to performing a task for the first time, at least an estimate of the operating procedures should be available.

Example situations for which operating procedures are required for a drillship operation are those for the selection of on-line position reference sensors as a function of operations going on in the area of moonpool. These procedures are important because the hydrophones of the acoustic position reference system are normally located in the moonpool area. Therefore, if there is any operation going on in the moonpool area which may interrupt the position information from the acoustic system, the operator should be aware of it and transfer the control system to a back-up sensor such as the taut wire.

Two moonpool area operations which can significantly affect the acoustic system are mud dumping, and loading and unloading support vessels. In the case of mud dumping, the mud contains just enough trapped air to interfere with a hydrophone receiving the subsea-beacon signal. Likewise, the propeller wash from the support vessel when its stern is toward the drillship contains sufficient trapped air to interrupt the acoustic transmission if it passes between the beacon and the hydrophone.

Both of these effects are localized so that there is a temptation to instruct everyone to be careful and leave the DP system on acoustics. But inevitably, the support vessel captain is careless or the sea current carries the aerated mud and the information from the acoustic system is interrupted. Therefore, a much safer procedure is to transfer the system to the back-up sensor until the mud is thoroughly dispersed or the supply vessel is gone or tied firmly to the side of the drillship with her screws off.

A certain amount of support equipment is required to keep the DP system operational. This is critical for a system that is intended to operate in remote areas and continuously for hundreds of consecutive days around the clock. The first item of equipment is a proper set of spare parts. With most modern electronic equipment modular design makes possible sparing at the electronic card level. Thus, by replacing a defective electronics card, the system can be restored to total operational status.

The replacement cycle for a defective card by properly trained personnel with sufficient experience is only a few minutes duration if the failure is on one of the electronics cards. Following the card replacement, the maintenance personnel can either return the card to the DP control system supplier's service depot or factory for repair or repair the card on board the vessel. The former is generally recommended by the supplier.

When the failed card is returned to the supplier, he has either already shipped a replacement spare card or will ship one immediately. Since spares can only be stocked in limited quantities because of the cost of the electronic cards, replacement of a spare can be critical.

Therefore, the spare replacement procedure of the supplier and the location of his nearest service depot are important considerations.

In addition to spares, maintenance equipment is required to trouble-shoot electronic problems and repair equipment. This maintenance equipment includes the standard electronic tools and test equipment such as a voltmeter and an oscilloscope.

Another item of equipment which is not maintenance equipment, but acts to maintain or improve the performance of the dynamic position-ing system over the life of the system is an event-recording system. In the course of the operation of the dynamic positioning system, events may occur which should, if possible, be avoided in the future.

However, during the event and the time immediately following it, the operating personnel are far too busy to accurately describe any more than the general symptoms of the event. Thus, a recording sys-tem that captures the events is an asset that pays for itself over the life of the system by making possible post-event analysis and correc-tive action.

If required, another value of the event-recording system is the col-lection of objective performance data from which the vessel capabilities can be accurately determined. These capabilities can be used to market the vessel to potential customers or as design inputs to the next-gen-eration vessel.

There are numerous other operational considerations.

REFERENCES

1. M. J. Morgan, "Dynamic Positioning of Drillships and Future Semi-submersibles," Offshore North Sea Conference Paper T-1/i-2, Stavanger, 1974.

2. H. Van Calcar and M. J. Morgan, "Dynamic Positioning Today," Oceanology In-ternational Conference, 1975.

3. J. S. Sargent and P. N. Cowgill, "Design Considerations for Dynamically Posi-tioned Utility Vessels," OTC 2633, 1976.

4. John T. Meyers, Editor, *Handbook of Ocean and Underwater Engineering*, Mc-Graw-Hill Book Co., 1969.

5. Roy Pearson, "Hardware Mechanization for a Safe and Reliable Dynamic Position-ing System," Safety and Reliability of Dynamic Positioning Systems, DNV Symposium, Oslo, Norway, November 17–18, 1975.

6. *Rules for Building and Classing Offshore Mobile Drilling Units*, American Bureau of Shipping, 1973.

7. *Tentative Rules for the Construction and Classification of Dynamic Positioning Systems for Ships and Mobile Offshore Units*, Det Norske Vertias, RM-7-76, Rev. 3, 17 November, 1976.

8. F. B. Williford and A. Anderson, "Dynamic Stationed-Drilling SEDCO 445," OTC 1882, 1973.

9. Denis T. Skinner and D. S. Hammett, "SEDCO 445 — Drilling Without Anchors in 2000-Feet-Plus Water Depth," OTC 2151, 1975.

10. Bernard C. Duval, "Exploratory Drilling on the Canadian Continental Shelf, Labrador Sea," OTC 2155, 1975.

11. C. L. Collier and D. S. Hammett, "Dynamic Stationed Drill Ship SEDCO 445," OTC 2034, 1974.

12. Darrell Sims, "Acoustic Re-entry and the Deep Sea Drilling Project," OTC 1390, 1971.

10

Extensions and Projections

10.1 Introduction

This is the final chapter of this book, but the use of dynamic position-ing and computers on offshore vessels has only begun. The trend, in fact, is being accelerated by the explosive expansion of offshore projects and their increasing demands for more sophisticated equip-ment.

The technologies needed to configure the sophisticated equipment are also in a rapidly expanding period when new, more powerful products appear on the market nearly every day. This chapter at-tempts to examine the future of the following areas:

- Dynamic positioning in new offshore applications
- Expanded uses of the dynamic positioning computing system
- Developments of technologies which will improve existing products and make possible new products to solve new problems

The field of dynamic positioning and computer systems in offshore applications is developing so rapidly that some of the projections dis-cussed here may well exist by the time this book becomes available. The presentation in this chapter is by no means exhaustive, but it will give some insight into the future and the potential of dynamic positioning and computers.

10.2 Extending the use of dynamic positioning

Chapter 1 described the various offshore missions for which dy-namic positioning is either required or highly desirable. This contribu-tion is increased efficiency and responsiveness to mission requirements or reduced cost of operation. To date, dynamic positioning has not been applied to all the offshore applications listed in Chapter 1.

Table 10-1 contains a recap of the offshore applications for which dynamic positioning has been used. Table 10-1 also contains the name of a vessel that has been used in the given application and a reference paper about the application, if one is available. The definition of dynamic positioning used in Table 10-1 is the one emphasized in this book, i.e., the system is a closed-loop feedback system which contains sensors, an automatic controller implemented in a computer, thrusters, and a power system.

TABLE 10-1
RECAP OF DYNAMIC POSITIONING APPLICATIONS

Applications	Vessel(s)	Reference papers
Sea bed sampling and exploration	Eureka, Glomar Challenger	2
Exploratory drilling	SEDCO 445	3
Cable laying	USS Naubuc	
Diving support	Arctic Surveyor	
Dynamic mooring assistance	Saipem Due	1
Salvage	Kattenturm	
Production field support	Arctic Surveyor, Seaway Falcon, Kattenturm	

There are other applications which have used a vessel with a thruster or a mooring system and manual control for positioning which have been termed dynamically positioned. In this discussion these applications are not included.

Dynamic positioning, as defined here, has now been installed and tested on many different types of vessels. The largest concentration of installations has been in ship-shaped hulls, both large and small. A dynamic positioning system has also been installed on a catamaran and just recently on a semisubmersible.[4] While each vessel configuration imposes special requirements on the dynamic positioning system there does not appear to be any hull shape for which dynamic positioning cannot be installed.

Future applications of dynamic positioning certainly include deep-ocean mining. In this application, the mining vessel or mother ship requires dynamic positioning so that it can more efficiently move along a preset track. Mining support vessels may require some form of positioning control so that they can off-load the mother ship as she continues to move in good or rough weather.

Dynamic positioning will certainly be applied to offshore support functions where a vessel must hold station near another fixed or floating platform. Already the offshore hydrocarbon industry is moving from the exploratory phase to the production phase and large offshore fields are now being developed. During such a development of an offshore oil and gas field, there are large numbers of installation, maintenance, and repair jobs throughout the field. In many of these fields, mooring the support vessel in the field with its labryinth of interconnecting flow lines is prohibited, if not at least highly discouraged.

As a result, the support vessel sent out to install, maintain, or repair some subsea element in the production field must use thrusters to maintain position during the performance of the work task.

In milder weather a coordinated thruster control system, using a single joystick control lever, can provide sufficiently accurate positioning control for the support vessel if the vessel is working between production field platforms.

However, the support vessel operator who wants to work in heavier weather and nearer to the platforms needs a dynamic positioning system with an automatic controller. For this reason, the support-vessel market will continue to grow, especially when deep water production becomes a reality.

The dynamic positioning of tankers during loading and unloading operations from offshore terminals may become both economical and necessary as water depths increase. Likewise, dynamic positioning of tankers will increase the weather windows in which loading and unloading can be performed. This is often a serious limitation in the use of tankers and offshore terminals, especially during the winter months.

The same system could also be used to improve the accuracy of the low-speed steerage of a tanker coming into a port, thereby reducing the chance of a damaging oil spill.

Since tankers are large, they require a significant amount of thruster horsepower to control their position and heading. Unfortunately, the standard propulsion system of a tanker, although very powerful, is not as useful for position and heading control because of the slowness of its response to changes. Even though most modern tankers possess bow thrusters to improve their low-speed steerage or maneuverability, these thrusters are typically too small for dynamic positioning.

As a result, nearly the entire thruster system of the tanker must be modified or replaced with one more suitable for dynamic positioning. Additionally, a power system to support the new thruster system is required. The cost of such modifications to an already expensive tanker is more than most owners are willing to spend.

It seems reasonable that at some point a pipe-laying vessel will be dynamically positioned. This action will be the result of a requirement for a pipeline from an offshore field in several thousand feet of water or a special pipe-laying technique which cannot tolerate the time required to maintain the presently used multipoint mooring system.

Either of these requirements makes the currently used multipoint mooring system unusable and are likely the only reasons to abandon the present position control system.[5] However, as is already occurring, the multipoint mooring systems will be augmented by a thruster system. These combination systems will use the thrusters to statically or dynamically off-load the mooring system.[1]

There are two major drawbacks to the totally dynamically positioned pipe-laying vessel. The first is the required horsepower to maintain proper position control, navigation along the pipeline route, and the tension in the pipe. For the third-generation pipe-laying vessel, the horsepower required for total dynamic positioning is significant because of the size of the vessel (typically 40,000 to 50,000 tons); the pipe-laying weather requirement (sea state 6 or more); and the constant pipe tension of 100,000 lb or more which transposes into several thousand horsepower.

As a result, valuable space is required for the thruster and supporting power system. However, a multipoint mooring system of 12 to 14 winches requires considerable space for its installation and power to drive the winches as the vessel moves along the pipeline route. Similarly, there is a cost trade-off between a multipoint mooring system and a thruster system which in water depths less than 1,000 or 2,000 ft favors the mooring system.

The second drawback to the all dynamically positioned pipe-laying vessel is the resistance to change in the pipe-laying industry. This industry has grown up using the multipoint mooring system to hold the vessel in position while the pipe is welded together and to pull the vessel along the pipeline route as the pipe is laid on the sea bed.

484 DYNAMIC POSITIONING OF OFFSHORE VESSELS

Therefore, a change to a position control system from one firmly attached to the bottom in a dozen or more places with huge anchors is significant. The change not only involves a "way-of-thinking" but also a major revision in operation, training, and maintenance procedures. Confidence in the security of the dynamic positioning system must equal that of the mooring system. This confidence is not easy to instill, as the operators of moored drillships who have changed to DP drillships will attest.

Beyond these current and projected uses of dynamic positioning there are undoubtedly others. Like so many other technologies, the more dynamic positioning is used, the more applications it will inspire. With the ever-increasing use of the oceans, dynamic positioning is here to stay and should become as common as classic marine-navigation techniques are today.

10.3   Extended uses of the DP computing system

Because of the capabilities of digital computers in computational speed, memory capacity, and software systems are increasing, it is possible to use the computing system of the DP system for other purposes if the computing system uses digital computers. In this section some potential uses are discussed.

Chapter 6 described the management of the power system, a major use of the digital computing system of the DP system. This use is very critical on larger dynamically positioned vessels where the power system is composed of multiple power-generating units, feeding a common bus which in turn feeds the various users. With such a power system, management is important to achieve the maximum system economy. Likewise, power system management can result in a real cost savings by reducing the number of personnel required to operate the system.

Even more important, power system management minimizes the likelihood of a total power system shutdown caused by a failure or overload. In the midst of heavy weather and a critical point in the mission of the vessel this feature alone may justify the expense of such a system.

The usage of the computing system is naturally extended to any kind of task requiring numerical computations or information storage. Included in the list of such tasks on board an offshore vessel are

- Stability and load calculation
- Inventory control
- Payroll

Also, special computer programs which assist in performing the mission of the vessel can be run on the computing system of the DP system. For example, on a drillship, computer programs exist which can be used to improve the drilling operation; on a pipe-laying vessel programs exist which predict the shape and stresses in the pipe.

Likewise, if real-time data is recorded on the vessel for post-data analysis of any event or for performance analysis, the on-board computing system can be used to process and format this data for on-board use. A list of typical parameters for a data recording system for a dynamic positioning system is given in Table 10-2. Without this capability the recorded data must be shipped back to a shore-based computer system for processing.

These programs are assumed to be preprogrammed and checked out so that they require only to be loaded into the DP computing system and they are ready to use. Programs can also be created, corrected (debugged), and then executed in a time-share manner. With this capability, the DP computing system becomes a general-purpose computing center, limited only by its memory and access time to execute the lower priority programs.

The only restriction on using the computing system for these other computing operations is that they are treated as lower priority than

TABLE 10-2

TYPICAL PARAMETERS RECORDED FOR A DP
SYSTEM

1. X-Y position from all position reference sensors
2. Heading
3. Wind speed and direction
4. Thruster commands, feedback, and status
5. Position and heading commands
6. Modes and alerts
7. Environmental parameters if measured
   a. Wave height
   b. Sea current
   c. Temperature
8. Riser angle

the control of the DP system or the management of the power system. Thus, if a higher-priority program is scheduled to be executed, the processing of the lower-priority program is interrupted.

Computer software systems exist which schedule programs on a priority and time-available basis. These systems when combined with modern high-speed digital computers can provide effective multitask time-share systems. Fig. 10–1 illustrates the multitask software scheduler concept.

In order to provide the expanded computational capacity, peripherals must be added to the DP computing system. The list of peripherals includes mass storage devices, high-speed printers, program loaders, and auxiliary displays and keyboards. In the category of mass-storage devices, there are magnetic tape recorders, magnetic cartridge discs, and magnetic drums. Magnetic tape has been used quite successfully for recording data, but either a disc or a drum is required to support a time-share system where programs and data are loaded at the same time that the CPU is performing other tasks.

Magnetic tape is also used to load programs, but only before the computing operation has started. Similarly, discs and drums are used to record data. However, magnetic tape is a more convenient data-storage medium from the standpoint of handling, storing, transporting, and cost.

High-speed printing capability is available with standard computer line printers. Program loading can be performed with either the mass storage devices or lower-speed devices such as paper tape readers, cartrifiles, or cassettes. Auxiliary displays and keyboards are usually composed of the same alphanumeric CRT display used in the DP system control console and a general-purpose full alphanumeric keyboard. For convenience the auxiliary displays and keyboards are usually not located near the DP control console and may be in a separate room and have a desk-top display and keyboard unit for convenient operator use. These additions to the DP computing system are illustrated in Fig. 10–2.

With the addition of the above-mentioned peripherals, the central processing unit (CPU) of the computing system requires features which allow it to transfer data in and out of the computer in the fastest, most efficient manner as well as features to increase its computational speed. High-speed data transfers between the CPU and its peripherals are achieved with Direct Memory Access (DMA), and special hard-

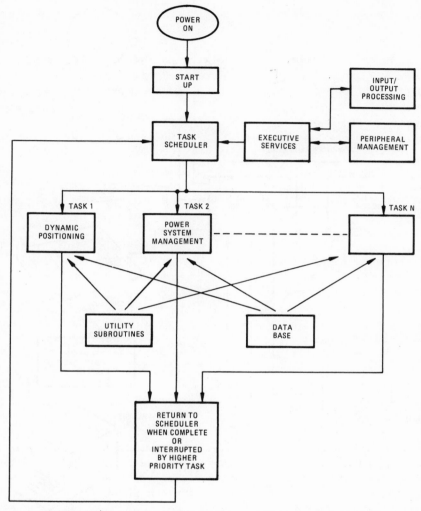

Figure 10-1. Multi-Task Software Program Scheduling

ware arithmetic packages increase the computational speed of the CPU.

Naturally, as more operations are performed in the CPU and more complex program-control software is created to give a powerful multi-

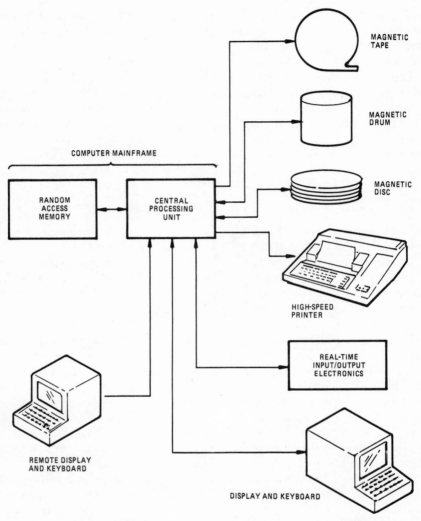

Figure 10-2. Block Diagram of Expanded DP Computing System

task time-share computing system, more directly-addressable memory is required to support the CPU.

Currently, high-speed minicomputers, suitable for dynamic positioning control and power management, are standardly configurable with 65,536 words of 16-bit random access memory. Since DP and power management use between 30,000 and 40,000 words of 16-bit random access memory, there is a sizable quantity of memory available for other data processing purposes.

There are definite benefits to be gained by extending the use of the DP computing system to other computational tasks. However, there is a limit to the amount of additional computational loads that the DP computing system can manage. The first limit is most likely computer time. With a time limit, there is either not sufficient time to execute the low-priority programs or so little time that the user cannot get a useful turn around on his computing job. As a result, the user stops using the DP computing system and the investment for the added capability is wasted.

The alternative to the central computing system, centered in the DP computers, is an independent computer center for the non-DP data processing. The distributed processing approach can also simplify the software system required to provide multitasking on a priority and time-available basis which, of course, is a cost-savings. In fact, with the present trend toward reduced digital computer hardware costs and the ever-increasing software costs due to the skilled labor involved in creating software, there is a very strong argument for adding computing hardware wherever there is a potential simplification in the software.

Then, if the transfer of data between computers is required or is desirable, high-speed inter-computer links can be installed. The distributed computer processing approach can also have the added advantage of dispersing the computing hardware so that a large single area is not required and the DP operators are not distracted during critical times. With such a multicomputer operation, commonality of computing equipment should be a basic requirement.

Then the maintenance and handling of spare parts is simplified and in case of serious failure in the DP computing system, which is not covered by a spare, the lower priority computing system can supply the needed equipment until it can be resupplied.

10.4   Future technological developments

Change in technology is commonplace. This is certainly true in the offshore industry where considerable energies are being applied to develop the offshore resources needed to power the world's industries.

A further impetus in the offshore industry is the nature of the problems and the environment in which they occur.

Techniques developed and used onshore are, in many cases, only conceptually valid in the offshore environment. Therefore, new techniques and tools for offshore projects have to be developed.

One good example of the offshore industry's demand for new technology to solve a well-developed onshore application is drilling for gas and petroleum. The basic drilling apparatus, i.e., the derrick, the drill bit, drilling pipe, etc., is applicable to offshore drilling. However, that is about all because such items as a suitable offshore platform, a marine riser, a blow-out prevention stack, and motion compensation system had to be added to the drilling equipment.

Then, as water depths have increased, the nature of the platform has changed from a jack-up rig or flat-bottom barge to a dynamically positioned drillship or semisubmersible.

One of the next offshore applications to go through a development process is mining for valuable minerals. The first phase, and the only phase to receive any attention in the near future, is subsea mining for minerals residing directly on the sea bed. Since these minerals are located in very deep water in mid-ocean areas, there are many problems to be solved before this form of mining is technically and economically feasible. In fact, the investments required for such operations will be so large that few single organizations will be able to attack this area.

As a result, consortiums will be the future offshore miners, bringing with them their special area of expertise, e.g., mineral collection and transport, vessel design and construction, and offshore operational methods.

A comprehensive discussion of future technological developments for the entire offshore industry is far beyond the scope of this book. As might be expected, this section concentrates on technological developments which are going to affect dynamic positioning. Furthermore, these discussions are limited to the sensors and controller elements of the dynamic positioning system.

Future developments in the fields of thrusters and power system equipment are beyond the authors specific area of experience and expertise. There will certainly be developments in these two critical elements of the DP system.

In the area of sensors for the dynamic positioning system, the major changes are going to be in the position-reference sensors. Currently, the sensors for heading and wind do not appear to need further development. Position-reference sensors do not require further development either. There are many local area position reference sensors which satisfy the current dynamic positioning requirements.

Their only apparent weakness is vulnerability to some form of interference. Therefore, there will be continuing efforts to develop the "perfect" local area position reference sensor or an integrated system which combines the best features from a suite of independent position-reference sensors.

The other area of development in the position reference sensors for dynamic positioning applications is in extending the area of position measurement to a truly global system with accuracies approaching the current local area sensors.

Of the position reference systems on the horizon, satellite navigation is probably the only true global navigation system candidate. Even then, a network of satellites as planned for the NAVSTAR[6] is required, each with a significantly improved positioning accuracy.

Another global or large-area position reference system candidate is the integrated navigation system which is formed around an inertial platform.[7] The inertial platform gives the system the short-term position reference accuracy and the other members of the integrated system provide the long-term positioning accuracy. The satellite navigation system is a candidate for the long-term position reference sensor member of the integrated system.

To achieve the desired results, the quality of the inertial platform may have to be on the level of those presently used for military purposes. One such platform is the electrostatically suspended gyro (ESG) system which is currently not available for civilian applications.[8]

Developments in the area of the dynamic positioning controller will continue to be dominated by technological developments in digital systems. With each day, new digital system components come on the market, each offering an improved product or the potential for increased capability.

In certain areas of the present controller, however, technological developments are not required because the current level of technology is sufficient. For example, the minicomputer used for the central processing element does not require further improvement to perform the computations for the dynamic positioning system.

Increased minicomputer capacity is not required to add many auxiliary functions to the DP computing system as is discussed in the previous section. Improved mass-memory systems will improve such things as reliability and processing speed over the currently available magnetic discs, drums, and tape drives. Thus, for systems using the DP controller for multiple computational purposes, the mention of bubble and charge-coupled memories will be heard.

New operator displays and interfaces will certainly appear in the DP system of the future. The currently-used black-and-white alphanumeric CRT displays and push buttons will be replaced by color graphic CRT displays with touch-panel switches and other man-machine interactive devices. This technology is available, so its introduction is only a matter of time. Some DP controller system supplier will use it in order to outdo his competition. As is the case with many new product features, their introduction is more a function of competition than required performance.

The same phenomenon will occur in the case of the minicomputer. All minicomputer suppliers continue to change their products to keep up with, to be slightly ahead of, or to be slightly different than their competitors. Presently, a minicomputer line that has not been completely replaced in five to ten years is unusual and, most likely, out of business. As a result, for a system for a ship whose lifetime may span 20 or more years, even providing spare parts for the computing system may be difficult, if not impossible.

The one area of the dynamic positioning controller which should see a significant change is in the transmission of data to and from the computing system. Currently, data is transmitted to and from the DP computing system in analog form on separate hard-wire conductors. With each cable there is a cost of installation and maintenance.

For the larger DP systems with many thrusters and sensors plus other functions requiring data in the DP computing system these costs are significant. Additionally, the quality of the data can be degraded by some form of electromagnetic coupling of noise or interference onto the electrical conductors.

Two developing technologies will improve the data-transmission situation. The first is not new, but it has yet to be applied to many marine applications. This technology is serial multiplexing where data are transmitted in digital message formats over the same transmission path. The single-path reduces cabling costs and the digital message format can give noise immunity.

Obviously, with serial-data transmission, time becomes important. If the transmission rate of the digital data system is too slow, all the

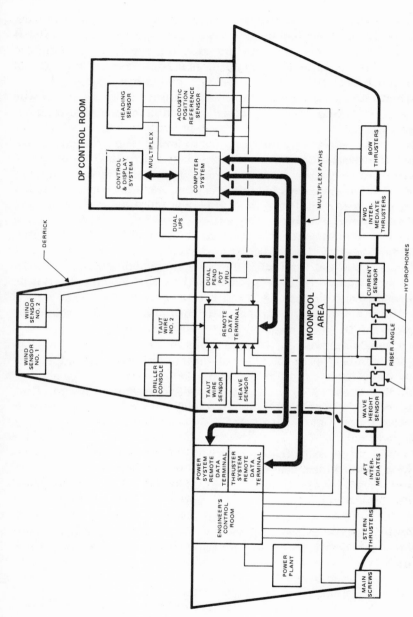

Figure 10-3. Dynamically Positioned Ship with Digital Data Multiplexing

required data will not reach its destination on time. However, with the currently available technology, data-transmission rates of 250,000 to 1,000,000 bits per second are possible. Therefore, even if a data message is 50 bits in length (a data message is normally 30 to 35 bits in length), 5,000 to 20,000 messages per second are possible. This is more than enough to do a combined dynamic positioning and power management system whose combined input and output requirements are less than 2,000 messages.

To be most effective, however, multiplexing requires that data sources and destinations be sufficiently concentrated. Then, short local cable runs can be made to remote input and output interface terminals which are connected to the central computing system with a single conductor or multiple conductors for purposes of redundancy.

For example, a dynamically positioned drillship with a combined DP and power management system has concentrations of data in at least three areas of the vessel. These areas include the stern of the vessel, the moonpool, and the DP room as shown schematically in Fig. 10–3. The data inputs in the stern of the vessel are from the power and thruster control stations. From the moonpool area comes sensor and drilling inputs. The control console furnishes the data inputs and outputs in the DP room area.

The second improvement to data transmission on offshore vessels is the use of fiber optics in place of the metallic signal conductors.[9] Since fiber optics use light as a transmission carrier, they give a significant improvement in immunity to electromagnetic sources of interference and a very wide transmission bandwidth. With the wide bandwidth much higher data transmission rates are possible, if they become necessary.

Ultimately, the cost of fiber optics will also reach a competitive level with current electrical transmission techniques.

So the future of the devices required to perform the control of dynamic positioning systems is only going to improve. Like many other product areas which use digital technology, the cost of the hardware will decrease. However, with more complex digital systems, additional software costs occur, generally driving the prices for the systems higher.

REFERENCES

1. J. S. Sargent and M. J. Morgan, "Augmentation of a Mooring System Through Dynamic Positioning," OTC 2064, 1974.

2. J. R. Graham and J. A. Reed, II, "Glomar Challenger-Newest Concept in Deep Sea Drilling," Society of Petroleum Engineers of AIME Conference, Paper SPE 2180, 1968.

3. D. S. Hammett, "SEDCO 445—Dynamic Stationed Drillship," OTC 1626, 1972.

4. C. Sims, et. al. "SEDCO 709—First Dynamically Stationed Semisubmersible," Multipart Series in *Ocean Industry* in 1976 and 1977.

5. M. J. Morgan, "Position Control of Pipelay Barges," ASME Session of ACD 1975 Winter Annual Meeting, 1975.

6. L. L. Booda, "NAVSTAR: Three Dimensional Satellite Positioning with Ten Meter Accuracy," *Sea Technology,* March 1977, pp 19–21.

7. L. E. Ott, "A Totally Integrated Navigation System," OTC 2461, 1976.

8. P. E. Hall, "Gimballed Electrically Suspended Gyro Aircraft Navigation System," 9th Joint Services Data Exchange for Inertial Systems, 19 November, 1975.

9. J. McDermott, "Telephone and Data Transmission Via Fiber Optics is Being Pushed," *Electrical Design,* April 12, 1976.

Appendix A

Idealized Maximum Required Thruster Horsepower

In this appendix the angle with respect to the bow where the maximum horsepower is required for a given set of force-moment and power-thrust functions is derived for the condition of constant current and wind velocity. Additionally the relative angle between the wind and current, to yield the maximum, is determined. The derivations are performed for fixed-axis and steerable thruster configurations.

The following angular force-moment functions and power-thrust relationship are assumed:

$$F_{XA} = C_{XA}\ V_A^2 \cos \alpha_A = K_{XA} \cos \alpha_A$$
$$F_{YA} = C_{YA}\ V_A^2 \sin \alpha_A = K_{YA} \sin \alpha_A$$
$$M_{ZA} = C_{ZA}\ V_A^2 \sin 2\alpha_A = K_{ZA} \sin 2\alpha_A$$
$$F_{XC} = C_{XC}\ V_C^2 \cos \alpha_C = K_{XC} \cos \alpha_C$$
$$F_{YC} = C_{YC}\ V_C^2 \sin \alpha_C = K_{YC} \sin \alpha_C$$
$$M_{ZC} = C_{ZC}\ V_C^2 \sin 2\alpha_C = K_{ZC} \sin 2\alpha_C$$
$$P = K_P F_T$$

where:

F_{XA}, F_{YA}, M_{ZA} = aerodynamic force-moment functions
F_{XC}, F_{YC}, M_{ZC} = hydrodynamic force-moment functions
α_A = the angle of attack of the wind with respect to the bow
α_C = the angle of attack of the current with respect to the bow
P = power
F_T = thrust

For a symmetrical set of identical fixed-axis thrusters with sufficient thrust capability not to be saturated, the following expression gives the total power used for a given θ_A and θ_C:

$$P_{total} = K_P \{(K_{XA} \cos \alpha_A + K_{XC} \cos \alpha_C) + (K_{YA} \sin \alpha_A + K_{YA} \sin \alpha_C)\} \quad (A1)$$

With a linear power-thrust relationship, moment does not consume any power because the thrust command to one thruster is increased by an equal amount to the decrease in another thruster. Collecting terms in Equation A1 yields:

496

$$P_{total} = K_p \{K_A \sin (\theta_A + \alpha_A) + K_C \sin (\theta_c + \alpha_c)\} \qquad (A2)$$

where:

$$K_A = \sqrt{K_{XA}^2 + K_{YA}^2}$$

$$\theta_A = \arctan \left[\frac{K_{XA}}{K_{YA}}\right]$$

$$K_C = \sqrt{K_{XC}^2 + K_{YC}^2}$$

$$\theta_C = \arctan \left[\frac{K_{XC}}{K_{YC}}\right]$$

For a ship-shape hull θ_A and θ_C can be as small as 20 degrees which means the peak of the power required for either the wind and current will be at 70 degrees from the bow. For a semisubmersible, K_X and K_Y for the wind and for the current are more nearly equal with K_Y being generally slightly larger than K_X. As a result, the peak of the power demand is slightly greater than 45 degrees from the bow.

If the wind and current are considered in the same direction, i.e., $\alpha_A = \alpha_C$, then the peak power occurs at the following angle from the bow:

$$\tan \alpha = \frac{K_A \cos \theta_A + K_C \cos \theta_C}{K_A \sin \theta_A + K_C \sin \theta_C} \qquad (A3)$$

If θ_A is approximately equal to θ_C, then Equation A3 becomes:

$$\tan \alpha = \frac{1}{\tan \theta_A}$$

which implies the peak power occurs at approximately:

$$\alpha = 90 - \theta_A \qquad (A4)$$

and the peak power is approximately:

$$P_{peak} = K_P (K_A + K_C) \qquad (A5)$$

For symmetrical pattern of steerable thrusters with sufficient thrust capability not to be saturated, the following expression gives the approximate total power used for a given α_A and α_C:

$$P = K_P \{(K_{XA} \cos \alpha_A + K_{XC} \cos \alpha_C)^2 + (K_{YA} \sin \alpha_A + K_{YC} \sin \alpha_C)^2\}^{1/2} \qquad (A6)$$

where the moment command is insignificant compared to the x-axis and y-axis commands.

For the semisubmersible with K_{XA} approximately equal to K_{YA} and K_{XC} approximately equal to K_{YC}, Equation A6 can be expressed as follows:

$$P = K_P \{K_{YA}^2 + K_{YC}^2 + 2K_{YA} K_{YC} \cos (\alpha_A - \alpha_C)\}^{1/2} \qquad (A7)$$

If the wind and the current are in the same direction, i.e., $\alpha_A = \alpha_C$, then the power used is a constant at all angles from the bow and is equal to:

$$P_{peak} = K_P (K_{YA} + K_{YC}) \tag{A8}$$

A comparison of the peak power for steerables, Equation A8, and for fixed-axis propellers for a semisubmersible, Equation A5, shows that the power for the fixed-axis propellers is approximately 1.4 times larger than for the steerable propellers.

Appendix B

A Method of Approximating the Wave Spectrum

One method of approximating the wave spectrum is with the summation of a finite number of sinusoids as expressed in the following equation:

$$A_w(\omega) = \sum_{i=1}^{m} A_i \sin(\omega_i t + \theta_i)$$

where:

A_i = the amplitude of the ith sinusoid
ω_i = the radian frequency of the ith sinusoid
θ_i = the phase angle of the ith sinusoid

In this appendix a method is presented for the selection of the amplitude, the phase angle, and the frequency of the individual sinusoids. The method uses equal partitioning of the wave energy spectrum and, therefore, the amplitude of the various sinusoids are equal.

A common model of the regular wave spectrum is the modified Bretschneider spectrum which is given as follows:[1]

$$S_w(\omega) = \frac{A}{\omega^5} \exp\left[-\frac{B}{\omega^4}\right] \tag{B1}$$

where:

$A = 2760\ H_s^2/\widetilde{T}_s^4$
$B = 690/\widetilde{T}_s^4$
ω = radian frequency
H_s = the significant wave height
$\widetilde{T}_s$ = the average period of the significant waves.

The integral of the wave spectrum is a measure of the energy in the waves between the integration limits. The integral of Equation B1 is given as follows:

$$E = \int_{\omega}^{\infty} S_w(\omega)\ d\omega = \frac{A}{4B}\left[1 - \exp\left(-\frac{B}{\omega^4}\right)\right] \tag{B2}$$

or:

499

$$E = H_s^2 \left(1 - \exp\left\{ -\frac{690}{(\omega T)^4} \right\} \right) \qquad (B3)$$

which yields a total energy equal to the square of the significant wave height, H_s for $\omega = 0$ in Equation B3.

For a given value of E, E_i, which is less than E, a unique frequency can be determined as follows:

$$\omega_i = \frac{1}{\tilde{T}_s} \left\{ \frac{690}{\ln \dfrac{1}{(1 - \delta E_i)}} \right\}^{1/4} \qquad (B4)$$

where $\delta E_i = \dfrac{E_i}{E}$. When the energy spectrum is divided into an integer number of parts for the purpose of approximating the spectrum for a dynamic simulation, the boundary frequencies between segments can be computed from Equation B4 by defining:

$$\delta E_i = \frac{M - i}{M} \qquad (B5)$$

where:

M $\doteq$ the number of equal parts into which spectrum is divided
i = 1, 2, . . . , M − 1

Substituting Equation B5 into Equation B4 yields the following expression:

$$\omega_i = \frac{1}{\tilde{T}_s} \left[\frac{690}{\ln \left(\dfrac{M}{i} \right)} \right]^{1/4} \qquad (B6)$$

Equation B6 gives the frequencies at the boundaries of the equal energy segments. The frequency used to simulate an energy segment is generally not the boundary frequencies. Instead the segment frequency is selected as some value between the two boundary frequencies. One such value is the frequency corresponding to one-half of the energy of the segment. The corresponding energy segment, δE_i, for Equation B4 is given as:

$$\delta E(j) = \frac{2M - 2j + 1}{2M} \qquad (B7)$$

where j = 1, 2, 3, . . . , M. With the substitution of Equation B7 into Equation B4, the simulation frequency for each energy segment can be computed from the following expression:

$$\omega_j = \frac{1}{\tilde{T}_s} \left[\frac{690}{\ln \left(\dfrac{2M}{2j - 1} \right)} \right]^{1/4} \qquad (B8)$$

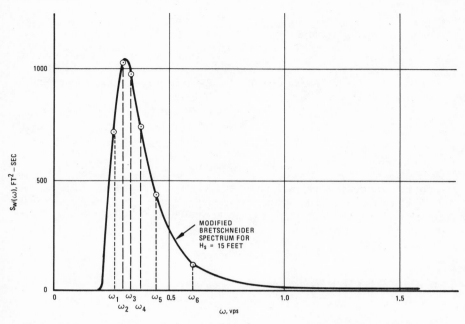

Figure B-1. Simulation Frequencies for a 15-Foot Wave Spectrum

or in terms of significant wave height:

$$\omega_j = \frac{1}{K_s \sqrt{H_s}} \left[\frac{690}{\ln \left\{ \dfrac{2M}{2j-1} \right\}} \right]^{1/4} \tag{B9}$$

since $\widetilde{T}_s = K_s \sqrt{H_s}$ for a fully developed sea. For the modified Bretschneider spectrum $K_s = 4.05$,[1] which results in the following expression:

$$\omega_j = \frac{1.265}{\sqrt{H_s}} \left[\ln \left(\frac{2M}{2j-1} \right) \right]^{-1/4} \tag{B10}$$

As mentioned, the representative function of each energy segment of the wave spectrum is a pure sinusoid with a constant frequency. The frequency is given by Equation B9. The amplitude of the pure sinusoid is determined by the following relationship:

$$A_i = 1.414 \sqrt{\frac{E}{M}} = \frac{H_s}{\sqrt{\dfrac{M}{2}}} \tag{B11}$$

since $E = H_s^2$.

As an example, the amplitude and frequencies of the sinusoids to simulate a regular wave spectrum with a 15-ft significant wave height using six energy segments or sinusoids are as follows:

$$\text{Amplitude, } A_i = 8.66 \text{ ft}$$
$$\omega_1 = 0.26 \text{ rps}$$
$$\omega_2 = 0.301 \text{ rps}$$
$$\omega_3 = 0.338 \text{ rps}$$
$$\omega_4 = 0.381 \text{ rps}$$
$$\omega_5 = 0.446 \text{ rps}$$
$$\omega_6 = 0.601 \text{ rps}$$

where the frequencies are carried out to three decimal places to increase the time before the simulated waveform repeats itself. These frequencies are illustrated in Fig. B1 with respect to the spectrum that they are representing.

REFERENCE

1. W. H. Michel, "Sea Spectra Simplified," *Marine Technology,* January, 1968.

Appendix C

Commonly Used Constants and Conversion Factors

Constants

Freshwater mass density .... 1.9384 slugs/ft³
1.000 grams/ml

Saltwater mass density .... 1.9905 slugs/ft³
1.027 grams/ml

Air mass density .... 0.00237 slugs/ft³
0.00122 grams/ml

Standard atmospheric pressure at sea level.............. 14.696 lb/in²

Gravitational constant .... 32.172 ft/sec²
9.80616 meter/sec²

π 3.141593

Base of Napierian Logs (ϵ).................................... 2.7183

Conversion Factors

Length:

1 inch .... 2.54 centimeters

1 foot .... 12 inches
0.3048 meters
1/6 fathom

1 meter .... 39.37 inches
3.2808 feet

1 statute mile .... 5280 feet
0.869 nautical miles

Mass:

1 slug .... 32.174 pounds

1 kilogram .... 2.2046 pounds

1 metric ton .... 2204.6 pounds
1000 kilogram

1 long ton .... 2240 pounds
1.12 short tons
1.016 metric tons

Speed:

1 foot/second .... 1.09728 kilometers/hour
                                                          0.68182 statute miles/hour
                                                          0.59248 knot
                                                          0.3048 meter/second

1 statute mile per hour............................. 1.4667 feet per sec.
                                                          0.869 knot
                                                          0.447 meter/sec

1 knot .... 1.8545 feet per sec.
                                                          1.151 miles per hour
                                                          0.5144 meters per sec.

1 meter per sec... 3.281 feet per sec.
                                                          2.237 mph
                                                          1.944 knots

Force:

1 poundal... 32.174 pounds
                                                          0.1383 newtons

1 kilogram (force)...................................... 9.807 newtons
                                                          1 kiloponds

1 pound (force)... 4.448 newtons

Power:

1 horsepower .... 550 foot-lb/sec.
                                                          746 watts

1 horsepower (metric) .... 542.5 foot-lb/sec.

1 horsepower (metric) .... 735.5 watts

1 foot-lb/sec... 1.356 watts

Appendix D

Wind and Sea Parameters for Fully Arisen Sea

WIND AND SEA PARAMETERS FOR FULLY ARISEN SEA

WIND AND SEA SCALE FOR FULLY ARISEN SEA

Sea state	Description	Beaufort wind force	Description	Range knots	Wind velocity knots	Wave height, ft — Average	Signifi-cant	Average 1/10 highest	Significant range of periods, sec	I max, period of maximum energy of spectrum	f̄ average period	T average wave-lenght	Minimum fetch, nmi	Minimum duration, hr
0	like a mirror	0	Calm	Less than	0									
	Ripples with the appearance of scales are formed but without foam crests	1	Light airs	1-3	2	0.05	0.08	0.10	up to 12 sec	0.7	0.5	10 in	5	18 min
1	Small wavelets, still short but more pronounced, crests have a glassy appearance, but do not break	2	Light breeze	4-6	5	0.18	0.29	0.37	0.4- 2.8	2.0	1.4	6.7 ft	8	39 min
	Large wavelets, crests begin to break. Foam of glassy appearance. Perhaps scattered white horses	3	Gentle breeze	7-10	8.5	0.6	1.0	1.2	0.8- 5.0	3.4	2.4	20	9.8	1.7 hr
2					10	0.88	1.4	1.8	1.0- 6.0	4	2.9	27	10	2.4
	Small waves, becoming larger, fairly frequent white horses	4	Moderate breeze	11-16	12	1.4	2.2	2.8	1.0- 7.0	4.8	3.4	40	18.	3.8
3					13.5	1.8	2.9	3.7	1.4- 7.6	5.4	3.9	52	24.	4.8
					14	2.0	3.3	4.2	1.5- 7.8	5.6	4.0	59	28.	5.2
					16	2.9	4.6	5.8	2.0- 8.8	6.5	4.6	71	40.	6.6
4	Moderate waves, taking a more pronounced long form, many white horses are formed (chance of some spray)	5	Fresh breeze	17-21	18	3.8	6.1	7.8	2.5-10.0	7.2	5.1	90	55.	8.3
					19	4.3	6.9	8.7	2.8-10.6	7.7	5.4	99	65.	9.2
					20	5.0	8.0	10	3.0-11.1	8.1	5.7	111	75.	10
5	Large waves begin to form; the white foam crests are more extensive everywhere (probably some spray)	6	Strong breeze	22-27	22	6.4	10.	13	3.4-12.2	8.9	6.3	134	100.	12.
					24	7.9	12.	16	3.7-13.5	9.7	6.8	160	130.	14
					24.5	8.2	13.	17	3.8-13.6	9.9	7.0	164	140.	15
					26	9.6	15	20	4.0-14.5	10.5	74.0	188	180.	17
6	Sea heaps up and white foam from breaking waves begins to be blown in streaks along the direction of the wind (spindrift begins to be seen)	7	Moderate gale	28-33	28	11.	18.	23	4.5-15.5	11.3	7.9	212	230	20.
					30	14	22.	28	4.7-16.7	12.1	8.6	250	280.	23.
					30.5	14.	23.	29	4.8-17.0	12.4	8.7	258	290.	24.
					32	16.	26.	33.	5.0-17.5	12.9	9.1	285	340.	27.

Sea state	Description	Beaufort force	Name	Wind (knots) range	Wind (knots)								
7	Moderately high waves of greater length; edges of crests break into spindrift. The foam is blown in well-marked streaks along the direction of the wind. Spray affects its visibility	8	Fresh gale	34-40	34	19.	30.	38.	5.5-18.5	13.6	9.7	322.	420.	30.
					36	21.	35.	44.	5.8-19.7	10.3	10.3	363.	500.	34.
					37	23.	37.	46.7	6.-20.5	14.9	10.5	376.	530.	37.
					38	25.	40.	50.	6.2-20.8	15.4	10.7	392.	600.	38.
					40	28.	45.	58.	6.5-21.7	16.1	11.4	444.	710.	42.
8	High waves. Dense streaks of foam along the direction of the wind. Sea begins to roll. Visibility affected	9	Strong gale	41-47	42	31.	50.	64.	7.-23	17.0	12.0	492.	830.	47.
					44	36.	58.	73.	7.-24.2	17.7	12.5	534.	960.	52.
					46	40.	64.	81.	7.-25	18.6	13.1	590.	1,110.	57.
9	Very high waves with long overhanging crests. The resulting foam is in great patches and is blown in dense white streaks along the direction of the wind. On the whole, the surface of the sea takes a white appearance. The rolling of the sea becomes heavy and shocklike. Visibility is affected	10	Whole gale*	48-55	48	44.	71.	90.	7.5-26.	19.4	13.8	650.	1,250.	63.
					50	49.	78.	99.	7.5-27.	20.2	14.3	700.	1,420.	69.
					51.5	52.	83.	106.	8.-28.2	20.8	14.7	736.	1,560.	73.
					52	54.	87.	110.	8.-28.5	21.0	14.8	750.	1,610.	75.
					54	59.	95.	121.	8.-29.5	21.8	15.4	810.	1,800.	81.
	Exceptionally high waves (small and medium-sized ships might for a long time be lost to view behind the waves). The sea is completely covered with long white patches of foam lying along the direction of the wind. Everywhere the edges of the wave crests are blown into froth. Visibility affected	11	Storm*	56-63	56	64.	103.	130.	8.5-31.	22.6	16.3	910.	2,100.	88.
					59.5	73.	116.	148.	10.-32.	24.	17.0	985.	2,500.	101.
	Air filled with foam and spray. Sea completely white with driving spray. Visibility very seriously affected	12	Hurricane*	64-71	>64	>80.‡	>128.‡	>164.‡	10.-(35.)	(26)	(18)	?	?	?

*For Hurricane winds (and often whole gale and storm winds) required durations and fetches are rarely attained. Seas are therefore not fully arisen.

†A heavy box around this value means that the values tabulated are at the center of the Beaufort range.

‡For such high winds, the seas are confused. The wave crests blow off, and the water and the air mix.

ORIGINAL SOURCE: W. A. McEwen and A. H. Lewis, "Encyclopedia of Nautical Knowledge," p. 483, Cornell Maritime Press, Cambridge, Md., 1953. "Manual of Seamanship," pp. 717-718, vol. II, Admiralty, London, H. M. Stationary Office, 1952. Pierson, Neumann, James, "Practical Methods for Observing and Forecasting Ocean Waves," New York University College of Engineering, 1953.
THIS BOOK'S SOURCE: John J. Meyers, Editor, "Handbook of Ocean and Underwater Engineering", McGraw-Hill Book Co., 1969, pp. 11-98 and 11-99.

Index